BUILDING HYBRID APPLICATIONS IN THE CLOUD

Building Hybrid Applications in the Cloud

Scott Densmore
Alex Homer
Masashi Narumoto
John Sharp
Hanz Zhang

978-1-62114-012-2

This document is provided "as-is". Information and views expressed in this document, including URL and other Internet Web site references, may change without notice.

Some examples depicted herein are provided for illustration only and are fictitious. No real association or connection is intended or should be inferred.

This document does not provide you with any legal rights to any intellectual property in any Microsoft product. You may copy and use this document for your internal, reference purposes.

© 2012 Microsoft. All rights reserved.

Microsoft, Active Directory, BizTalk, Hotmail, MSDN, SharePoint, SQL Azure, Visual C#, Visual Studio, Windows, Windows Azure, Windows Live, and Windows PowerShell are trademarks of the Microsoft group of companies. All other trademarks are property of their respective owners.

Contents

Contents		v
Foreword		xi
Preface		xiii
	Who This Book Is For	xiv
	Why This Book Is Pertinent Now	xiv
	How This Book Is Structured	xv
	What You Need to Use the Code	xvi
	Who's Who	xvii
	Where to Go for More Information	xviii
Acknowledgments		xix
1	**The Trey Research Scenario**	**1**
	Integrating with the Cloud	1
	The Challenges of Hybrid Application Integration	2
	The Trey Research Company	4
	Trey Research's Strategy	5
	The Orders Application	5
	The Original On-Premises Orders Application	6
	The Windows Azure Hybrid Application	7
	How Trey Research Tackled the Integration Challenges	10
	Staged Migration to the Cloud	12
	Technology Map of the Guide	12
	Summary	13
	More Information	14

2 Deploying the Orders Application and Data in the Cloud — 15
Scenario and Context — 15
Deploying the Application and Data to the Cloud — 17
- Choosing the Location for Data — 17
  - Deploy All of the Data in the Cloud — 18
  - Keep All Data On-premises — 18
  - Deploy Some of the Data in the Cloud — 19
- How Trey Research Chose the Location for Deploying Data — 19
  - Customer Data — 20
  - Product Data — 20
  - Order Data — 20
  - Audit Log Data — 21
- Choosing the Data Storage Mechanism — 21
  - Windows Azure Storage — 21
  - SQL Azure — 22
  - Alternative Database System or Custom Repository — 23
- How Trey Research Chose a Storage Mechanism for Data — 23
  - Encrypting Data Stored in Windows Azure Storage and Databases — 23

Synchronizing Data across Cloud and On-Premises Locations — 24
- Choosing a Data Synchronization Solution — 24
  - SQL Azure Data Sync — 24
  - Microsoft Sync Framework — 25
  - A Custom or Third Party Synchronization Solution — 25
- How Trey Research Chose the Data Synchronization Solution — 26
- How Trey Research Uses SQL Azure Data Sync — 26

Implementing a Reporting Solution for Cloud-Hosted Data — 29
- Choosing a Reporting Solution — 29
  - SQL Server Reporting Services — 30
  - SQL Azure Reporting Service — 30
  - A Custom or Third Party Reporting Solution — 31
- How Trey Research Chose the Reporting Solution — 31
- How Trey Research Uses the SQL Azure Reporting Service — 31
- How Trey Research Makes Reporting Data Available to External Partners — 32

Summary — 36
More Information — 37

3 Authenticating Users in the Orders Application — 39
Scenario and Context — 39
Authenticating Visitors to the Orders Application — 42
- Choosing an Authentication Technique — 42
  - ASP.NET Forms Authentication — 42
  - Claims-Based Authentication with Microsoft Active Directory Federation Service — 42

        Claims-Based Authentication with Windows Azure Access
        Control Service   43
        Claims-Based Authentication with ACS and ADFS   44
        Combined Forms and Claims-Based Authentication   45
    How Trey Research Chose an Authentication Technique   45
    How Trey Research Uses ACS and ADFS to Authenticate
    Visitors   45
        Access Control Service Configuration   47
        Handling Multiple User IDs   48
        Authentication Implementation   48
        Authentication with Windows Identity Foundation   49
        ASP.NET Request Validation   52
        Visitor Authentication and Authorization   53
        The Custom Logon Page   54
        Using a Custom Authorization Attribute   55
        Customer Details Storage and Retrieval   56
    Authenticating Access to Service Bus Queues and Topics   60
    Summary   61
    More Information   61

**4 Implementing Reliable Messaging and Communications
with the Cloud   63**
    Scenario and Context   63
    Communicating with Transport Partners   67
        Choosing a Communications Mechanism   68
            Electronic Data Interchange (EDI)   68
            Web Services (Push Model)   68
            Web Services (Pull Model)   69
            Windows Azure Storage Queues   69
            Windows Azure Service Bus Queues   70
            Windows Azure Service Bus Topics and Subscriptions   71
        How Trey Research Communicates with Transport Partners   71
            Sending Messages to a Service Bus Queue Asynchronously   75
            Receiving Messages from a Service Bus Queue and
            Processing Them Asynchronously   77
            Sending Messages to a Service Bus Topic   84
            Subscribing to a Service Bus Topic   88
            Receiving Messages from a Topic and Processing Them
            Asynchronously   90
            Implementing Adapters and Connectors for Translating
            and Reformatting Messages   91
            Correlating Messages and Replies   93
            Securing Message Queues, Topics, and Subscriptions   94
            Securing Messages   97

	Sending Orders to the Audit Log	100
	Choosing a Mechanism for Sending Orders to the Audit Log	100
	How Trey Research Sends Orders to the Audit Log	101
	Verifying Orders to Ensure Regulatory Compliance	104
	Choosing Where to Host the Compliance Application	105
	How Trey Research Hosted the Compliance Application	105
	Summary	107
	More Information	107
5	**Processing Orders in the Trey Research Solution**	**109**
	Scenario and Context	109
	Processing Orders and Interacting with Transport Partners	111
	How Trey Research Posts Messages to a Topic in a Reliable Manner	112
	Recording the Details of an Order	114
	Sending an Order to a Service Bus Topic from the Orders Application	117
	How Trey Research Decouples the Order Process from the Transport Partners' Systems	131
	Receiving and Processing an Order in a Transport Partner	132
	Acknowledging an Order or Indicating that it has Shipped in a Transport Partner	135
	Receiving Acknowledgement and Status Messages in the Orders Application	139
	Summary	141
	More Information	141
6	**Maximizing Scalability, Availability, and Performance in the Orders Application**	**143**
	Scenario and Context	143
	Controlling Elasticity in the Orders Application	144
	Choosing How to Manage Elasticity in the Orders Application	144
	Do Not Scale the Application	144
	Implement Manual Scaling	145
	Implement Automatic Scaling using a Custom Service	145
	Implement Automatic Scaling using the Enterprise Library Autoscaling Application Block	146
	How Trey Research Controls Elasticity in the Orders Application	146
	Hosting the Autoscaling Application Block	147
	Defining the Autoscaling Rules	148

	Managing Network Latency and Maximizing Connectivity to the Orders Application	152
	Choosing How to Manage Network Latency and Maximize Connectivity to the Orders Application	152
	Build a Custom Service to Redirect Traffic	152
	Use Windows Azure Traffic Manager to Route Customers' Requests	153
	How Trey Research Minimizes Network Latency and Maximizes Connectivity to the Orders Application	154
	Optimizing the Response Time of the Orders Application	156
	Choosing How to Optimize the Response Time of the Orders Application	156
	Implement Windows Azure Caching	156
	Configure the Content Delivery Network	157
	How Trey Research Optimizes the Response Time of the Orders Application	158
	Defining and Configuring the Windows Azure Cache	158
	Synchronizing the Caches and Databases in the Orders Application	159
	Retrieving and Managing Data in the Orders Application	159
	Implementing Caching Functionality for the Products Catalog	160
	Instantiating and Using a ProductsStoreWithCache Object	164
	Summary	167
	More Information	167
7	**Monitoring and Managing the Orders Application**	**169**
	Scenario and Context	169
	Monitoring Services, Logging Activity, and Measuring Performance	170
	Choosing a Monitoring and Logging Solution	171
	Windows Azure Diagnostics	172
	Enterprise Library Logging Application Block	172
	Third Party Monitoring Solution	173
	Custom Logging Solution	173
	How Trey Research Chose a Monitoring and Logging Solution	174
	How Trey Research Uses Windows Azure Diagnostics	174
	Selecting the Data and Events to Record	175
	Configuring the Diagnostics Mechanism	176
	Implementing Trace Message Logging and Specifying the Level of Detail	177
	Writing Trace Messages	179
	Transferring Diagnostics Data from the Cloud	181

Deployment and Management	184
Choosing Deployment and Management Solutions	184
Windows Azure Management Portal	184
Windows Azure Service Management REST API and Windows Azure SDK	185
Windows Azure PowerShell Cmdlets	185
How Trey Research Chose Deployment and Management Solutions	185
How Trey Research Deploys and Manages the Orders Application	186
Configuring Windows Azure by Using the Service Management Wrapper Library	186
Configuring Windows Azure by Using the Built-in Management Objects	188
Summary	190
More Information	190

Appendix A: Replicating, Distributing, and Synchronizing Data — 193

Use Cases and Challenges	193
Replicating Data across Data Sources in the Cloud and On-Premises	194
Synchronizing Data across Data Sources	199
Cross-Cutting Concerns	201
Data Access Security	201
Data Consistency and Application Responsiveness	201
Integrity and Reliability	202
Windows Azure and Related Technologies	202
Replicating and Synchronizing Data Using SQL Azure Data Sync	203
Guidelines for Configuring SQL Azure Data Sync	203
Guidelines for Using SQL Azure Data Sync	211
SQL Azure Data Sync Security Model	220
Implementing Custom Replication and Synchronization Using the Sync Framework SDK	221
Replicating and Synchronizing Data Using Service Bus Topics and Subscriptions	222
Guidelines for Using Service Bus Topics and Subscriptions	223
More Information	227

Appendix B: Authenticating Users and Authoring Requests — 229

Uses Cases and Challenges	230
Authenticating Public Users	230
Authenticating Corporate Users and Users from Partner Organizations	230

Authorizing User Actions	231
Authorizing Service Access for Non-Browser Clients	231
Authorizing Access to Service Bus Queues	232
Authorizing Access to Service Bus Relay Endpoints	232
Cross-Cutting Concerns	232
Security	232
Responsiveness	233
Reliability	233
Interoperability	233
Claims-Based Authentication and Authorization Technologies	233
Federated Authentication	234
An Overview of the Claims-Based Authentication Process	235
Authorizing Web Service Requests	236
Windows Identity Foundation	237
Windows Azure Access Control Service	238
ACS and Unique User IDs	239
Windows Azure Service Bus Authentication and Authorization	239
Client Authentication	240
Service Bus Tokens and Token Providers	243
Service Bus Endpoints and Relying Parties	243
Authorization Rules and Rule Groups	244
More Information	244

Appendix C: Implementing Cross-Boundary Communication — 245

Uses Cases and Challenges	245
Accessing On-Premises Resources From Outside the Organization	246
Accessing On-Premises Services From Outside the Organization	246
Implementing a Reliable Communications Channel across Boundaries	247
Cross-Cutting Concerns	248
Security	248
Responsiveness	248
Interoperability	249
Windows Azure Technologies for Implementing Cross-Boundary Communication	249
Accessing On-Premises Resources from Outside the Organization Using Windows Azure Connect	251
Guidelines for Using Windows Azure Connect	251
Windows Azure Connect Architecture and Security Model	253

Limitations of Windows Azure Connect	255
Accessing On-Premises Services from Outside the Organization Using Windows Azure Service Bus Relay	256
Guidelines for Using Windows Azure Service Bus Relay	256
Guidelines for Securing Windows Azure Service Bus Relay	264
Guidelines for Naming Services in Windows Azure Service Bus Relay	267
Selecting a Binding for a Service	268
Windows Azure Service Bus Relay and Windows Azure Connect Compared	270
Implementing a Reliable Communications Channel across Boundaries Using Service Bus Queues	271
Service Bus Messages	271
Guidelines for Using Service Bus Queues	272
Guidelines for Sending and Receiving Messages Using Service Bus Queues	283
Guidelines for Securing Service Bus Queues	286
More Information	287

Appendix D: Implementing Business Logic and Message Routing across Boundaries — 289

Use Cases and Challenges	289
Separating the Business Logic from Message Routing	290
Routing Messages to Multiple Destinations	291
Cross-Cutting Concerns	291
Security	291
Reliability	291
Responsiveness and Availability	291
Interoperability	291
Windows Azure Technologies for Routing Messages	292
Separating the Business Logic from Message Routing Using Service Bus Topics and Subscriptions	292
Guidelines for Using Service Bus Topics and Subscriptions to Route Messages	293
Limitations of Using Service Bus Topics and Subscriptions to Route Messages	304
Routing Messages to Multiple Destinations Using Service Bus Topics and Subscriptions	304
Guidelines for Using Service Bus Topics and Subscriptions to Route Messages to Multiple Destinations	304
Limitations of Using Service Bus Topics and Subscriptions to Route Messages to Multiple Destinations	308
Security Guidelines for Using Service Bus Topics and Subscriptions	308
More Information	308

Appendix E: Maximizing Scalability, Availability, and Performance — 309

- Requirements and Challenges — 310
  - Managing Elasticity in the Cloud — 310
  - Reducing Network Latency for Accessing Cloud Applications — 311
  - Maximizing Availability for Cloud Applications — 312
  - Optimizing the Response Time and Throughput for Cloud Applications — 312
- Windows Azure and Related Technologies — 313
  - Managing Elasticity in the Cloud by Using the Microsoft Enterprise Library Autoscaling Application Block — 314
    - How the Autoscaling Application Block Manages Role Instances — 315
    - Constraint Rules — 316
    - Reactive Rules — 316
    - Actions — 316
    - Guidelines for Using the Autoscaling Application Block — 317
  - Reducing Network Latency for Accessing Cloud Applications with Windows Azure Traffic Manager — 318
    - How Windows Azure Traffic Manager Routes Requests — 319
    - Using Monitoring Endpoints — 321
    - Windows Azure Traffic Manager Policies — 321
    - Guidelines for Using Windows Azure Traffic Manager — 322
    - Guidelines for Using Windows Azure Traffic Manager to Reduce Network Latency — 323
    - Limitations of Using Windows Azure Traffic Manager — 323
  - Maximizing Availability for Cloud Applications with Windows Azure Traffic Manager — 324
    - Guidelines for Using Windows Azure Traffic Manager to Maximize Availability — 326
  - Optimizing the Response Time and Throughput for Cloud Applications by Using Windows Azure Caching — 327
    - Provisioning and Sizing a Windows Azure Cache — 327
    - Implementing Services that Share Data by Using Windows Azure Caching — 329
    - Updating Cached Data — 331
    - Implementing a Local Cache — 334
    - Caching Web Application Session State — 335

Caching HTML Output	335
Guidelines for Using Windows Azure Caching	336
Limitations of Windows Azure Caching	346
Guidelines for Securing Windows Azure Caching	347
More Information	347

Appendix F: Monitoring and Managing Hybrid Applications — 349

Use Cases and Challenges	350
Measuring and Adjusting the Capacity of Your System	350
Monitoring Services to Detect Performance Problems and Failures Early	351
Recovering from Failure Quickly	352
Logging Activity and Auditing Operations	352
Deploying and Updating Components	353
Cross-Cutting Concerns	353
Performance	353
Security	353
Windows Azure and Related Technologies	354
Monitoring Services, Logging Activity, and Measuring Performance in a Hybrid Application by Using Windows Azure Diagnostics	355
Guidelines for Using Windows Azure Diagnostics	356
Guidelines for Securing Windows Azure Diagnostic Data	360
Deploying, Updating, and Restoring Functionality by Using the Windows Azure Service Management API and PowerShell	360
Guidelines for using the Windows Azure Service Management API and PowerShell	361
Guidelines for Securing Management Access to Windows Azure Subscriptions	363
More Information	364

Index — 367

Foreword

The first platform-as-a-service cloud capabilities to be released by Microsoft as a technical preview were announced on May 31, 2006 in form of the "Live Labs" Relay and Security Token services (see *http://blogs.msdn.com/b/labsrelay/archive/2006/05/31/612288.aspx*), well ahead of the compute, storage, and networking capabilities that are the foundation of the Windows Azure platform. In the intervening years, these two services have changed names a few times and have grown significantly, both in terms of capabilities and most certainly in robustness, but the mission and course set almost six years ago for the Windows Azure Service Bus and the Windows Azure Access Control Service has remained steady: Enable Hybrid Solutions.

We strongly believe that our cloud platform – and also those that our competitors run – provides businesses with a very attractive alternative to building and operating their own datacenter capacity. We believe that the overall costs for customers are lower, and that the model binds less capital. We also believe that Microsoft can secure, run, and manage Microsoft's server operating systems, runtime, and storage platforms better than anyone else. And we do believe that the platform we run is more than ready for key business workloads. But that's not enough.

From the start, the Microsoft cloud platform, and especially the Service Bus and Access Control services, was built recognizing that "moving to the cloud" is a gradual process and that many workloads will, in fact, never move into the cloud. Some services are bound to a certain location or a person. If you want to print a document, the end result will have to be a physical piece of paper in someone's hand. If you want to ring an alarm to notify a person, you had better do so on a device where that person will hear it. And other services won't "move to the cloud" because they are subjectively or objectively "perfectly fine" in the datacenter facilities and on their owner's existing hardware – or they won't move because regulatory or policy constraints make that difficult, or even impossible.

However, we did, and still do, anticipate that the cloud value proposition is interesting for corporations that have both feet solidly on the ground in their own datacenters. Take the insurance business as an example. Insurance companies were some of the earliest adopters of Information Technology. It wouldn't be entirely inaccurate to call insurance companies (and banks) "datacenters with a consumer service counter." Because IT is at the very heart of their business operations (and has been there for decades) and because business operations fall flat on the floor when that heart stops beating, many of them run core workloads that are very mature; and these workloads run on systems that are just as mature and have earned their trust.

Walking into that environment with a cloud value proposition is going to be a fairly sobering experience for a young, enthusiastic, and energetic salesperson. Or will it be? It turns out that there are great opportunities for leveraging the undeniable flexibility of cloud environments, even if none of the core workloads are agile and need to stay put. Insurance companies spend quite a bit of energy (and money) on client acquisition, and some of them are continuously present and surround us with advertising. With the availability of cloud computing, it's difficult to justify building up dedicated on-premises hardware capacity to run the website for a marketing campaign – if it weren't for the nagging problem that the website also needs to deliver a rate-quote that needs to be calculated by the core backend system and, ideally, can close the deal right away.

But that nagging problem would not be a problem if the marketing solution was "hybrid" and could span cloud and the on-premises assets. Which is exactly why we've built what we started building six years ago.

A hybrid application is one where the marketing website scales up and runs in the cloud environment, and where the high-value, high-touch customer interactions can still securely connect and send messages to the core backend systems and run a transaction. We built Windows Azure Service Bus and the "Service Bus Connect" capabilities of BizTalk Server for just this scenario. And for scenarios involving existing workloads, we offer the capabilities of the Windows Azure Connect VPN technology.

Hybrid applications are also those where data is spread across multiple sites (for the same reasons as cited above) and is replicated and updated into and through the cloud. This is the domain of SQL Azure Data Sync. And as workloads get distributed across on-premises sites and cloud applications beyond the realms of common security boundaries, a complementary complexity becomes the management and federation of identities across these different realms. Windows Azure Access Control Service provides the solution to this complexity by enabling access to the distributed parts of the system based on a harmonized notion of identity.

This guide provides in-depth guidance on how to architect and build hybrid solutions on and with the Windows Azure technology platform. It represents the hard work of a dedicated team who collected good practice advice from the Windows Azure product teams and, even more importantly, from real-world customer projects. We all hope that you will find this guide helpful as you build your own hybrid solutions.

Thank you for using Windows Azure!

Clemens Vasters
Principal Technical Lead and Architect
Windows Azure Service Bus

Preface

Modern computing frameworks and technologies such as the Microsoft .NET Framework, ASP.NET, Windows Communication Foundation, and Windows Identity Framework make building enterprise applications much easier than ever before. In addition, the opportunity to build applications that you deploy to the cloud using the Windows Azure™ technology platform can reduce up-front infrastructure costs, and reduce ongoing management and maintenance requirements.

Most applications today are not simple; they may consist of many separate features that are implemented as services, components, third-party plug-ins, and other systems or resources. Integrating these items when all of the components are hosted locally in your datacenter is not a trivial task, and it can become even more of a challenge when you move your applications to a cloud-based environment.

For example, a typical application may use web and worker roles running in Windows Azure, store its data in a SQL Azure™ technology database, and connect to third-party services that perform tasks such as authenticating users or delivering goods to customers. However, it is not uncommon for an application to also make use of services exposed by partner organizations, or services and components that reside inside the corporate network which, for a variety of reasons, cannot be migrated to the cloud.

Applications such as this are often referred to as *hybrid applications*. The issues you encounter when building them, or when migrating parts of existing on-premises applications to the cloud, prompt questions such as "How can I integrate the various parts across network boundaries and domains so that all of the parts can work together to implement the complete application?" and "How do I maximize performance and availability when some parts of the application are located in the cloud?"

This guide focuses on the common issues you will encounter when building applications that run partly in the cloud and partly on-premises, or when you decide to migrate some or all elements of an existing on-premises application to the cloud. It focuses on using Windows Azure as the host environment, and shows how you can take advantage of the many features of this platform, together with SQL Azure, to simplify and speed the development of these kinds of applications.

Windows Azure provides a set of infrastructure services that can help you to build hybrid applications. These services, such as Service Bus Security, Messaging, Caching, Traffic Manager, and Azure Connect, are the main topics of this guide. The guide demonstrates scenarios where these services are useful, and shows how you can apply them in your own applications.

This guide is based on the experiences of a fictitious corporation named Trey Research who evolved their existing on-premises application to take advantage of Windows Azure. The guide does not cover the individual migration tasks, but instead focuses on the way that Trey Research utilizes the services exposed by Windows Azure and SQL Azure to manage interoperability, process control, performance, management, data synchronization, and security.

Who This Book Is For

This book is the third volume in a series on Windows Azure. Volume 1, *Moving Applications to the Cloud on Windows Azure*, provides an introduction to Windows Azure, discusses the cost model and application life cycle management for cloud-based applications, and describes how to migrate an existing ASP.NET application to the cloud. Volume 2, *Developing Applications for the Cloud on Windows Azure*, discusses the design considerations and implementation details of applications that are designed from the beginning to run in the cloud. It also extends many of the areas covered in Volume 1 to provide information about more advanced techniques that you can apply in Windows Azure applications.

This third volume in the series demonstrates how you can use the powerful infrastructure services that are part of Windows Azure to simplify development; integrate the component parts of a hybrid application across the cloud, on-premises, and third-party boundaries; and maximize security, performance scalability, and availability.

This guide is intended for architects, developers, and information technology (IT) professionals who design, build, or operate applications and services that run on or interact with the cloud. Although applications do not need to be based on the Microsoft® Windows® operating system to operate in Windows Azure, this book is written for people who work with Windows-based systems. You should be familiar with the Microsoft .NET Framework, the Microsoft Visual Studio® development system, ASP.NET MVC, and the Microsoft Visual C#® development language.

Why This Book Is Pertinent Now

Software designers, developers, project managers, and administrators are increasingly recognizing the benefits of locating IT services in the cloud to reduce infrastructure and ongoing data center runtime costs, maximize availability, simplify management, and take advantage of a predictable pricing model. However, it is common for an application to contain some components or features that cannot be located in the cloud, such as third-party services or sensitive data that must be maintained onsite under specialist control.

Applications such as this require additional design and development effort to manage the complexities of communication and integration between components and services. To prevent these complexities from impeding moving applications to the cloud, Windows Azure is adding a range of framework services that help to integrate the cloud and on-premises application components and services. This guide explains how these services can be applied to typical scenarios, and how to use them in applications you are building or migrating right now.

How This Book Is Structured

This is the road map of the guide.

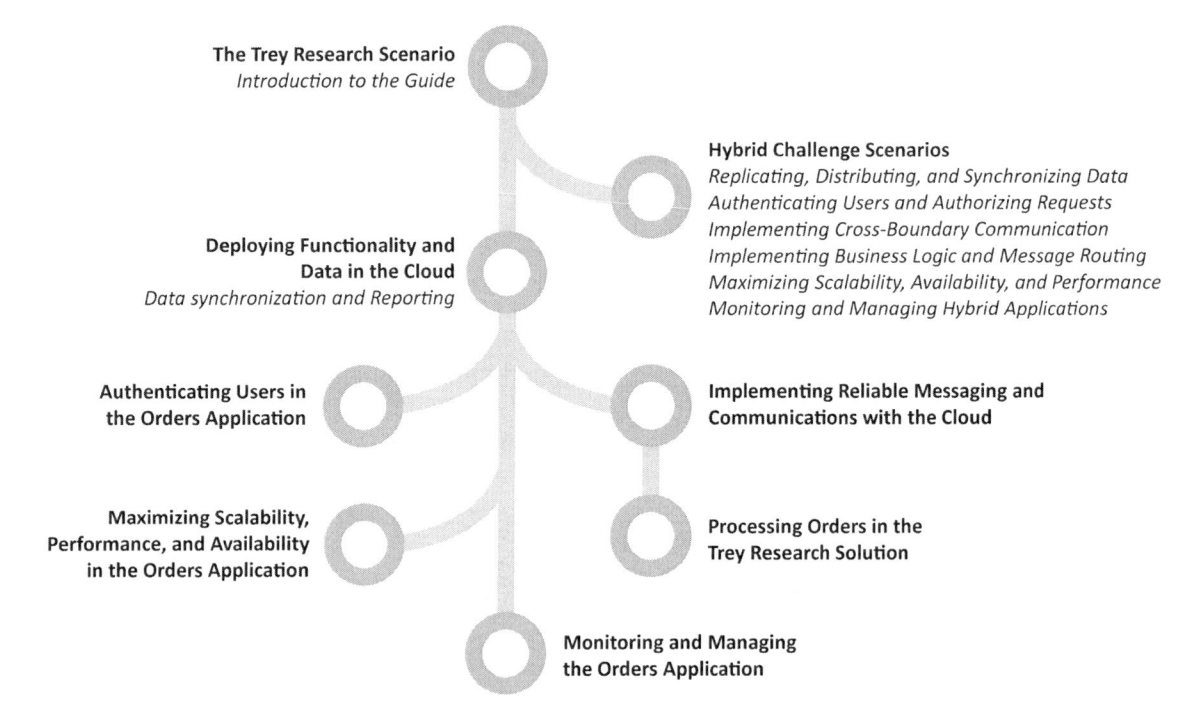

The Trey Research Scenario
Introduction to the Guide

Hybrid Challenge Scenarios
Replicating, Distributing, and Synchronizing Data
Authenticating Users and Authorizing Requests
Implementing Cross-Boundary Communication
Implementing Business Logic and Message Routing
Maximizing Scalability, Availability, and Performance
Monitoring and Managing Hybrid Applications

Deploying Functionality and Data in the Cloud
Data synchronization and Reporting

Authenticating Users in the Orders Application

Implementing Reliable Messaging and Communications with the Cloud

Maximizing Scalability, Performance, and Availability in the Orders Application

Processing Orders in the Trey Research Solution

Monitoring and Managing the Orders Application

Chapter 1, "The Trey Research Scenario" provides an introduction to Trey Research and its plan for evolving the on-premises Orders application into a hybrid application. It also contains overviews of the architecture and operation of the original on-premises application and the completed hybrid implementation to provide you with context for the remainder of the guide.

Chapter 2, "Deploying the Orders Application and Data in the Cloud" discusses the techniques and technologies Trey Research considered for deploying the application and the data it uses to the cloud, how Trey Research decided which data should remain on-premises, and the deployment architecture that Trey Research decided would best suit its requirements. The chapter also explores technologies for synchronizing the data across the on-premises and cloud boundary, and how business intelligence reporting could still be maintained.

Chapter 3, "Authenticating Users in the Orders Application" describes the technologies and architectures that Trey Research examined for evolving the on-premises application from ASP.NET Forms authentication to use claims-based authentication when deployed as a hybrid application.

Chapter 4, "Implementing Reliable Messaging and Communications with the Cloud" describes the technologies that Trey Research investigated for sending messages across the on-premises and cloud boundary, and the solutions it chose. This includes the architecture and implementation for sending messages to partners in a reliable way, as well as to on-premises services.

Chapter 5, "Processing Orders in the Trey Research Solution" describes the business logic that Trey Research requires to securely and reliably process customers' orders placed by using the Orders website. This logic includes directing messages to the appropriate partner or service, receiving acknowledgements, and retrying operations that may fail due to transient network conditions.

Chapter 6, "Maximizing Scalability, Availability, and Performance in the Orders Application" describes how Trey Research explored techniques for maximizing the performance of the Orders application by autoscaling instances of the web and worker roles in the application, deploying the application in multiple datacenters, and improving data access performance through caching.

Chapter 7, "Monitoring and Managing the Orders Application" describes the techniques that Trey Research examined and chose for monitoring and managing the Orders application. These techniques include capturing diagnostic information, setting up and configuring the Windows Azure services, and remotely managing the application configuration and operation.

While the main chapters of this guide concentrate on Trey Research's design process and the choices it made, the *"Hybrid Challenge Scenarios"* appendices focus on a more generalized series of scenarios typically encountered when designing and building hybrid applications. Each appendix addresses one specific area of challenges and requirements for hybrid applications described in Chapter 1, "The Trey Research Scenario," going beyond those considered by the designers at Trey Research for the Orders application. In addition to the scenarios, the appendices provide more specific guidance on the technologies available for tackling each challenge. The appendices included in this guide are:

- Appendix A - Replicating, Distributing, and Synchronizing Data
- Appendix B - Authenticating Users and Authorizing Requests
- Appendix C - Implementing Cross-Boundary Communication
- Appendix D - Implementing Business Logic and Message Routing across Boundaries
- Appendix E - Maximizing Scalability, Availability, and Performance
- Appendix F - Monitoring and Managing Hybrid Applications

The information in this guide about Windows Azure, SQL Azure, and the services they expose is up to date at the time of writing. However, Windows Azure is constantly evolving and new capabilities and features are frequently added. For the latest information about Windows Azure, see "What's New in Windows Azure" and the Windows Azure home page at http://www.microsoft.com/windowsazure/.

What You Need to Use the Code

These are the system requirements for running the scenarios:

- Microsoft Windows 7 with Service Pack 1 or later (32 bit or 64 bit edition), or Windows Server 2008 R2 with Service Pack 1 or later
- Microsoft Internet Information Server (IIS) 7.0
- Microsoft .NET Framework version 4.0
- Microsoft ASP.NET MVC Framework version 3
- Microsoft Visual Studio 2010 Ultimate, Premium, or Professional edition with Service Pack 1 installed
- Windows Azure SDK for .NET (includes the Visual Studio Tools for Windows Azure)
- Microsoft SQL Server or SQL Server Express 2008
- Windows Identity Foundation

- Microsoft Enterprise Library 5.0 (required assemblies are included in the source code download)
- Windows Azure Cmdlets (install the Windows Azure Cmdlets as a Windows PowerShell® snap-in, this is required for scripts that use the Azure Management API)
- Sample database (scripts are included in the Database folder of the source code)

*You can download the sample code from **http://wag.codeplex.com/releases/**. The sample code contains a dependency checker utility you can use to check for prerequisites and install any that are required. The dependency checker will also install the sample databases.*

Who's Who

This book uses a sample application that illustrates integrating applications with the cloud. A panel of experts comments on the development efforts. The panel includes a cloud specialist, a software architect, a software developer, and an IT professional. The delivery of the sample application can be considered from each of these points of view. The following table lists these experts.

	**Bharath** is a cloud specialist. He checks that a cloud-based solution will work for a company and provide tangible benefits. He is a cautious person, for good reasons. *"Implementing hybrid applications for the cloud can be a challenge, but the many services and features offered by Windows Azure can help you to resolve these issues quickly and easily".*
	**Jana** is a software architect. She plans the overall structure of an application. Her perspective is both practical and strategic. In other words, she considers the technical approaches that are needed today and the direction a company needs to consider for the future. *"It's not easy to balance the needs of the company, the users, the IT organization, the developers, and the technical platforms we rely on."*
	**Markus** is a senior software developer. He is analytical, detail-oriented, and methodical. He's focused on the task at hand, which is building a great cloud-based application. He knows that he's the person who's ultimately responsible for the code. *"For the most part, a lot of what we know about software development can be applied to the cloud. But, there are always special considerations that are very important."*
	**Poe** is an IT professional who's an expert in deploying and running applications in the cloud. Poe has a keen interest in practical solutions; after all, he's the one who gets paged at 03:00 when there's a problem. *"Running applications in the cloud that are accessed by thousands of users involves some big challenges. I want to make sure our cloud apps perform well, are reliable, and are secure. The reputation of Trey Research depends on how users perceive the applications running in the cloud."*

If you have a particular area of interest, look for notes provided by the specialists whose interests align with yours.

Where to Go for More Information

There are a number of resources listed in text throughout the book. These resources will provide additional background, bring you up to speed on various technologies, and so forth. For your convenience, there is a bibliography online that contains all the links so that these resources are just a click away.

You can find the bibliography at: *http://msdn.microsoft.com/en-us/library/hh968447.aspx*.

Acknowledgments

The IT industry has been evolving, and will continue to evolve at a rapid pace; and with the advent of the cloud computing, the rate of evolution is accelerating significantly. Back in January 2010, when we started work on the first guide in this series, Windows Azure offered only a basic set of features such as compute, storage and database. Two years later, as we write this guide, we have available many more advanced features that are useful in a variety of scenarios.

Meanwhile, general acceptance and use of cloud computing by organizations has also been evolving. In 2010, most of the people I talked to were interested in the cloud, but weren't actually working on real projects. This is no longer the case. I'm often impressed by the amount of knowledge and experience that customers have gained. There's no doubt in my mind that industry as a whole is heading for the cloud.

However, transition to the cloud is not going to happen overnight. Most organizations still have a lot of IT assets running in on-premises datacenters. These will eventually be migrated to the cloud, but a shift to the next paradigm always takes time. At the moment we are in the middle of a transition between running everything on-premises and hosting everything in the cloud. "Hybrid" is a term that represents the application that positions its architecture somewhere along this continuum. In other words, hybrid applications are those that span the on-premises and cloud divide, and which bring with them a unique set of challenges that must be addressed. It is to address these challenges that my team and I have worked hard to provide you with this guide.

The goal of this guide is to map Windows Azure features with the specific challenges encountered in the hybrid application scenario. Windows Azure now offers a number of advanced services such as Service Bus, Caching, Traffic Manager, Azure Connect, SQL Azure Data Sync, VM Role, ACS, and more. Our guide uses a case study of a fictitious organization to explain the challenges that you may encounter in a hybrid application, and describes solutions using the features of Windows Azure that help you to integrate on-premises and the cloud.

As we worked with the Windows Azure integration features, we often needed to clarify and validate our guidelines for using them. We were very fortunate to have the full support of product groups and other divisions within Microsoft. First and foremost, I want to thank the following subject matter experts: Clemens Vasters, Mark Scurrell, Jason Chen, Tina Stewart, Arun Rajappa, and Corey Sanders. We relied on their knowledge and expertise in their respective technology areas to shape this guide. Many of the suggestions raised by these reviewers, and the insightful feedback they provided, have been incorporated into this guide.

The following people were also instrumental in providing technical expertise during the development of this guide: Kashif Alam, Vijaya Alaparthi, Matias Woloski, Eugenio Pace, Enrique Saggese, and Trent Swanson (Full Scale 180). We relied on their expertise to validate the scenario as well as to shape the solution architecture.

I also want to extend my thanks to the project team. As the technical writers, John Sharp (Content Master) and Alex Homer brought to the project both considerable writing skill and expertise in software engineering. Scott Densmore, Jorge Rowies (Southworks), Alejandro Jezierski (Southworks), Hanz Zhang, Ravindra Mahendravarman (Infosys Ltd.), and Ravindran Paramasivam (Infosys Ltd.) served as the development and test team. By applying their expertise with Windows Azure, exceptional passion for technology, and many hours of patient effort, they developed the sample code.

I also want to thank RoAnn Corbisier and Richard Burte (ChannelCatalyst.com, Inc.) for helping us to publish this guide. I relied on their expertise in editing and graphic design to make this guide accurate, as well as interesting to read.

The visual design concept used for this guide was originally developed by Roberta Leibovitz and Colin Campbell (Modeled Computation LLC) for "A Guide to Claims-Based Identity and Access Control." Based on the excellent responses we received, we decided to reuse it for this book. The book design was created by John Hubbard (eson). The cartoon faces were drawn by the award-winning Seattle-based cartoonist Ellen Forney.

Many thanks also go out to the community at our CodePlex website. I'm always grateful for the feedback we receive from this very diverse group of readers.

Masashi Narumoto
Senior Program Manager – patterns & practices
Microsoft Corporation
Redmond, January 2012

1 The Trey Research Scenario

This guide focuses on the ways that you can use the services exposed by Windows Azure™ technology platform, and some other useful frameworks and components, to help you integrate applications with components running in the cloud to build hybrid solutions. A *hybrid application* is one that uses a range of components, resources, and services that may be separated across datacenter, organizational, network, or trust boundaries. Some of these components, resources, and services may be hosted in the cloud, though this is not mandatory. However, in this guide, we will be focusing on applications that have components running in Windows Azure.

The guide is based on the scenario of a fictitious company named Trey Research that wants to adapt an existing application to take advantage of the opportunities offered by Windows Azure. It explores the challenges that Trey Research needed to address and the architectural decisions Trey Research made.

Integrating with the Cloud

Using the cloud can help to minimize running costs by reducing the need for on-premises infrastructure, provide reliability and global reach, and simplify administration. It is often the ideal solution for applications where some form of elasticity or scalability is required.

> Hybrid applications make use of resources and services that are located in different physical or virtual locations; such as on-premises, hosted by partner organizations, or hosted in the cloud. Hybrid applications represent a continuum between running everything on-premises and everything in the cloud. Organizations building hybrid solutions are most likely to position their architectures somewhere along this continuum.

It's easy to think of the cloud as somewhere you can put your applications without requiring any infrastructure of your own other than an Internet connection and a hosting account; in much the same way as you might decide to run your ASP.NET or PHP website at a web hosting company. Many companies already do just this. Applications that are self-contained, so that all of the resources and components can be hosted remotely, are typical candidates for the cloud.

But what happens if you cannot relocate all of the resources for your application to the cloud? It may be that your application accesses data held in your own datacenter where legal or contractual issues limit the physical location of that data, or the data is so sensitive that you must apply special security policies. It could be that your application makes use of services exposed by other organizations, which may or may not run in the cloud. Perhaps there are vital management tools that integrate with your application, but these tools run on desktop machines within your own organization.

In fact there are many reasons why companies and individuals may find themselves in the situation where some parts of an application are prime targets for cloud hosting, while other parts stubbornly defy all justification for relocating to the cloud. In this situation, to take advantage of the benefits of the cloud, you can implement a hybrid solution by running some parts in the cloud while other parts are deployed on-premises or in the datacenters of your business partners.

THE CHALLENGES OF HYBRID APPLICATION INTEGRATION

When planning to move parts of an existing application from on-premises to the cloud, it is likely that you will have concerns centered on issues such as communication and connectivity. For example, how will cloud-based applications call on-premises services, or send messages to on-premises applications? How will cloud-based applications access data in on-premises data stores? How can you ensure that all instances of the application running in cloud datacenters have data that is up-to-date?

In addition, moving parts of an application to the cloud prompts questions about performance, availability, management, authentication, and security. When elements of your application are now running in a remote location, and are accessible only over the Internet, can they still work successfully as part of the overall application?

It is possible to divide the many challenges into separate areas of concern. This helps you to identify them more accurately, and discover the solutions that are available to help you to resolve them. The areas of concern typically consist of the following:

> Self-contained applications are often easy to locate in the cloud, but complex applications may contain parts that are not suitable for deployment to the cloud.

It is often helpful to divide the challenges presented by hybrid applications into distinct categories that focus attention on the fundamental areas of concern.

- **Deploying functionality and data to the cloud.** It is likely that you will need to modify the code in your existing on-premises applications to some extent before it, and the data it uses, can be deployed to the cloud. At a minimum you will need to modify the configuration, and you may also need to refactor the code so that it runs in the appropriate combination of Windows Azure web and worker roles. You must also consider how you will deploy data to the cloud; and handle applications that, for a variety of reasons, may not be suitable for deploying to Windows Azure web and worker roles.
- **Authenticating users and authorizing requests.** Most applications will need to authenticate and authorize visitors, customers, or partners at some stage of the process. Traditionally, authentication was carried out against a local application-specific store of user details, but increasingly users expect applications to allow them to use more universal credentials; for example, existing accounts with social network identity providers such as Windows Live® ID, Google, Facebook, and Open ID. Alternatively, the application may need to authenticate using accounts defined within the corporate domain to allow single sign on or to support federated identity with partners.
- **Cross-boundary communication and service access.** Many operations performed in hybrid applications must cross the boundary between on-premises applications, partner organizations, and applications hosted in Windows Azure. Service calls and messages must be able to pass through firewalls and Network Address Translation (NAT) routers without compromising on-premises security. The communication mechanisms must work well over the Internet and compensate for lower bandwidth, higher latency, and less reliable connectivity. They must also protect the contents of messages, authenticate senders, and protect the services and endpoints from Denial of Service (DoS) attacks.
- **Business logic and message routing.** Many hybrid applications must process business rules or workflows that contain conditional tests, and which result in different actions based on the results of evaluating these rules. For example, an application may need to update a database, send the order to the appropriate transport and warehouse partner, perform auditing operations on the content of the order (such as checking the customer's credit limit), and store the order in another database for accounting purposes. These operations may involve services and resources located both in the cloud and on-premises.
- **Data synchronization.** Hybrid applications that run partly on-premises and partly in the cloud, run in the cloud and use on-premises data, or run wholly in the cloud but in more than one datacenter, must synchronize and replicate data between locations and across network boundaries. This may involve synchronizing only some rows and columns, and you may also want to perform translations on the data.
- **Scalability, performance, and availability.** While cloud platforms provide scalability and reliability, the division of parts of the application across the cloud/on-premises boundary may cause performance issues. Bandwidth limitations, the use of chatty interfaces, and the possibility of throttling in Windows Azure may necessitate caching data at appropriate locations, deploying additional instances of the cloud-based parts of the application to handle varying load and to protect against transient network problems, and providing instances that are close to the users to minimize response times.

- **Monitoring and management.** Companies must be able to effectively manage their remote cloud-hosted applications, monitor the day-to-day operation of these applications, and have access to logging and auditing data. They must also be able to configure, upgrade, and administer the applications, just as they would if the applications were running in an on-premises datacenter. Companies also need to obtain relevant and timely business information from their applications to ensure that they are meeting current requirements such as Service Level Agreements (SLAs), and to plan for the future.

To help you meet these challenges, Windows Azure provides a comprehensive package of cloud-based services, management tools, and development tools that make it easier to build integrated and hybrid applications. You can also use many of these services when the entire application is located within Windows Azure, and has no on-premises components.

The Trey Research Company

Trey Research is a medium sized organization of 600 employees, and its main business is manufacturing specialist bespoke hardware and electronic components for sale to research organizations, laboratories, and equipment manufacturers. It sells these products over the Internet through its Orders application. As an Internet-focused organization, Trey Research aims to minimize all non-central activities and concentrate on providing the best online service and environment without being distracted by physical issues such as transport and delivery. For this reason, Trey Research has partnered with external companies that provide these services. Trey Research simply needs to advise a transport partner when an order is received into manufacturing, and specify a date for collection from Trey Research's factory. The transport partner may also advise Trey Research when delivery to the customer has been made.

The developers at Trey Research are knowledgeable about various Microsoft products and technologies, including the .NET Framework, ASP.NET MVC, SQL Server®, and the Microsoft Visual Studio® development system. The developers are also familiar with Windows Azure, and aim to use any of the available features of Windows Azure that can help to simplify their development tasks.

> The services exposed by Windows Azure are useful for both integrating on-premises applications with the cloud, and for applications that run entirely in the cloud.

> *The Orders application is just one of the many applications that Trey Research uses to run its business. Other on-premises applications are used to manage invoicing, raw materials, supplier orders, production planning, and more. However, this guide is concerned only with the Orders application and how it integrates with other on-premises systems such as the main management and monitoring applications.*

Trey Research's Strategy

Trey Research was an early adopter of cloud-based computing and Windows Azure; it has confirmed this as the platform for new applications and for extended functionality in existing applications. Trey Research hopes to minimize on-premises datacenter costs, and is well placed to exploit new technologies and the business opportunities offered by the cloud.

Although they are aware of the need to maintain the quality and availability of existing services to support an already large customer base, the managers at Trey Research are willing to invest in the development of new services and the modification of existing services to extend their usefulness and to improve the profitability of the company. This includes planning ahead for issues such as increased demand for their services, providing better reporting and business information capabilities, improving application performance and availability, and handling additional complexity such as adding external partners.

The Orders Application

Trey Research's Orders application enables visitors to place orders for products. It is a web application that has evolved over time to take advantage of the benefits of cloud-based deployment in multiple datacenters in different geographical locations, while maintaining some essential services and applications within the on-premises corporate infrastructure. This is a common scenario for many organizations, and it means that solutions must be found to a variety of challenges. For example, how will the application connect cloud-based services with on-premises applications in order to perform tasks that would normally communicate over a corporate datacenter network, but most now communicate over the Internet?

In Trey Research's case, some vital functions connected with the application are not located in the cloud. Trey Research's management and operations applications and some databases are located on-premises in their own datacenter. The transport and delivery functions are performed by separate transport partners affiliated to Trey Research. These transport partners may themselves use cloud-hosted services, but this has no impact on Trey Research's own application design and implementation.

> The developers at Trey Research use the latest available technologies: Visual Studio 2010, ASP.NET MVC 3.0, and .NET Framework 4. Over time they have maintained and upgraded the Orders application using these technologies.

The Original On-Premises Orders Application

When Trey Research originally created the Orders application it ran entirely within their own datacenter, with the exception of the partner services for transport and delivery. The application was created as two separate components: the Orders application itself (the website and the associated business logic), and the suite of management and reporting applications.

In addition, the public Orders web application would need to be able to scale to accommodate the expected growth in demand over time, whereas the management and reporting applications would not need to scale to anything like the same extent. Trey Research proposed to scale the management and reporting applications as demand increases by adding additional servers to an on-premises web farm in their datacenter. Figure 1 shows the application running on-premises.

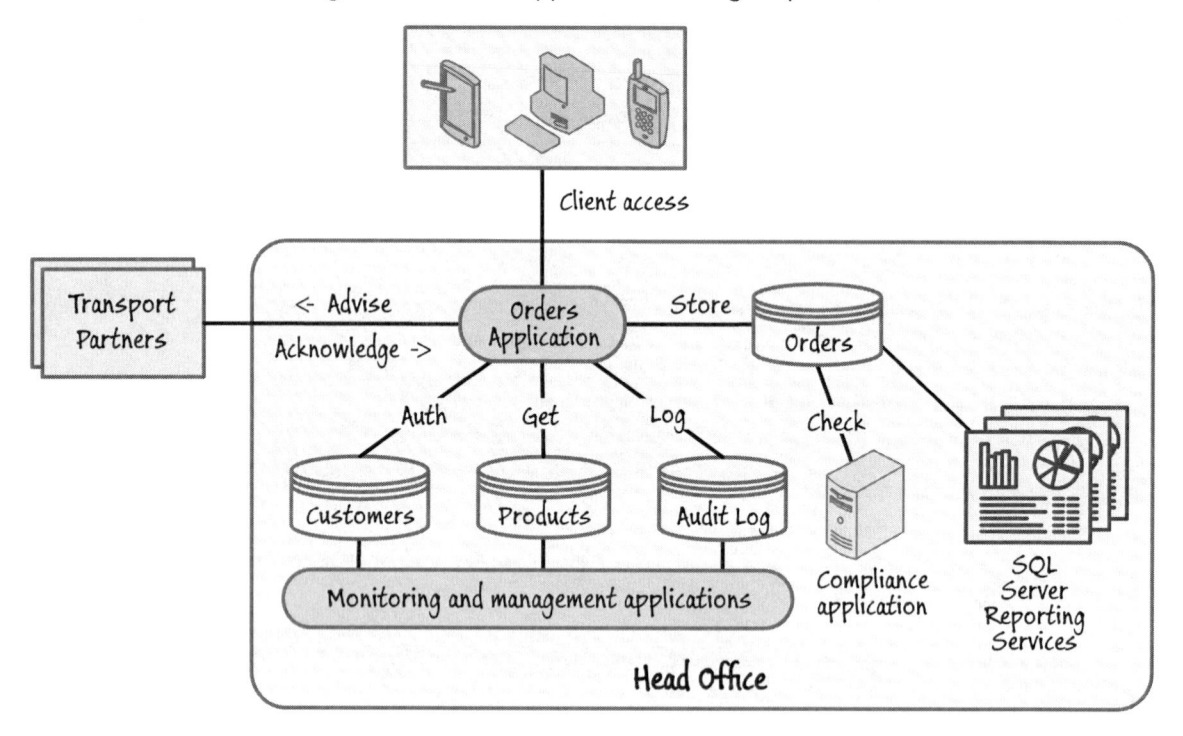

Figure 1
High-level overview of the Trey Research Orders application running on-premises

As you can see in Figure 1, the Orders application accesses several databases. It uses ASP.NET Forms authentication to identify customers and looks up their details in the Customers table using a unique user ID. It obtains a list of the products that Trey Research offers from the Products table in the database, and stores customer orders in the Orders table. The Audit Log table in the on-premises database holds a range of information including runtime and diagnostic information, together with details of notable orders such as those over a specific total value. Managers can obtain business information from the Orders table by using SQL Server Reporting Services.

The Orders application sends a message to the appropriate transport partner when a customer places an order. Currently, Trey Research has two transport partners: one for local deliveries in neighboring states and one for deliveries outside of the area. This message indicates the anticipated delivery date and packaging information for the order (such as the weight and number of packages). The transport partner may send a message back to the Orders application after the delivery is completed so that the Orders database table can be updated.

Due to the nature of the products Trey Research manufactures, it must also ensure that it meets legal requirements for the distribution of certain items, particularly for export to other countries and regions. These requirements include keeping detailed records of the sales of certain electronic components that may be part of Trey Research's products, and hardware items that could be used in the manufacture of munitions. Analyzing the contents of orders is a complex and strictly controlled process accomplished by a legal compliance application from a third party supplier, and it runs on a specially configured server.

Finally, Trey Research uses separate applications to monitor the Orders application, manage the data it uses, and perform general administrative tasks. These monitoring and management applications interact with Trey Research's corporate systems for performing tasks such as invoicing and managing raw materials stock, but these interactions are not relevant to the topics and scenarios of this guide.

The Windows Azure Hybrid Application

With the availability of affordable and reliable cloud hosting services, Trey Research decided to investigate the possibility of moving the application to Windows Azure.

Applications that run across the cloud and on-premises boundary may use web, worker, and virtual machine roles hosted in one or more Windows Azure data centers; SQL Azure™ technology platform databases in the same or different data centers; third-party remote services built using Windows or other technologies; and on-premises resources such as databases, services, and file shares. Integrating and communicating between these resources and services is not a trivial task, especially when there are firewalls and routers between them.

In addition, applications should be designed and deployed in such a way as to be scalable to meet varying loads, robust so that they are available at all times, secure so that you have full control over who can access them, and easy to manage and monitor.

One of the most immediate concerns when evolving applications to the cloud is how you will expose internal services and data stores to your cloud-based applications and services.

Figure 2 shows a high-level view of the architecture Trey Research implemented for their hybrid application. Although Figure 2 may seem complicated, the Orders application works in much the same way as when it ran entirely on-premises. You will see more details about the design decisions and implementation of each part of the application in subsequent chapters of this guide.

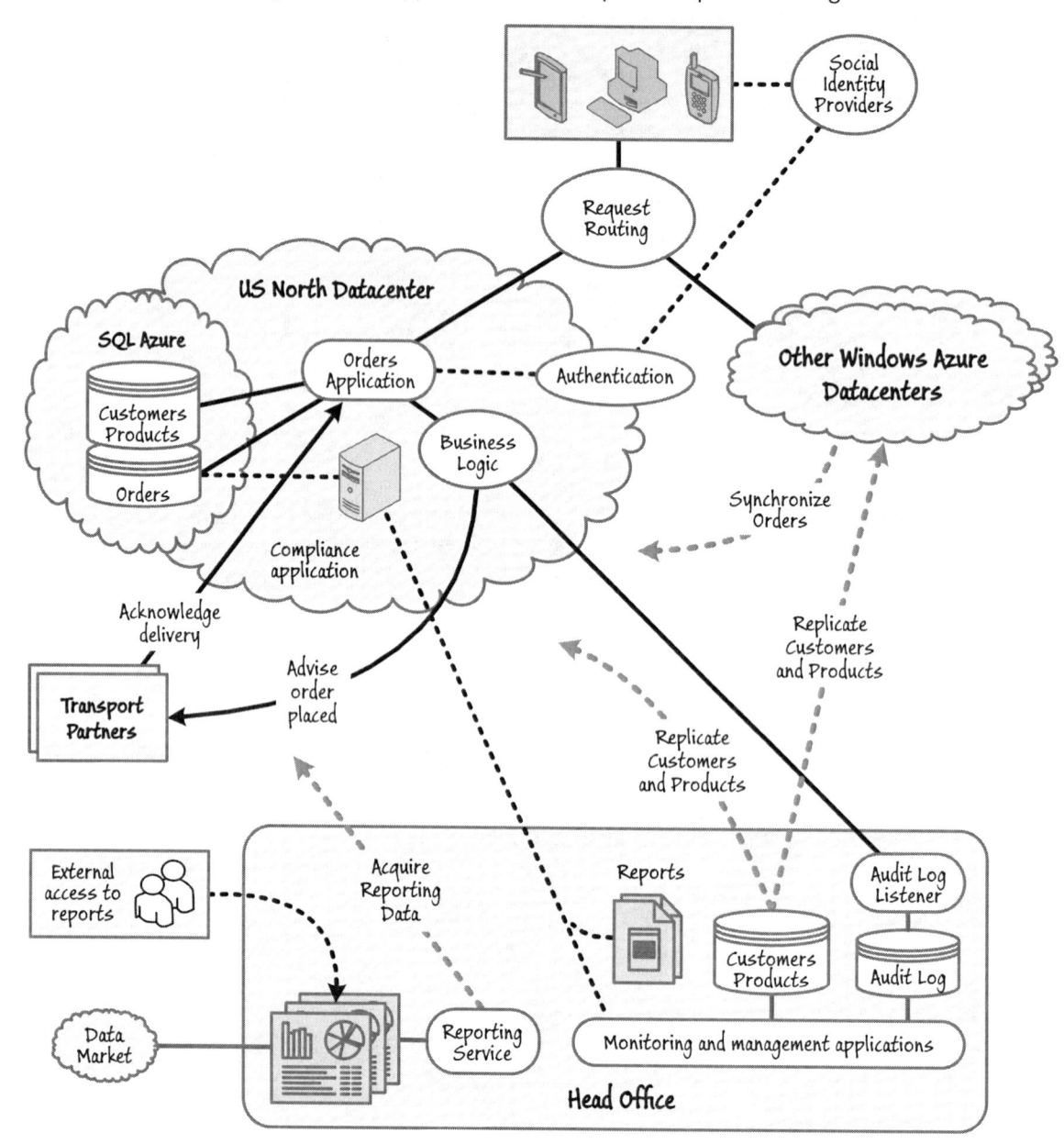

Figure 2
High-level overview of the Trey Research Orders application running in the cloud

Here is a brief summary of the features shown in Figure 2:
- Customer requests all pass through Windows Azure Traffic Manager, which redirects the customer to the instance of the Orders application running in the closest datacenter, based on response time and availability.
- Instead of using ASP.NET Forms authentication, customers authenticate using a social identity provider such as Windows Live ID, Yahoo!, or Google. Windows Azure Access Control Service (ACS) manages this process, and returns a token containing a unique user ID to the Orders application. The Orders application uses this token to look up the customer details in the Customers and Products tables of the database running in a local SQL Azure datacenter.
- New customers can register with Trey Research and obtain an account for using the Orders application. (Registration is performed as an out-of-band operation by the Head Office accounting team, and this process is not depicted in Figure 2.) When a customer has been provisioned within Trey Research's on-premises customer management system, the account details are synchronized between the Customers table held in the on-premises database and SQL Azure in all the datacenters. This enables customers to access the application in any of the global datacenters Trey Research uses.

  After the initial deployment, Trey Research decided to allow customers to edit some of their details, such as the name, billing address, and password (but not critical data such as the user's social identity information) using the application running in the cloud. These changes are be made to the local SQL Azure database, and subsequently synchronized with the on-premises data and SQL Azure in the other datacenters. You will see how this is done in Chapter 2, "Deploying the Orders Application and Data in the Cloud." However, the example application provided with this guide works in a different way. It allows you to register only by using the cloud application. This is done primarily to avoid the need to configure SQL Data Sync before being able to use the example application.

- The Orders application displays a list of products stored in the Products table. The Products data is kept up to date by synchronizing it from the master database located in the head office datacenter.
- When a customer places an order, the Orders application:
  - Stores the order details in the Orders table of the database in the local SQL Azure datacenter. All orders are synchronized across all Windows Azure datacenters so that the order status information is available to customers irrespective of the datacenter to which they are routed by Traffic Manager.
  - Sends an order message to the appropriate transport partner. The transport company chosen depends on the type of product and delivery location.
  - Sends any required audit information, such as orders over a specific total value, to the on-premises management and monitoring application, which will store this information in the Audit Log table of the database located in the head office datacenter.

- The third-party compliance application running in a virtual machine role in the cloud continually validates the orders in the Orders table for conformance with legal restrictions and sets a flag in the database table on those that require attention by managers. It also generates a daily report that it stores on a server located in the head office datacenter.
- When transport partners deliver the order to the customer they send a message to the Orders application (running in the datacenter that originally sent the order advice message) so that it can update the Orders table in the database.
- To obtain management information, the on-premises Reporting application uses the Business Intelligence features of the SQL Azure Reporting service running in the cloud to generate reports from the Orders table. These reports can be combined with data obtained from the Data Market section of Windows Azure Marketplace to compare the results with global or local trends. The reports are accessible by specific external users, such as remote partners and employees.

How Trey Research Tackled the Integration Challenges

This guide shows in detail how the designers and developers at Trey Research evolved the Orders application from entirely on-premises architecture to a hybrid cloud-hosted architecture. To help you understand how Trey Research uses some of the technologies available in Windows Azure and SQL Azure, Figure 3 shows them overlaid onto the architectural diagram you saw earlier in this chapter.

Keep in mind that, for simplicity, some of the features and processes described here are not fully implemented in the example we provide for this guide, or may work in a slightly different way. This is done to make it easier for you to install and configure the example, without requiring you to obtain and configure Azure accounts in multiple data centers, and for services such as SQL Azure Data Sync and SQL Reporting.

THE TREY RESEARCH SCENARIO   11

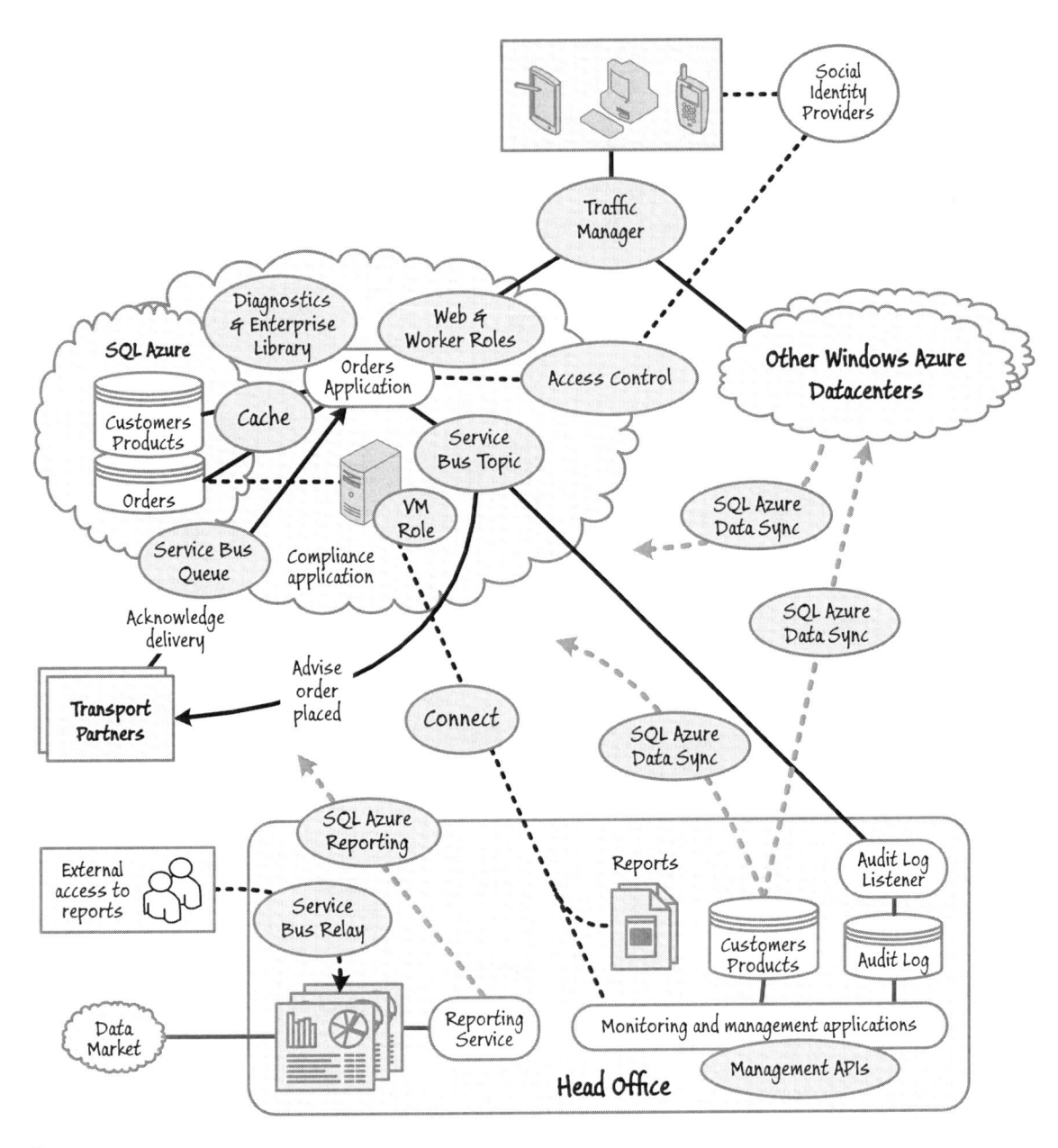

FIGURE 3
Technology map of the Trey Research Orders application running in the cloud

The information in this guide about Windows Azure, SQL Azure, and the services they expose is up to date at the time of writing. However, Windows Azure is constantly evolving and adding new capabilities and features. For the latest information about Windows Azure, see "What's New in Windows Azure" on MSDN.

> Staged or partial migration of existing on-premises applications to Windows Azure hybrid applications is not straightforward, and can require considerable effort and redesign to maintain security, reliability, and performance when communication channels cross the Internet. However, in large applications the effort required may be worthwhile compared to the complexity of a single-step migration.

Staged Migration to the Cloud

When converting an existing solution into a hybrid application, you may consider whether to carry out a staged approach by moving applications and services one at a time to the cloud. While this seems to be an attractive option that allows you to confirm the correct operation of the system at each of the intermediate stages, it is not always the best approach.

For example, the developers at Trey Research considered moving the web applications into Windows Azure web roles and using a connectivity solution such as the Windows Azure Connect service to allow the applications to access on-premises database servers. This approach introduces latency that will have an impact on the web application responsiveness, and it will require some kind of caching solution in the cloud to overcome this effect. It also leaves the application open to problems if connectivity should be disrupted.

Another typical design Trey Research considered was using Windows Azure Service Bus Relay to enable cloud-based applications to access on-premises services that have not yet moved to the cloud. As with the Windows Azure Connect service, Windows Azure Service Bus Relay depends on durable connectivity; application performance may suffer from the increased latency and transient connection failures that are typical on the Internet.

However, applications that are already designed around a Service Oriented Architecture (SOA) are likely to be easier to migrate in stages than monolithic or closely-coupled applications. It may not require that you completely redesign the connectivity and communication features to suit a hybrid environment, though there may still be some effort required to update these features to work well over the Internet if they were originally designed for use over a high-speed and reliable corporate network.

Technology Map of the Guide

The following chapters of this guide discuss the design and implementation of the Trey Research's hybrid Orders application in detail, based on a series of scenarios related to the application. The table below shows these scenarios, the integration challenges associated with each one, and the technologies that Trey Research used to resolve these challenges.

Chapter	Challenge	Technologies
Chapter 2, "Deploying the Orders Application and Data in the Cloud"	Deploying functionality and data to the cloud. Data synchronization.	SQL Azure SQL Azure Data Sync SQL Azure Reporting Service Windows Azure Marketplace (Data Market) Service Bus Relay
Chapter 3, "Authenticating Users in the Orders Application"	Authenticating users and authorizing requests in the cloud.	Windows Azure Access Control Service Windows Identity Framework Enterprise Library Transient Fault Handling Application Block
Chapter 4, "Implementing Reliable Messaging and Communications with the Cloud"	Cross-boundary communication and service access.	Windows Azure Connect service Service Bus Queues Service Bus Topics and Rules
Chapter 5, "Processing Orders in the Trey Research Solution"	Business logic and message routing.	Service Bus Queues Service Bus Topics and Rules
Chapter 6, "Maximizing Scalability, Availability, and Performance in the Orders Application"	Scalability, performance, and availability.	Windows Azure Caching service Windows Azure Traffic Manager Enterprise Library Autoscaling Application Block
Chapter 7, "Monitoring and Managing the Orders Application"	Monitoring and management.	Windows Azure Diagnostics Windows Azure Management REST APIs Windows Azure Management Cmdlets

Some of the features and services listed here (such as Windows Azure virtual machine role, Windows Azure Connect service, and Windows Azure Traffic Manager) were still prerelease or beta versions at the time of writing. For up to date information, see the Microsoft Windows Azure home page at http://www.microsoft.com/windowsazure/. In addition, this guide does not cover ACS in detail. ACS is discussed in more depth in "Claims Based Identity & Access Control Guide" (see http://claimsid.codeplex.com/), which is part of this series of guides on Windows Azure.

Summary

This chapter introduced you to hybrid applications that take advantage of the benefits available from hosting in the cloud. Cloud services provide a range of opportunities for Platform as a Service (Paas) and Infrastructure as a Service (IaaS) deployment of applications, together with a range of built-in features that can help to resolve challenges you may encounter when evolving an existing application to the cloud or when building new hybrid applications that run partially on-premises and partially in the cloud.

This chapter also introduced you to Trey Research's online Orders application, and provided an overview of how Trey Research evolved it from an entirely on-premises application into a hybrid application where some parts run in the cloud, while maintaining other parts in their on-premises datacenter. Finally, this chapter explored the final architecture of the Orders application so that you are familiar with the result.

The subsequent chapters of this guide drill down into the application in more detail, and provide a great deal more information about choosing the appropriate technology, how Trey Research implemented solutions to the various challenges faced, and how these solutions could be extended or adapted to suit other situations.

You'll see how Trey Research modified its application to work seamlessly across on-premises and cloud locations, and to integrate with external partner companies (whose applications may also be running on-premises or in the cloud), using services exposed by Windows Azure and SQL Azure.

More Information

All links in this book are accessible from the book's online bibliography available at: *http://msdn.microsoft.com/en-us/library/hh968447.aspx*.

- For the latest information about Windows Azure, see "What's New in Windows Azure" at *http://msdn.microsoft.com/en-us/library/windowsazure/gg441573*.
- The website for this series of guides at *http://wag.codeplex.com/* provides links to online resources, sample code, Hands-on-Labs, feedback, and more.
- The portal with information about Microsoft Windows Azure is at *http://www.microsoft.com/windowsazure/*. It has links to white papers, tools, and many other resources. You can also sign up for a Windows Azure account here.
- Find answers to your questions on the Windows Azure Forum at *http://social.msdn.microsoft.com/Forums/en-US/category/windowsazureplatform*.
- Eugenio Pace, a principal program manager in the Microsoft patterns & practices group, is creating a series of guides on Windows Azure, to which this documentation belongs. To learn more about the series, see his blog at *http://blogs.msdn.com/eugeniop*.
- Masashi Narumoto is a program manager in the Microsoft patterns & practices group, working on guidance for Windows Azure. His blog is at *http://blogs.msdn.com/masashi_narumoto*.
- Scott Densmore, lead developer in the Microsoft patterns & practices group, writes about developing applications for Windows Azure on his blog at *http://scottdensmore.typepad.com/*.
- Steve Marx's blog is at *http://blog.smarx.com/* is a great source of news and information on Windows Azure.
- Code and documentation for the patterns & practice Windows Azure Guidance project is available on the Codeplex Windows Azure Guidance site at *http://wag.codeplex.com/*.
- Comprehensive guidance and examples on Windows Azure Access Control Service is available in the patterns & practices book *"A Guide to Claims–based Identity and Access Control,"* also available online at *http://claimsid.codeplex.com/* and on MSDN at *http://msdn.microsoft.com/en-us/library/ff423674.aspx*.

2 Deploying the Orders Application and Data in the Cloud

The first stage in moving parts of the Orders system to the cloud as elements of a hybrid application required the designers at Trey Research to consider how to deploy these pieces in Windows Azure™ technology platform. Windows Azure offers several options for deployment of application functionality, and a wide range of associated services that Trey Research can take advantage of when designing and building hybrid applications.

In this chapter, you will see how Trey Research addressed the challenges associated with deploying the key elements of the Orders application to the cloud, and how the designers integrated the application with the services provided by Windows Azure and the SQL Azure™ technology platform.

Scenario and Context

In the original implementation of the Orders application, the components and services it uses ran on-premises and accessed data stored in local SQL Server databases in Trey Research's datacenter. You saw the architecture and a description of the original on-premises system in Chapter 1, "The Trey Research Scenario." Trey Research had to decide how to segregate the functionality, the types of Windows Azure roles to use, and how this might architecture affects the security, performance, and reliability of the application.

In addition, the designers had to consider where and how to host the data used by the application when some parts of the application are located remotely and communication must cross the Internet, and how to maintain the ability to produce business reports from that data.

When they examined the existing Orders application with a view to moving some parts to Windows Azure, it soon became clear that the management and reporting part of the application, which does not need to scale to the same extent as the public website, should remain on premises. This allowed Trey Research to more closely control the aspects of the application that require additional security and which, for logistical reasons, they felt would be better kept within their own datacenter. However, Trey Research wished to make some non-confidential elements of the reporting data available to trusted partners for use in their own systems.

The public section of the application could easily be deployed to the cloud as it was already effectively a separate application, and is the part of the application that will be required to scale most over time to meet elastic demand. This allowed Trey Research to take full advantage of the cloud in terms of reliability, availability, security, lower running costs, reduced requirements for on-premises infrastructure, and the capability to scale up and down at short notice to meet peaks in demand.

There are other advantages to hosting in Windows Azure that served to make a strong case for moving the public parts of the Orders application to the cloud. These include the ability to deploy it to multiple datacenters in different geographical locations to provide better response times and to maximize availability for customers. By using Windows Azure Traffic Manager, requests to the application are automatically routed to the instance that will provide the best user experience. Traffic Manager also handles failed instances by rerouting requests to other instances.

In addition, Trey Research were able to take advantage of the built-in distributed data caching feature for transient data used by the public website, the claims-based authentication service for easily implementing federated authentication, the connectivity features for secure communication and service access across the cloud/on-premises boundary, the capabilities for data synchronization, a comprehensive cloud-based reporting system, and the availability of third party components and frameworks to simplify development.

Figure 1 shows a high-level view of the way that Trey Research chose to segregate the parts of the application across the cloud and on-premises boundary.

> Taking advantage of available components, services, frameworks, and features designed and optimized for the cloud simplifies both the design and development of cloud-based applications.

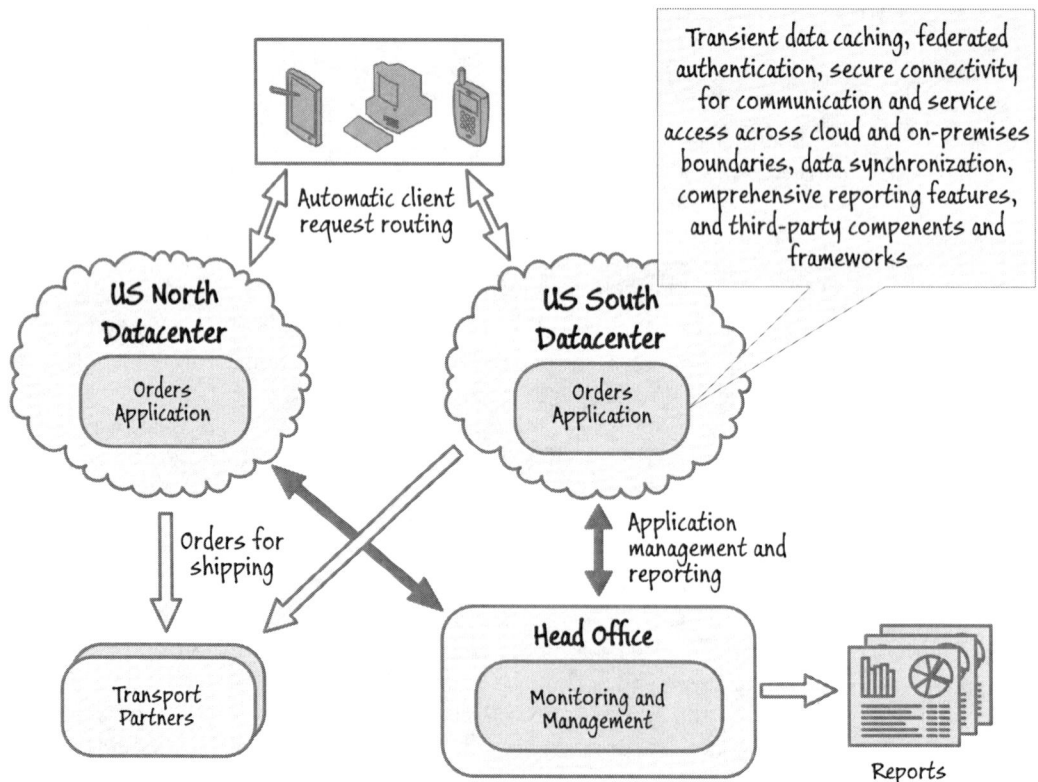

FIGURE 1
A high-level view of the segregation across the cloud and on-premises boundary

In this chapter you will see how the designers at Trey Research chose where to locate the data the application uses, how they implemented a synchronization mechanism that ensures that the relevant data is available and consistent in all of the locations where it is required, and how they maintain comprehensive business intelligence reporting capabilities. These decisions required the designers to consider the options available, and the tradeoffs that apply to each one.

Deploying the Application and Data to the Cloud

The Orders application is a website, and so the designers at Trey Research realized that this could easily be deployed in Windows Azure as a web role. Deploying multiple instances of the web role allows the website to scale to meet demand, and ensures that it provides the availability and reliability that Trey Research requires. Background processing tasks, which occur after a customer places an order, are handed off to a worker role. Trey Research can deploy multiple instances of the worker role to handle the varying load as customers place orders in the website.

The Orders website requires access to several items of data as it runs. This data includes the list of products that customers can order, the list of customers so that the application can authenticate visitors and access information about them, the orders that customers place at the website, and auditing and runtime logging information. The designers at Trey Research needed to decide where and how to locate each of these items, and also identify the appropriate storage mechanism for this data.

> You write new web applications or adapt existing web applications for deployment to Windows Azure in a very similar manner to that you would follow if you were building items for local deployment in your own datacenter. However, there are some aspects that differ, such as session state management, data storage, and configuration.

Choosing the Location for Data

All elements of a hybrid application, whether they are located on-premises, in the cloud, or at a partner location, are likely to need to access data. A fundamental part of the design of a hybrid application is locating this data in the appropriate places to maximize efficiency and performance, while maintaining security and supporting any replication and synchronization requirements. Typically, data should be located as close as possible to the applications and components that use it. However, this is not always advisable, or possible, depending on individual circumstances.

The major decision is whether to locate data remotely (such as in the cloud or at a partner location), or to keep it on-premises. The Orders application uses four types of data:

- Customer information, including sensitive data such as credit limits and payment information. This includes personally identifiable information (PII) and must be protected to the highest extent possible.
- Product information such as the product catalog, prices, and details. Trey Research manufactures all products to order, and so there is no stock level data.
- Order information, including full details of orders placed by customers and delivery information.
- Audit log information, such as events and exceptions raised by the application and details of orders over a total value of $10,000. This data may contain sensitive information that must be fully secured against access by non-administrative staff.

The designers at Trey Research considered three options for locating the data used by the Orders application. They could deploy all of the data in the cloud, keep all of the data on-premises, or deploy some in the cloud while the rest remains on-premises.

Deploy All of the Data in the Cloud

Deploying all of the data in the cloud so that it is close to the Orders application can help to maximize performance and minimize response times, and removes the requirement to synchronize data between cloud and on-premises locations. It also allows Trey Research to take advantage of the scalability and performance of either Windows Azure storage or SQL Azure, both of which provide reliable, fast, and efficient data access for the application and make it easy to expand storage availability as required.

However, deploying all of the data in the cloud would mean head-office applications that require access to this data must do so over the Internet. This could cause users in the head office to encounter delays and failed connections due to occasional Internet networking and performance issues, and additional costs would be incurred for access to the data from the on-premises applications. In addition, the storage costs for deploying large volumes of data or multiple databases could be an issue, and there is still likely to be a requirement to synchronize the data between these deployments if the application is located in more than one datacenter.

Keep All Data On-premises

Keeping all of the data on-premises means that it can be secured and managed by Trey Research administrators and operations staff more easily, especially if most of the update operations are done by on-premises staff and other on-premises applications within the organization. This approach also allows Trey Research to ensure they comply with legal or regulatory limitations on the location and security of sensitive information. In addition, there is no requirement to migrate or deploy data to a remote location, and other operations such as backing up data are easier.

However, keeping all of the data on-premises means that remote applications and services in the cloud or at partner locations must access the data over the Internet, although this can be mitigated to some extend by the judicious use of caching. The designers at Trey Research also considered whether it would be possible to implement the required business logic so that it worked securely and reliably when remote applications and services must perform updates across the Internet in multiple databases.

Deploy Some of the Data in the Cloud

Deploying some of the data in the cloud and keeping the remainder on-premises provides several advantages. For example, data for applications and services that require fast and reliable access can be located in the cloud, close to the application or service that uses it, whereas data that is mostly accessed by head office applications can remain on-premises to provide fast and reliable access for these applications. In addition, data that is subject to legal or regulatory limitations regarding its storage location, or requires specific security mechanisms to be in place, can remain on-premises. Finally, data that does not need to scale can remain on-premises, saving hosting costs, whereas data that must scale can be located in Windows Azure storage or SQL Azure to take advantage of the scalability these services offer.

However, deploying some of the data in the cloud means that, where it is used in both cloud-hosted or on-premises applications, it will still need to be accessed across the Internet. A suitably secure and reliable connectivity mechanism will be required, and a data replication and synchronization solution will be necessary to ensure that data in all locations is consistent.

> Accessing data held on-premises from a cloud-hosted application is not usually the best approach due to the inherent network latency and reliability of the Internet. If you decide to follow this approach, you must consider using a robust caching mechanism such as Windows Azure Caching to minimize the impact of network issues.

How Trey Research Chose the Location for Deploying Data

After considering the options for where to deploy data, Trey Research made the following decisions for locating the information used by the Orders application.

Customer Data

Customer information is maintained by Trey Research's own operations staff in conjunction with the existing on-premises accounting system that Trey Research uses within its wider organization. Trey Research requires customers to register through the head office, and operators add customers to the on-premises database. Using the Orders application, it is planned that customers will be able modify some of their own information (this functionality is not yet implemented), but the application will not allow them to modify critical identity or other secure data. Customer data is likely to be relatively static and not change much over time.

Trey Research decided to keep the master Customer database on-premises to maximize security, and to maintain the existing capabilities of all the on-premises applications to interact with the data efficiently. However, customer data is also required by the Orders website to authenticate visitors and to accept orders from them. Therefore, to maximize performance and reliability, Trey Research decided to locate a replica of the customer data in the cloud, close to the Orders website.

This means that a bidirectional synchronization mechanism is required to ensure that updates to the customer data made by on-premises operators are replicated to all datacenters that host the Orders application, and changes made in the Orders application by customers to certain parts of their own data are replicated back to the master copy of the data held on-premises and out to the SQL Azure databases located in other datacenters.

Product Data

Product information is also maintained by Trey Research's operations staff. This data can only be updated on-premises in conjunction with the existing on-premises manufacturing processes and parts catalogs that Trey Research uses within its wider organization. Because there is no stock level information (all products are manufactured on-demand), the Product data is relatively static.

Trey Research decided to keep the master Product data on-premises to maintain the existing capabilities of all the on-premises applications to interact with the data efficiently. However, to maximize performance and reliability, Trey Research decided to locate a replica of some fields of the Product data (just the data required to list products, show product details, and accept orders) in the cloud, close to the Orders application. This means that a unidirectional synchronization mechanism is required to ensure that updates to the Product data made by on-premises operators are replicated to all datacenters that host the Orders application.

Order Data

Order information is generated by the Orders application running in the cloud, and cannot be edited elsewhere. The Orders application also reads Order data when displaying lists of current orders and delivery information to users. Unlike Customer and Product data, which is relatively static, Order data is highly dynamic because it changes as customer place orders and as they are shipped by the transport partners.

Trey Research decided that there was no requirement to locate Order data on-premises. Instead, Order data is stored only in the cloud, close to the Orders application. However, when the Orders application is deployed to more than one datacenter, bi-directional synchronization of the order data between datacenters ensures that customers see their order information if, due to an application failure (or when a user moves to a new geographical location), they are redirected to a different datacenter. The only issue with this decision is that Trey Research will no longer be able to use SQL Server Reporting Services to create business intelligence reports on the data directly. You will see how Trey Research resolved this issue later in this chapter, in the section "Choosing a Reporting Solution."

Audit Log Data

Audit log information is generated by the Orders application in response to events and exceptions raised by the application, and for orders over a total value of $10,000. It is also generated by other on-premises applications within Trey Research's organization, and so the Audit Log database is a complete repository for all application management and monitoring facilities.

Trey Research decided that, because the most intensive access to this data is from monitoring tools and administrative management applications, the data should remain on-premises. In addition, government regulations on the sale of some high-tech products that Trey Research manufactures means Trey Research must maintain full and accurate records of such sales and store these records locally. Keeping the Audit Log data, which may contain sensitive information about the application, on-premises also helps to ensure that it is fully secured within Trey Research's domain against access by unauthorized parties.

CHOOSING THE DATA STORAGE MECHANISM

Having decided that some of the data used by the Orders application will be hosted in Windows Azure, the designers at Trey Research needed to choose a suitable mechanism for storing this data in the cloud. The most common options are Windows Azure storage, SQL Azure or another database system, or a custom repository.

Windows Azure Storage

Windows Azure storage provides blob storage, table storage, and queues. Queues are typically used for passing information between roles and services, and are not designed for use as a persistent storage mechanism. However, Trey Research could use table storage or blob storage. Both of these are cost-effective ways of storing data.

Blob storage is ideal for storing unstructured information such as images, files, and other resources. Table storage is best suited to structured information. Table storage is very flexible and can be very efficient, especially if the table structure is designed to maximize query performance. It also supports geographical replication, so that access is fast and efficient from different client locations. Table storage is significantly cheaper than using a SQL Azure database.

However, table storage does not support the familiar SQL-based techniques for reading and writing data, and some of the standard relational database data types. Data is stored as collections of entities, which are similar to rows but each has a primary key and a set of properties. These properties consist of a name and a series of typed-value pairs. The designers at Trey Research realized that migrating an existing application that uses a SQL database to the cloud, and deciding to use Windows Azure table storage, meant that they would need to redesign their data model and rewrite some of the data access code. This would add cost and time to the migration process.

In addition, Windows Azure table storage does not support the concept of database transactions, although it does provide transacted access to a single table. Finally, data cannot be directly imported from a relational database system such as SQL Server into table storage. Trey Research would need to create or source tools to perform the translation and upload the data.

> *For more information about using Windows Azure table storage, see the section "Storing Business Expense Data in Windows Azure Table Storage" in Chapter 5 of the guide "Moving Applications to the Cloud."*

SQL Azure

SQL Azure is a high-performance database service that fully supports SQL-based operations, and can be used to create relational databases in the cloud. It is implemented by SQL Server instances installed in Microsoft datacenters.

SQL Azure offers much of the core functionality of a local SQL Server installation, and it delivers data to the application using the familiar SQL Server Tabular Data Stream (TDS) protocol. This architecture enables you to use the same .NET Framework data providers (such as **System.Data.SqlClient**) to connect to the database, and T-SQL to access and manipulate the data. SQL Azure is also compatible with existing connectivity APIs, such as the Entity Framework (EF), ADO.NET, and Open Database Connectivity (ODBC). Data can be updated using database transactions to ensure consistency.

These advantages mean that developers at Trey Research would not have to make major changes to the application code, and administrators could quickly and easily deploy the data to SQL Azure without needing to change the schema of the tables. Trey Research administrators and operators can manage SQL Azure databases through the Windows Azure Management Portal, and by using familiar tools such as SQL Server Management Studio and the Visual Studio database tools. A range of other tools for activities such as moving and migrating data, as well as command line tools for deployment and administration, are also available.

In addition, data synchronization across cloud-hosted and on-premises databases is easy to achieve through the Windows Azure Data Sync service or the Data Sync APIs. SQL Azure supports business intelligence reporting with the SQL Azure Reporting Service.

However, the designers at Trey Research also needed to consider that, while SQL Azure is very similar to SQL Server, certain concepts such as server-level controls or physical file management do not apply in an auto-managed environment such as SQL Azure. In addition, the subscription costs for SQL Azure are higher than those of Windows Azure storage.

Alternative Database System or Custom Repository

If your application currently uses a relational database system, or utilizes a custom repository to store its data, you may be able to migrate the data to SQL Azure easily—depending on the existing format of the data. Alternatively, if you use a database system other than SQL Server (such as Mongo DB, see *http://www.mongodb.org/*), you might be able to run this database system in the cloud using the Windows Azure worker role or VM role.

Using an existing database system or custom repository that already provides data for your application means that you will probably be able to use the same data access code as you employed on-premises. This is an advantage if developers are familiar with the mechanism you choose, and it can reduce the transition time and effort of learning a new system.

However, using an alternative database system or custom repository means that you must maintain this database or repository yourself. For example, you must install updated versions of the database management software or debug your own custom code. You may also have difficulty importing data or moving data to another data storage mechanism in the future.

How Trey Research Chose a Storage Mechanism for Data

Trey Research uses SQL Server to store data in their on-premises applications, including the original Orders application. The data formats and types, and the data access code, are all designed to work with SQL Server. Therefore, it made sense for Trey Research to choose SQL Azure as the data storage mechanism for the hybrid version of the Orders application. The additional cost compared to using Windows Azure table storage is partly mitigated by the savings in schema redesign and code development costs.

In addition, Trey Research wanted to be able to use database transactions and perform complex queries when working with data. Implementing code to achieve the equivalent functionality using Windows Azure table storage would require additional development time and incur subsequent additional costs. Administrators at Trey Research are also familiar with SQL Server, including the tools used to manage data, and are comfortable using systems based on SQL Server so working with SQL Azure does not require them to learn new paradigms.

Encrypting Data Stored in Windows Azure Storage and Databases

The designers at Trey Research realized that when moving data to the cloud, they must consider the level of protection required for that data, irrespective of the selected storage mechanism. Sensitive data, such as customers' passwords and credit card numbers, and PII such as addresses and telephone numbers, typically require higher levels of protection than data such as product lists.

At the time of writing, neither Windows Azure storage nor SQL Azure support built-in data encryption mechanisms. This means that the application is responsible for encrypting or decrypting sensitive data that requires an additional level of protection. Trey Research achieves this by using the standard cryptography algorithms exposed by the .NET Framework, or with other code libraries.

> *For information about encrypting data in Windows Azure, see "Crypto Services and Data Security in Windows Azure" in MSDN® Magazine and "Encrypting Data in Windows Azure Storage." For details of the security features of SQL Azure, see "Security Guidelines and Limitations (SQL Azure Database)."*

Synchronizing Data across Cloud and On-Premises Locations

The architecture Trey Research chose for the Orders application has some data located in the cloud in SQL Azure, and some data located on-premises. This means that the designers at Trey Research must consider how to synchronize data across these locations to ensure it is consistent.

Choosing a Data Synchronization Solution

The choice of data synchronization solution depends on both the type of data stores that hold the data and the requirements for consistency. For example, if data must always be consistent across different locations, the solution must detect and replicate changes to data in each location as soon as they occur. If the application can work successfully when data is eventually consistent, but may be stale for short periods, a scheduled synchronization process may be sufficient. The following sections of this chapter describe the options that Trey Research considered for synchronizing data in the Orders application.

SQL Azure Data Sync

If data is deployed to SQL Azure, the natural solution for synchronizing this data is to use SQL Azure Data Sync. This is a service that can synchronize data between on-premises SQL Server databases and one or more SQL Azure databases hosted in the cloud. SQL Azure Data Sync offers a variety of options for unidirectional and bi-directional synchronization.

Using SQL Azure Data Sync would mean that the developers at Trey Research wouldn't need to write any custom code because synchronization is configured and managed through the Windows Azure web portal. This helps to reduce the cost and time required to implement a solution compared to building a custom solution.

However, SQL Azure Data Sync works with only SQL Server and SQL Azure databases; it cannot be used if data is stored in Windows Azure storage or another database system. In addition, SQL Azure Data Sync imposes some restrictions on column data types and nullability that may necessitate changes to existing database schemas. SQL Azure Data Sync handles conflicting changes made in different databases by using one of a small number of predefined policies. It isn't possible to customize these policies, and SQL Azure Data Sync does not provide synchronization events that you can use to implement your own mechanism.

The designers at Trey Research also realized that in some scenarios synchronization requires two passes to complete; the data is moved to a hub database first (which may be one of the existing operational databases) and then to client databases. This means that, when there is more than one database synchronizing from the hub database, some instances of the data may be stale until the second synchronization pass occurs. However, when simply synchronizing one on-premises database to the SQL Azure hub database, all updates are applied during a single pass.

See "Appendix A - Replicating, Distributing, and Synchronizing Data" for more information about using SQL Azure Data Sync.

Microsoft Sync Framework

SQL Azure Data Sync uses the components of the Microsoft Sync Framework to perform data synchronization. The Sync Framework is a comprehensive synchronization platform that supports any data type, any data store, any transfer protocol, and any network topology. It is not confined to use with just SQL Server and SQL Azure databases.

If the developers at Trey Research needed more control over the synchronization process, they could use the components of the Sync Framework SDK directly in code. This has the advantage that the application could react to events, such as data being changed, and initiate synchronization. The application could also handle events occurring during the synchronization process to manage conflicts and errors, or to provide more traceability. Of course, it will also mean that the developers would have to write additional code to control the synchronization process, which would incur additional cost and time compared to using the SQL Azure Data Sync service.

> *For more information about the Sync Framework SDK, see "Microsoft Sync Framework Developer Center."*

A Custom or Third Party Synchronization Solution

If Trey Research decided not to use SQL Azure Data Sync or the Microsoft Sync Framework, the designers could have considered implementing a custom or third party solution for synchronizing data. In particular, where there are special requirements for synchronizing or replicating data, a custom mechanism might be a better choice than an off the shelf solution. For example, if Trey Research needed to carry out specific types of synchronization not supported by available third-party solutions or services, a custom mechanism that passes messages between services located on-premises and at each datacenter using Windows Azure Service Bus brokered messaging could have been be a good choice.

Messaging solutions are flexible and can be used across different types of data repository because the service that receives update messages can apply the update operations in the repository using the appropriate methods. Message-based replication and synchronization solutions are particularly suited to performing real-time updates, but this was not a requirement of the Orders application.

In addition, messaging solutions can expose more information about the synchronization process as it proceeds; for example, allowing developers to trace each data modification and handle conflicts or errors in an appropriate way. It is also possible to implement a solution that follows the principles of the Command Query Responsibility Segregation (CQRS) pattern by separating the queries that extract data from the commands that update the target data repository.

However, if you cannot locate a third party solution that provides the required features and can interface with your existing data stores, and you decide to create a custom solution, implementing, testing, and debugging this solution is likely to incur additional costs and require additional development time.

> *See "Appendix A - Replicating, Distributing, and Synchronizing Data" for more information about creating a custom message-based data synchronization solution.*

How Trey Research Chose the Data Synchronization Solution

The designers at Trey Research decided to use SQL Azure Data Sync as the replication and synchronization solution for the Orders application. All of the data is stored in either SQL Server on-premises or SQL Azure in the cloud, and so SQL Azure Data Sync will be able to access and synchronize all of the data as required. The saving in development cost and time compared to a custom solution compensated to some extent for the costs of using the SQL Azure Data Sync service.

How Trey Research Uses SQL Azure Data Sync

This section is provided for information only. For simplicity, the sample solution is deployed to a single datacenter and, therefore, is not configured to replicate and synchronize data across multiple datacenters.

Trey Research stores information about products, customers, and the orders that these customers have placed. Trey Research uses a combination of SQL Server running on-premises and SQL Azure hosted at each datacenter to manage the data required by the Orders application. Therefore Trey Research decided to implement data replication and synchronization in the Orders application.

The different types of information that Trey Research synchronizes are managed and maintained in different ways, specifically:

- Order data is maintained exclusively in the cloud by the Orders application using SQL Azure, and is synchronized between datacenters. This information is not propagated back to the on-premises database.

- Product information is maintained exclusively on-premises by using SQL Server, but the details required for placing orders are copied to each SQL Azure database at each datacenter on a periodic basis.

- New customers are registered on-premises and their details are added to the SQL Server database held at Head Office. These details are replicated out to SQL Azure at each datacenter, enabling a customer to log in and access the Orders application without the system requiring recourse to the Head Office. In the future, once an account has been created, the Orders application may enable certain customer information to be changed by a customer without requiring the intervention of the Head Office, and these changes will be made to the SQL Azure database located in whichever datacenter the customer is currently connected to (this functionality is not currently implemented, but Trey Research wished to deploy the Customers data to allow for this eventuality). These changes will then be subsequently propagated back to the Head Office, and also replicated out to the other datacenters.

Figure 2 shows the solution Trey Research adopted.

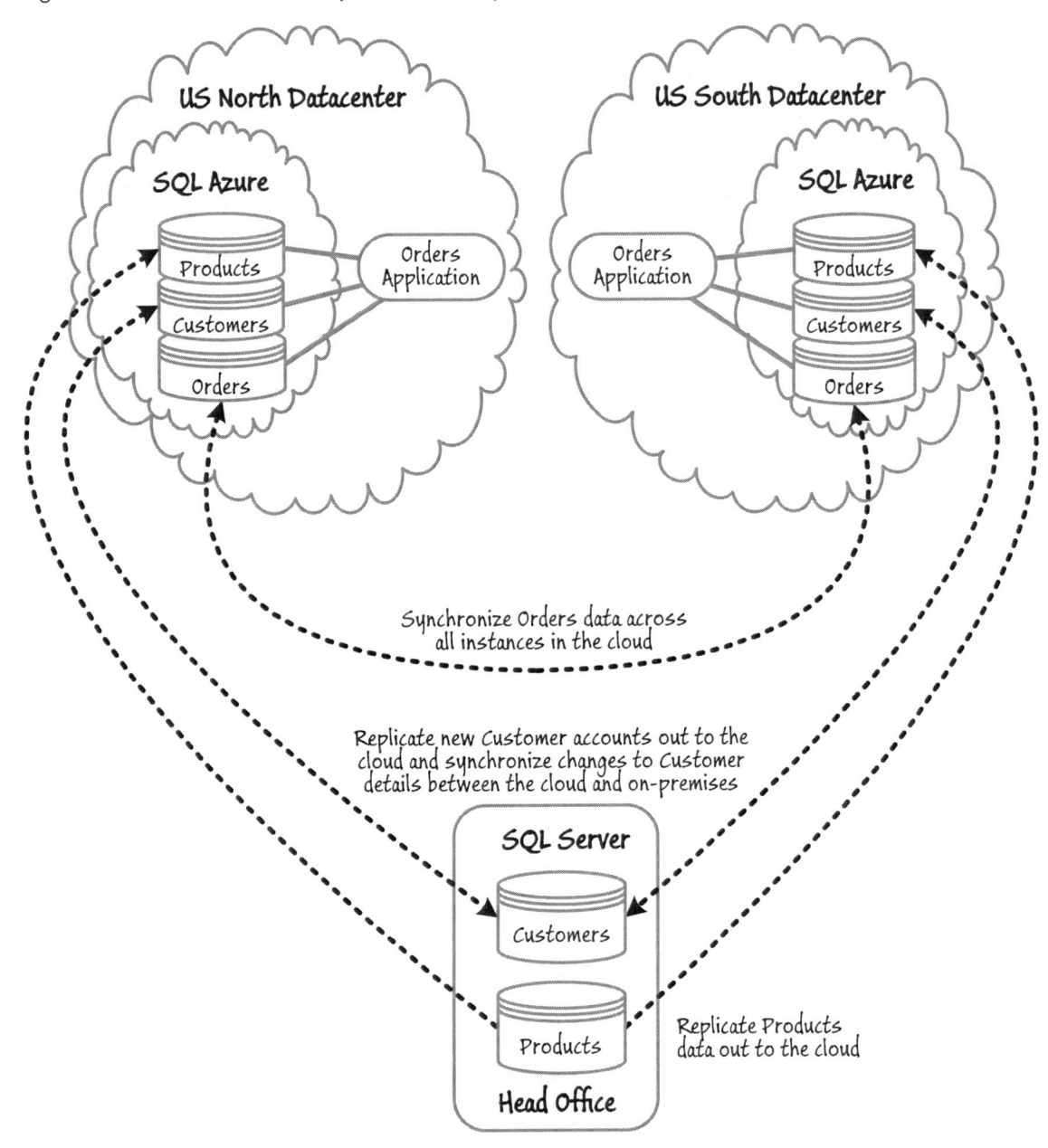

FIGURE 2
Data replication in the Trey Research Orders application

In this solution, the Product data is synchronized one way, from the on-premises database to the cloud. The Orders data is replicated bidirectionally between datacenters. The Customer data is also replicated bidirectionally, but including the on-premises database as well as those in the datacenters.

Figure 3 shows the physical implementation of these approaches based on SQL Azure Data Sync. This implementation uses three sync groups; each sync group defines a sync dataset and conflict resolution policy for each type of data (as described above, there are two overlapping sync groups for replicating customer details). The SQL Azure databases located in the US North Data Center also act as the synchronization hubs. This is the nearest datacenter to the head office (the Trey Research head office is located in Illinois), so selecting this location helps to reduce the network latency when synchronizing with the on-premises database.

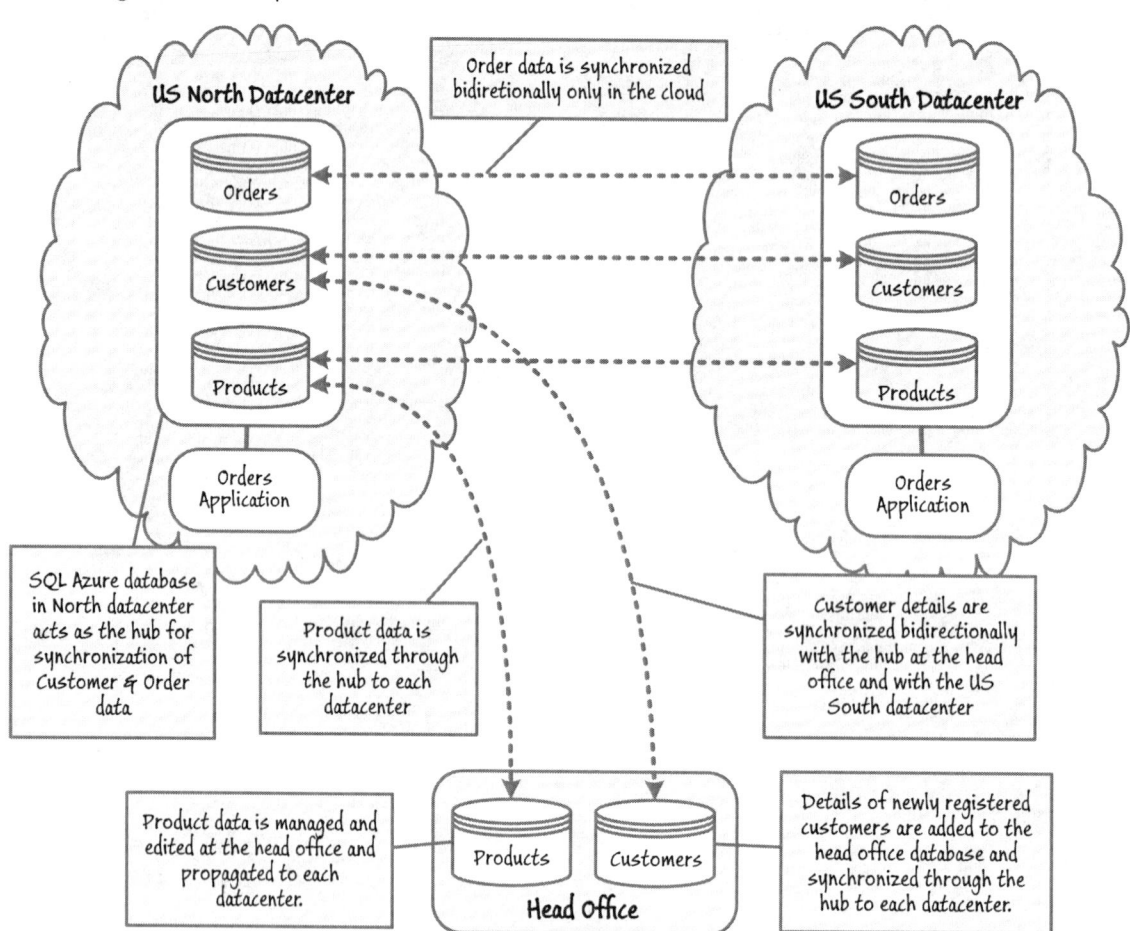

FIGURE 3
Physical implementation of data synchronization for Trey Research

The following table summarizes the configuration of each sync group implemented by Trey Research.

Name and Description	Location of Hub Database	Location of Member Databases and Replication Direction	DataSet	Conflict Resolution Policy
ProductsSyncGroup One way synchronization of product information from on-premises to the cloud	US North Data Center	Head Office Sync to the Hub US South Data Center Sync from the Hub	Product table (every column required to place an order)	Hub Wins
OrdersSyncGroup Bidirectional synchronization of order information between datacenters in the cloud	US North Data Center	US South Data Center Bidirectional	Order, OrderDetail, and OrderStatus tables (every column in each table)	Hub Wins
CustomersSyncGroup Bidirectional synchronization of customer information between on-premises and the cloud	US North Data Center	Head Office Bidirectional US South Data Center Bidirectional	Customer and ACSIdentity tables (every column in each table)	Hub Wins

*The information for each customer spans two tables; the **Customer** table which contains the public information about a customer, and the **ACSIdentity** table which contains the ACS tokens that identify each customer. This is required because, if more than one ACS instance is used to authenticate customers, some identity providers will return a different ACS user identifier from each instance. Therefore, the database must be able to associate more than one user ACS identifier with each customer record.*

*However, the sample solution provided with this guide only implements a single ACS identity for each user, stored in the **UserName** column in the **Customer** table in the TreyResearch database. The sample solution does not include the **ACSIdentity** table because ACS identifiers are stored in the **UserName** column in the **Customer** table.*

See Chapter 3, "Authenticating Users in the Orders Application," for more information about how Trey Research uses ACS to authenticate users.

Implementing a Reporting Solution for Cloud-Hosted Data

In some cases, moving functionality to the cloud will preclude you from using existing services. For example, in the original on-premises version of the Orders application, Trey Research used SQL Server to store all corporate data, and generated business intelligence reports by using SQL Server Reporting Services. However, when Trey Research moved the source data (the Orders database) to SQL Azure, the designers needed to consider whether running a reporting system on the on-premises network was still feasible. It is possible to run an on-premises reporting system and connect to a SQL Azure database over the Internet, but this approach is likely to consume large amounts of bandwidth and provide poor performance.

Choosing a Reporting Solution

The designers at Trey Research decided that they would need to find a better solution for creating business intelligence reports from the cloud-hosted Orders database. The following sections of this chapter describe the options for business intelligence reporting that Trey Research considered.

SQL Server Reporting Services

SQL Server Reporting Services is a feature of SQL Server that allows you to create attractive and comprehensive business intelligence reports from data stored in the database tables, or from a variety of other data sources. If you use SQL Server to store corporate information, SQL Server Reporting Services provides an easy-to-use solution for creating reports.

SQL Server Reporting Services can read data from relational, multidimensional, XML, and custom data sources, and generate reports in a variety of formats including web-oriented, page-oriented, and desktop application formats. It also supports publishing reports directly to a variety of targets, including a report server, SharePoint® services, file shares, internal archive stores, and Office applications.

SQL Server Reporting Services must connect to the data source to analyze the data. If the data source is remote to the application that uses the report, the process may generate considerable network traffic and may require a service that exposes the data for SQL Server Reporting Services to access. This may impact the security of the data source if it is located remotely from the application that uses the report.

SQL Azure Reporting Service

The SQL Azure Reporting Service is a service exposed by Windows Azure that can generate a range of customized business intelligence reports in common web and Microsoft Office-ready formats. If you use SQL Azure to store corporate information, the SQL Azure Reporting Service provides an easy-to-use solution for creating reports.

The SQL Azure Reporting Service runs in the same datacenter as the SQL Azure database, and so network traffic between SQL Azure Reporting Service and the application that displays the report is minimized. This means that it is likely to provide much faster report generation and better performance than connecting from an on-premises reporting application to the SQL Azure database.

> Wide-ranging and up to date information is vital in all companies for managing investment, planning resource usage, and monitoring business performance. The SQL Azure Reporting Service extends these capabilities to data hosted in SQL Azure.

However, the SQL Azure Reporting Service is a chargeable service and so incurs subscription costs, though the consequential reduction in data transfer costs to a reporting system located on-premises can help to compensate for this. It also avoids the high initial cost of an on-premises reporting system for organizations that do not already have such a system. Also consider that the SQL Azure Reporting Service offers lower interactivity and reduced capabilities compared to SQL Server Reporting Services and other high-end reporting solutions, although the variety of report formats is usually sufficient for the vast majority of requirements.

A Custom or Third Party Reporting Solution

It is possible to use any reporting package or create your own custom data analysis and reporting solution for a hybrid application provided that the solution you choose can access the data stored in your application's database or repository. Some organizations may require a custom solution that integrates with other applications and services, or wish to continue using an existing custom or third party application to create business intelligence reports.

Custom or third party reporting solutions may be closely tailored to the organization's reporting requirements, and this focus on specific areas of interest can provide faster report generation and overall performance compared to more generic solutions. Specially tailored third party solutions may be more appropriate for specialist types of applications, and may cost less than a more generalized solution.

Using an existing reporting solution can reduce the cost of migrating an application to the cloud. However, the reporting solution must be able to connect to the cloud-based data source without compromising the security of the data; this may require developers to create additional services to expose data. In addition, when the data source is remote to the application that uses the report, the process may generate considerable network traffic. Finally, an existing or third party solution may not offer the required variety of formats, or equivalent functionality, compared to SQL Server Reporting Services or the SQL Azure Reporting Service.

How Trey Research Chose the Reporting Solution

When Trey Research moved the Orders application to the cloud, the designers chose to locate the data generated by the application when customers place orders in SQL Azure. Before the move, Trey Research used SQL Server Reporting Services to generate business information from the Orders database.

The designers at Trey Research realized that one solution when the source data is located remotely would be to download all the orders data to an on-premises database, and continue to use SQL Server Reporting Services to analyze it. However, unless data synchronization occurs on a scheduled basis, which will incur additional cost, the data transfer operation will result in longer waiting times while reports are generated. This approach could also cause considerable network traffic over the Internet as data would be repeatedly queried to build the reports.

Instead, after moving the data to SQL Azure, it made more sense to adopt the business intelligence capabilities of the SQL Azure Reporting Service. This approach minimizes network traffic over the Internet, ensures that the most recent data is included in the reports without incurring additional delays, and can still generate the reporting information in a variety of formats.

How Trey Research Uses the SQL Azure Reporting Service

An on-premises reporting service application uses SQL Azure Reporting Service to collate the reports that the management applications running on-premises use. Trey Research extended the usefulness of the reports by combining the raw data available from SQL Azure Reporting with data streams exposed through Windows Azure Marketplace.

Figure 4 shows the architecture Trey Research implemented.

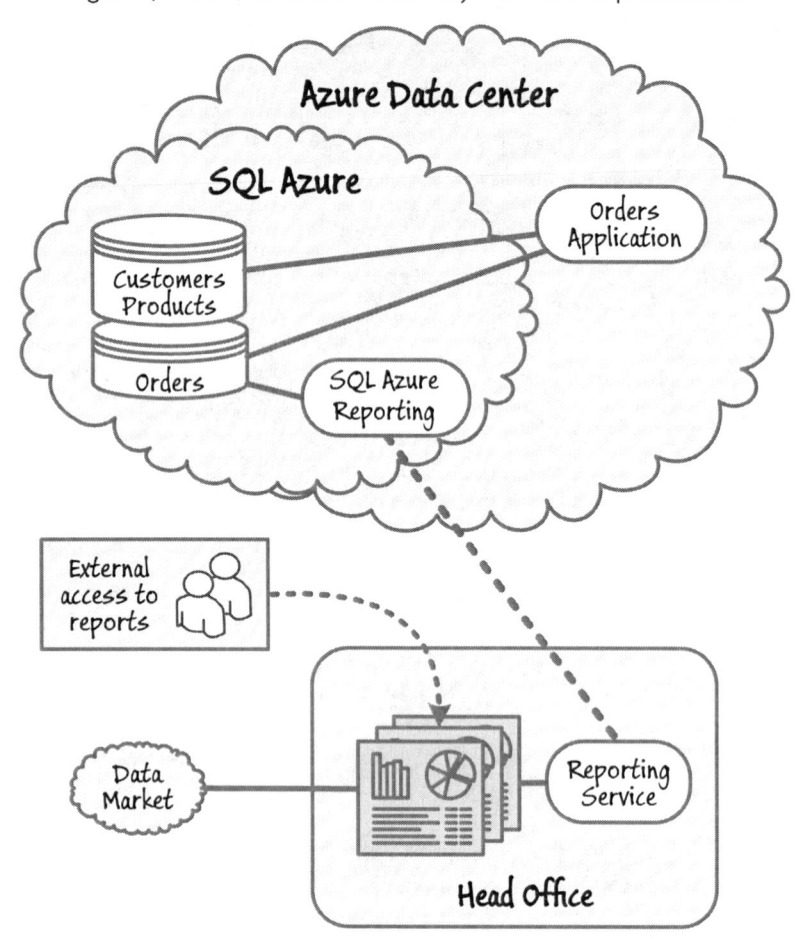

Figure 4
Creating business intelligence reports with SQL Azure Reporting Service

> *For simplicity of installation, the example application for this guide does not include an implementation of the SQL Azure Reporting Service. For more information about the SQL Azure Reporting Service, see "Business Analytics." For more information about incorporating external data feeds into your reports, see "One-Stop Shop for Premium Data and Applications."*

How Trey Research Makes Reporting Data Available to External Partners

Trey Research also decided to expose the report data to specific users who access the on-premises Reporting Service over the Internet through Service Bus Relay.

Trey Research chose to implement a trial version of this functionality while the organization gathered more details on the type of information that external partners should be able to view without intruding on the commercial confidentiality of the business. Therefore, in this first version, Trey Research simply published the total value of goods sold, broken down by reporting quarter or by region, as defined by the service contract available in the IOrdersStatistics.cs file in the Services folder of the **HeadOffice** project:

```C#
[ServiceContract]
public interface IOrdersStatistics
{
  [OperationContract]
  double SalesByQuarter(int quarter);

  [OperationContract]
  double SalesByRegion(string region);
}
```

Subsequent versions of this service may provide a more detailed breakdown of the sales data.

The HeadOffice application, which runs on-premises within Trey Research, hosts a WCF service called **OrderStatistics** that implements the **SalesByQuarter** and **SalesByRegion** operations. These operations simply retrieve the requested data from the underlying database and pass it back as the return value. The implementation of this service is available in the OrdersStatistics.cs file in the Services folder of the **HeadOffice** project.

The code that initiates the service is located in the **OpenServiceHost** method in the Global.asax.cs file also in the **HeadOffice** project.

> *The technique used by the sample application to start the **OrderStatistics** service is not necessarily applicable to all web applications, and only works in the sample application because the user is expected to explicitly start the web application that hosts the service. In other situations, it may be preferable to utilize the **Auto-Start** feature of Windows Server and IIS to initialize the service. For more information, see the topic "Auto-Start Feature."*

The service was made available to external partners and users through Windows Azure Service Bus Relay, using the service path **Services/RelayedOrdersStatistics**, and the application publishes the service through a TCP endpoint using a **netTcpRelayBinding** binding. The connection to the Service Bus is secured through ACS by using an authentication token; Chapter 3, "Authenticating Users in the Orders Application," provides more information about configuring ACS, and the section "Guidelines for Securing Windows Azure Service Bus Relay" in "Appendix C - Implementing Cross-Boundary Communication" includes guidance for authenticating and authorizing partners connecting to Service Bus Relay. The details of the Service Bus namespace, the service path, the security credentials, and the service configuration are stored in the web.config file of the project. Notice that the web application connects to the Service Bus by using the identity named **headoffice**; this identity is granted the privileges associated with the **Listen** claim, enabling the application to accept incoming requests but preventing it from performing other operations such as sending requests:

```xml
<?xml version="1.0" encoding="utf-8"?>
<configuration>
  ...
  <appSettings>
    ...
    <!-- ServiceBus config-->
    <add key="serviceBusNamespace"
      value="treyresearchscenario" />
    <add key="UriScheme" value="sb" />
    <add key="RelayServicePath"
      value="Services/RelayedOrdersStatistics" />
    ...
  </appSettings>
  ...
  <system.serviceModel>
    <services>
      <service name="HeadOffice.Services.OrdersStatistics">
        <endpoint behaviorConfiguration=
            "SharedSecretBehavior"
          binding="netTcpRelayBinding"
          contract=
            "HeadOffice.Services.IOrdersStatistics"
          name="RelayEndpoint"/>
      </service>
    </services>
    <behaviors>
      <endpointBehaviors>
        <behavior name="SharedSecretBehavior">
          <transportClientEndpointBehavior
            credentialType="SharedSecret">
            <clientCredentials>
              <sharedSecret issuerName="headoffice"
```

```xml
              issuerSecret="<data omitted>" />
            </clientCredentials>
          </transportClientEndpointBehavior>
          <serviceRegistrySettings
            discoveryMode="Public"
            displayName=
              "RelayedOrdersStatistics_Service"/>
        </behavior>
      </endpointBehaviors>
    </behaviors>
    ...
  </system.serviceModel>
</configuration>
```

Trey Research also built a simple command-line application to test the connectivity to the **Order-Statistics** service through Windows Azure Service Bus Relay, and verify that the **SalesByQuarter** and **SalesByRegion** operations function as expected. This application is available in the **ExternalDataAnalyzer** project. It is a WCF client that establishes a connection to the service by using the Service Bus APIs of the Windows Azure SDK together with the Service Model APIs of WCF. The connection to the Service Bus is secured by providing the appropriate authentication token. The endpoint definition for connecting to the service and the security credentials are all defined in the app.config file. Like the web application, the **ExternalDataAnalyzer** project connects to the Service Bus by using a specific identifier, **externaldataanalyzer**, which has been granted the privileges associated with the **Send** claim, enabling it to submit requests to the service but preventing it from performing other tasks such as listening for requests from other clients.

XML
```xml
<?xml version="1.0" encoding="utf-8"?>
<configuration>
  ...
  <system.serviceModel>
    ...
    <behaviors>
      <endpointBehaviors>
        <behavior name="SharedSecretBehavior">
          <transportClientEndpointBehavior
            credentialType="SharedSecret">
            <clientCredentials>
              <sharedSecret issuerName=
                "externaldataanalyzer"
                issuerSecret="<data omitted>" />
            </clientCredentials>
          </transportClientEndpointBehavior>
        </behavior>
      </endpointBehaviors>
    </behaviors>
```

```
    ...
    </system.serviceModel>
</configuration>
```

Figure 5 summarizes the structure and implementation details of the OrderStatistics service and the ExternalDataAnalyzer client application.

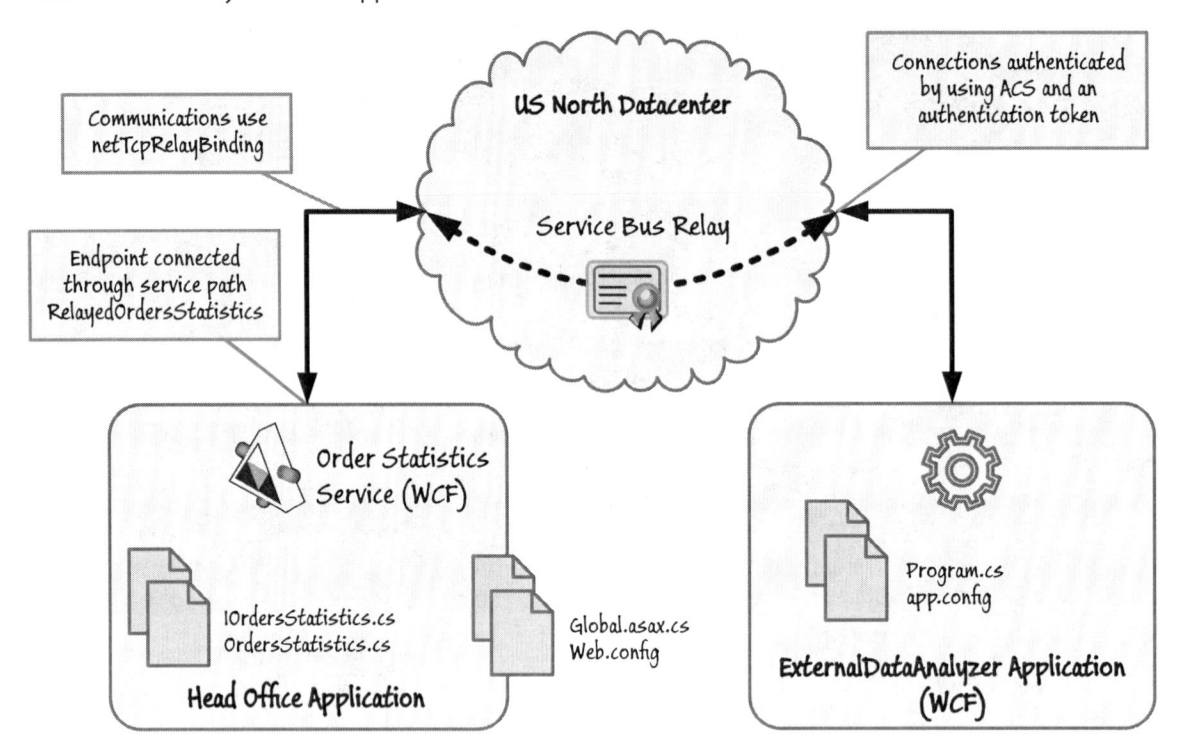

Figure 5
Structure of the **OrderStatistics** service and **ExternalDataAnalyzer** client application

Summary

This chapter concentrated on the deployment scenarios related to building applications where some parts run on-premises, some parts run in the cloud, and some parts are implemented by or for external partners. The topics in this chapter concern deployment challenges that Trey Research needed to tackle, such as locating data in the cloud or on-premises, synchronizing data across the different locations that are part of a hybrid solution, and generating business intelligence reports.

Because Trey Research used SQL Server in the original Orders application, deploying data to SQL Azure and using SQL Azure Data Sync to maintain consistency and replicate master data is a simple and natural way to migrate the application to the cloud.

Finally, having chosen to use SQL Azure to store data in the cloud, the SQL Azure Reporting Service is the obvious choice for implementing the business intelligence and reporting solution Trey Research requires, while the Service Bus Relay provides the ideal mechanism for publishing reporting data to partner organizations.

More Information

All links in this book are accessible from the book's online bibliography available at: *http://msdn.microsoft.com/en-us/library/hh968447.aspx*.

- Windows Azure features tour at *http://www.windowsazure.com/en-us/home/tour/overview/*.
- "Windows Azure Developer Center" at *http://www.windowsazure.com/en-us/develop/overview/*.
- "SQL Azure Data Sync Overview" at *http://social.technet.microsoft.com/wiki/contents/articles/sql-azure-data-sync-overview.aspx*.
- "SQL Azure Data Sync FAQ" at *http://social.technet.microsoft.com/wiki/contents/articles/sql-azure-data-sync-faq.aspx*.
- "SQL Azure Data Sync- Synchronize Data across On-Premise and Cloud (E2C)" at *http://channel9.msdn.com/Series/SQL-Azure-Data-Sync/SQL-Azure-Data-Sync-Synchronize-Data-across-On-Premise-and-Cloud-E2C*.
- "SQL Server to SQL Azure Synchronization using Sync Framework 2.1" at *http://blogs.msdn.com/b/sync/archive/2010/08/31/sql-server-to-sql-azure-synchronization-using-sync-framework-2-1.aspx*.
- "Business Analytics" (SQL Azure Reporting) at *http://www.windowsazure.com/en-us/home/tour/business-analytics/*.
- "One-Stop Shop for Premium Data and Applications" at *http://datamarket.azure.com/*.
- "Windows Azure AppFabric: An Introduction to Service Bus Relay" at *http://www.microsoft.com/en-us/showcase/details.aspx?uuid=395930db-6622-4a9f-8152-e0cb1fc5149c*.

3 Authenticating Users in the Orders Application

This chapter explores how Trey Research adapted the authentication and authorization implementation in the Orders application when they migrated parts of the application to the cloud.

The original on-premises Orders application used ASP.NET Forms authentication. All visitors to the site were required to log in using credentials allocated to them by Trey Research employees. These credentials were stored in a SQL Server® database.

When the designers at Trey Research were considering how to adapt the application to run in the cloud, they realized that they could take advantage of the Windows Azure™ technology platform Access Control Service (ACS) to implement a more flexible and user-friendly authentication approach for visitors, based on claims issued by trusted identity providers.

Scenario and Context

Most corporate applications and websites require secure and reliable authentication. Users establish their identity by providing a secret that only they and the authentication mechanism know. Typically this is a combination of a username and password, but it may instead be done using a smartcard, a biometric technique such as a fingerprint or iris scan, or some other approach. The important factor is that application designers must choose a mechanism that can uniquely identify each user, even if this identification consists only of establishing a unique user ID.

ASP.NET Forms authentication is a natural approach for authenticating visitors to a website created using ASP.NET, and for authorizing each visitor's actions as they access the website. It is flexible and easy to implement, and provides mechanisms for users to manage their credentials (such as changing their password). It was for these reasons that the designers at Trey Research originally used ASP.NET Forms authentication in the on-premises version of the Orders application.

> Forms authentication is the natural approach for authentication in ASP.NET applications, but it does not offer the flexibility of a claim-based approach for federated identity and single sign-on.

However, ASP.NET Forms authentication no longer meets the expectations of visitors who are becoming more used to authentication mechanisms that allow them manage their credentials themselves, and that enable them to use the same credentials to access multiple websites. For example, users that have a Windows Live® ID can set up and manage their account at Windows Live, and then use the same credentials for accessing websites such as the Hotmail® web-based email service, the MSDN® developer program, and other Microsoft and third-party websites.

In this situation, Windows Live acts as an identity provider by authenticating visitors and then confirming their identity to websites that trust Windows Live. This mechanism is usually referred to as *federated authentication*. It also supports the capability for single sign-on, where users sign on once and do not need to re-enter their credentials when accessing another website.

> *For more information about federated authentication, claims-based authentication, identity providers, and single sign-on see "Appendix B - Authenticating Users and Authorizing Requests" of this guide.*

The designers at Trey Research wanted to offer visitors the flexibility of using federated authentication, and decided to examine the options available for implementing it in the Orders application. In addition, they wanted to simplify access to the application for both Trey Research employees and employees at partner organizations (such as transport partners).

Figure 1 shows the existing authentication architecture for the on-premises Orders application. In this version of the application, customers are authenticated using ASP.NET Forms authentication (shown at **1** in the diagram) and are granted access the Orders application if their web browser presents the correct ASP.NET authentication token to the application (**2** in the diagram). Employees of partner organizations, such as transport partners, are authenticated in the same way (**3** and **4** in the diagram). Employees in the Trey Research head office use credentials specific to the Orders application to sign into the Orders application using the same ASP.NET Forms authentication mechanism used by customers and partner employees (**5** and **6** in the diagram). However, Trey Research employees also use internal applications, such as accounting and manufacturing applications, and they must be authenticated for these applications using different credentials that are stored in the on-premises Windows Active Directory® (**7** and **8** in the diagram).

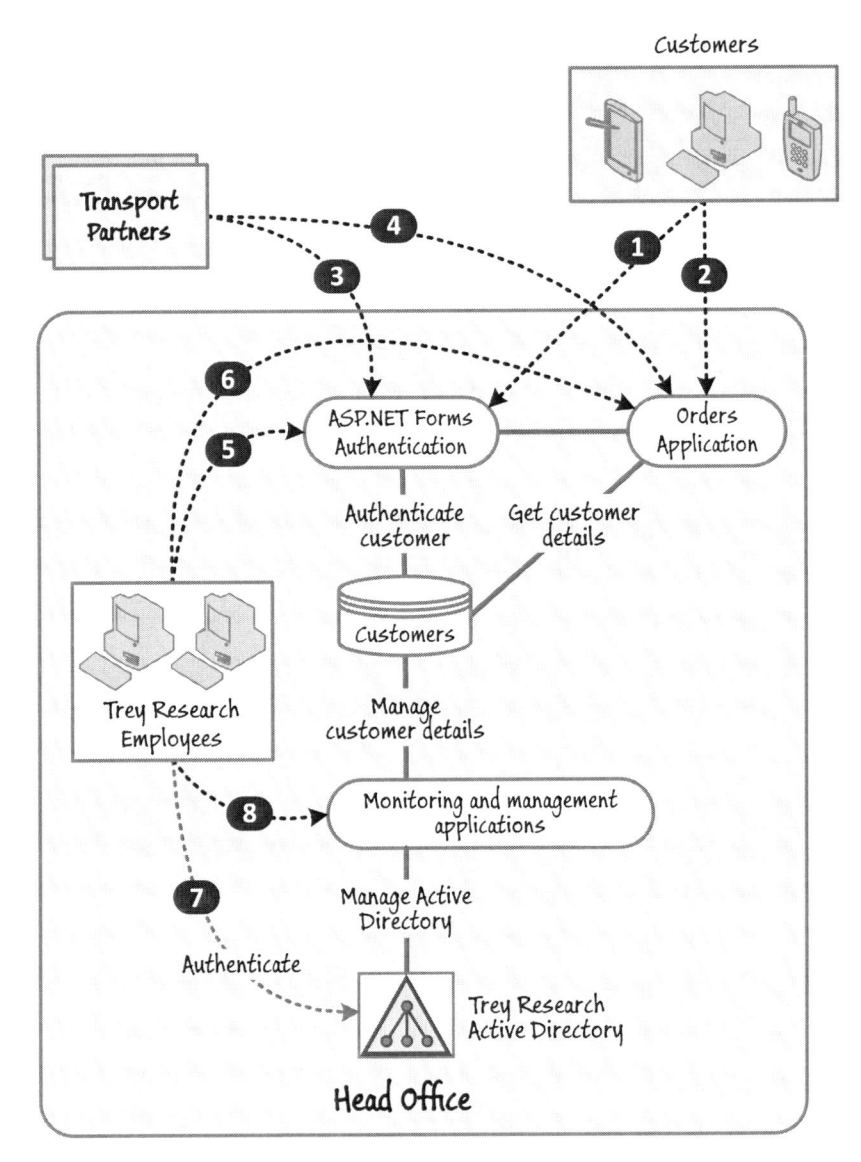

FIGURE 1
The original on-premises authentication mechanism at Trey Research

The following sections of this chapter describe how the designers at Trey Research evaluated and chose a more flexible authentication and authorization approach for visitors, partners, and Trey Research employees, and how they implemented this approach in the Orders application. The final section of this chapter discusses how ACS can also act as the authentication mechanism for Windows Azure Service Bus.

Authenticating Visitors to the Orders Application

The designers at Trey Research were aware of the growing use of federated identity on the Internet, and wanted to evaluate the options available for authentication in modern websites and web applications. The following sections of this chapter describe how they carried out this evaluation.

Choosing an Authentication Technique

The designers at Trey Research assessed common authentication techniques for the Orders application by considering the advantages and limitations of using claims-based authentication compared to ASP.NET Forms-based authentication, in conjunction with incorporating their existing user repository implemented by Windows Active Directory.

ASP.NET Forms Authentication

Forms authentication is a built-in feature of ASP.NET, and is quick and easy to implement in an ASP.NET website. There is a range of server controls that can be inserted into web pages to enable users to manage their account credentials, apply for an account, and sign in and sign out. The Orders application already used Forms authentication, and so using this same mechanism in the hybrid version of the application would not require any changes to the code.

Using ASP.NET Forms authentication would require the database containing the user's credentials to be available to the website, though having already decided to deploy the Customers data to SQL Azure™ technology platform this would not be an issue.

However, Trey Research had already decided that it wanted to offer more flexible options for visitors, Trey Research employees, and partners by allowing them to access the Orders website using their existing credentials rather than having to register new ones. Using ASP.NET Forms authentication means that users must have a separate account just for the Orders application.

Claims-Based Authentication with Microsoft Active Directory Federation Service

Claims-based authentication allows visitors to use an existing account that they have already established with an identity provider that Trey Research trusts. This approach requires the designers at Trey Research to decide which identity provider(s) they will trust. One option is to use Microsoft Active Directory Federation Service (ADFS) in conjunction with their existing Windows Active Directory. ADFS allows users to sign in over the Internet using credentials stored in Active Directory; they do not need to connect directly to the Active Directory domain network.

> ADFS acts as a Security Token Service (STS) that can authenticate users against Windows Active Directory and issue security tokens containing claims about the user. These claims may be just an authenticated user ID, or there may be a larger set of claims in the token that specify more information about the user such as their name, roles, email address, and more.

However, using ADFS would mean that Trey Research employees would need to create accounts for all visitors within their Windows Active Directory domain or in a separate Windows Active Directory domain dedicated to customers, which increases administrative requirements and may have unwelcome security implications. It would be an acceptable approach for authenticating a small and well-defined subset of users, such as employees at Trey Research and at partner organizations, but is not a practical approach for authenticating all visitors.

Claims-Based Authentication with Windows Azure Access Control Service

Trey Research can extend the capabilities for authenticating visitors by using ACS. This service provides a mechanism for visitors to sign in using their existing account credentials from a range of well-known social networks and identity providers.

The designers at Trey Research realized that using ACS would give visitors the flexibility required to use their existing identity and credentials, with a corresponding improvement in users' authentication experience. It also has the advantage that Trey Research would not need to manage the users' credentials, such as providing the facility to change a password. All features of identity management are the responsibility of the identity provider. Trey Research just needed to decide which identity providers it would trust to establish a user's identity.

However, there are some issues that Trey Research had to address when adopting this approach. Social identity providers typically return only claims containing a unique user identifier, and perhaps a user name or an email address, and so the application code must associate this identifier with a registered user stored in the Customers database table. In addition, Trey Research may need to consider how to migrate existing user accounts to use the claims-based approach. Using claims-based authentication will also require the developers at Trey Research to modify the application code to use the claims returned from ACS.

Trey Research also realized that using ACS with social identity providers does not provide a single sign-on solution for Trey Research employees and employees at Trey Research's partners. Both of these sets of employees sign into applications in their respective corporate networks using an account defined in their company's Active Directory (or other corporate credentials repository). However, they will be required to sign into the Orders application using an account defined by a social identity provider.

> ACS is also a STS that issues security tokens containing claims about the user, but it can authenticate users against a range of social identity providers such as Windows Live ID, Google, Facebook, and OpenID. ACS can also act as an identity provider; a feature that is used for authenticating access to Windows Azure Service Bus queues and topics.

Claims-Based Authentication with ACS and ADFS

The designers at Trey Research recognized that they could resolve the challenges encountered with the options presented in the previous sections by combining ACS with ADFS authentication techniques. Together ACS and ADFS can act as identity providers and token issuers (STSs) for accounts defined at social identity providers such as Windows Live ID, Google, Facebook, and OpenID, and accounts defined within Windows Active Directory.

All this requires is for ACS to be configured to trust the ADFS instances exposed by Trey Research and by partner organizations that wish to take part in the federated identity authentication realm. Users are directed to ACS for authentication. ACS sends a request to authenticate each user to the appropriate identity provider. This may be one of the configured social identity providers, or it may be one of the configured ADFS instances. When an ADFS instance receives a request for authentication, it looks up the identity in its local Windows Active Directory. After confirming the identity through one of the identity providers or ADFS instances, ACS returns a token containing any discovered claims to the application.

By combining ACS and ADFS, Trey Research enables visitors to the Orders application to sign in using a social identity or with an identity defined in one of the corporate Active Directory domains. Customers can use Windows Live ID, Google, Facebook, OpenID, and more; Trey Research employees can sign in with their corporate credentials defined within the Trey Research domain; and employees at partner organizations can sign in using corporate credentials defined within their own domain.

This means that Trey Research are no longer required to maintain authentication details for visitors or partner employees, and the details for Trey Research's own employee accounts are already managed by the network administrator in Active Directory. Consequently, there is no requirement for additional management of employee accounts.

The only real issues are the additional complexity of configuring ACS and ADFS to support this federated identity approach, and the loss of absolute control over the list of permitted users. For example, the administrator at partner companies will maintain the list of their employees that have access to the Orders application. While freed of the effort and responsibility for this task, Trey Research loses the ability to control which users do have access, and must trust the partner organization to grant only the appropriate privileges. This may be an issue if the permissions available to users depend on roles specified in the claims returned from the identity provider, such as "Manager" or "Administrator."

> Combining authentication through social identities and corporate directory mechanisms by using ACS offers flexibility and allows the application to support authentication for a broad set of users. It also makes it possible to use role-based claims to authorize user actions because role information is typically available from corporate directories, even though it is usually not available from social identity providers.

Combined Forms and Claims-Based Authentication

One final approach Trey Research considered was combining both a claims-based approach, as discussed above, with ASP.NET Forms authentication so that users can continue to use their existing Forms authentication credentials. This approach would mean administrators at Trey Research would not be required to migrate existing users to the claims-based approach. New and existing users would be able to use the application, with new users being provided with claims-based authentication credentials, and existing users could be encouraged to change over to using claims-based authentication credentials.

However, this means that developers and administrators at Trey Research would need to manage and maintain two incompatible systems. It would also be more difficult to perform common tasks such as generating a list of customers or managing the customer details held in the database. In addition, developers would still need to modify the application code to support both authentication techniques.

How Trey Research Chose an Authentication Technique

After considering all of the options, Trey Research decided to adopt a claims-based federated authentication approach based on combining ACS and ADFS. This approach offers the maximum flexibility for new customers, and supports single sign-on for all visitors; including Trey Research employees and employees at partner organizations.

The additional work and cost involved in also supporting Forms authentication was not considered to be worthwhile, however, and so Trey Research needed to decide how to manage the accounts of existing customers and employees. The developers discovered that they could add code to the application so that users signing in with a social or federated identity could be connected to an existing user account. This saves Trey Research from migrating the existing accounts from Forms authentication.

It also has the added benefit that customers and employees can have more than one identity linked to their registered account in the Customers table, and can therefore access the application when signed in to any of these identity providers. This offers an even better single sign-on experience for visitors. This implementation is described in more detail in the section "Handling Multiple User IDs," later in this chapter.

How Trey Research Uses ACS and ADFS to Authenticate Visitors

Trey Research authenticates visitors to the Orders website through claims-based authentication using ACS as the STS, three social identity providers (Windows Live ID, Yahoo!, and Google), and ADFS instances exposed by Trey Research itself and by its main partner organizations. Figure 2 shows a schematic view of the overall authentication cycle for different types of users and different identity providers.

46 CHAPTER THREE

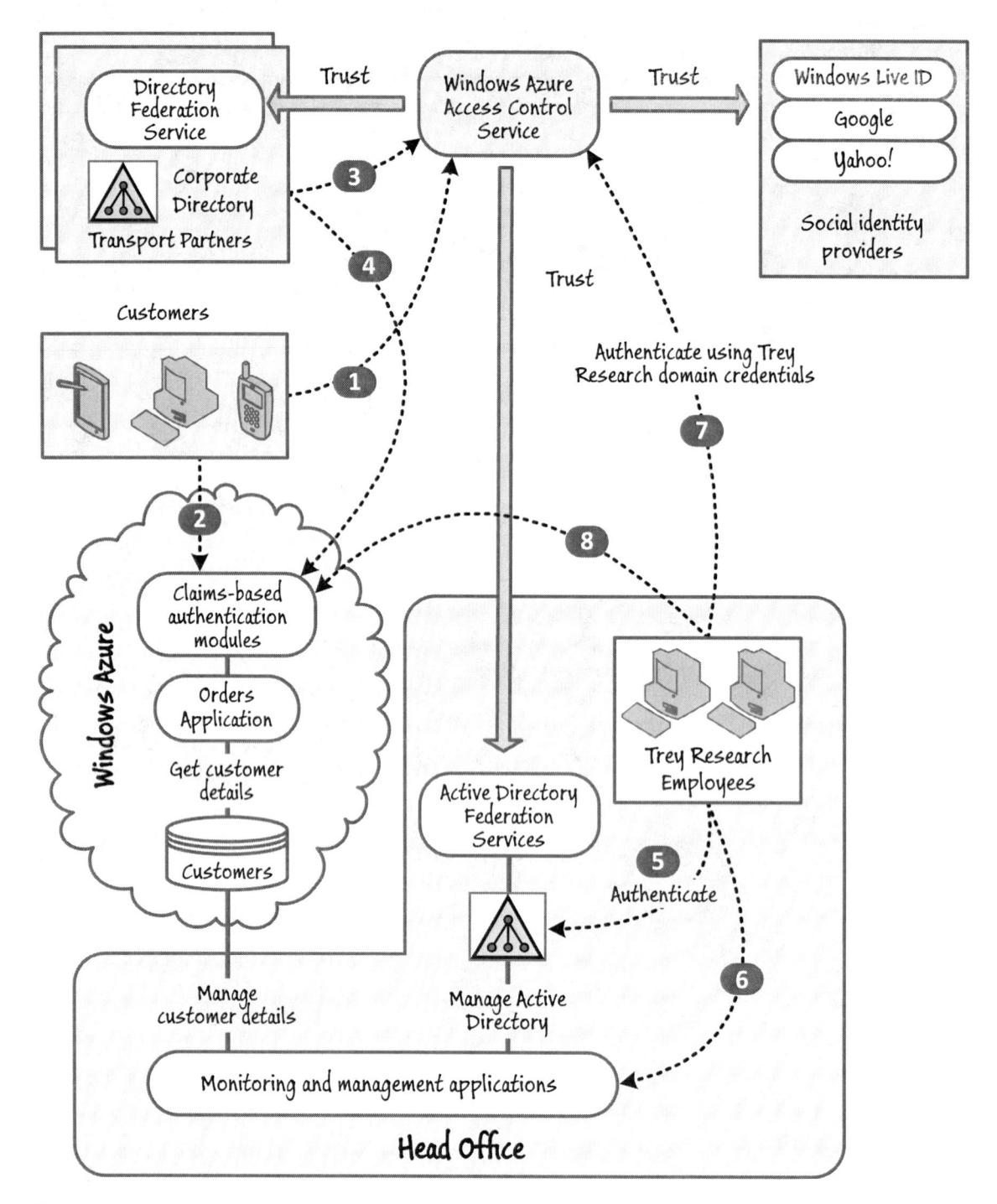

Figure 2
The authentication architecture and sequence in the hybrid Orders application

In this architecture, customers authenticate through ACS with their chosen social identity provider (shown at **1** in the diagram). These identity providers are configured in ACS, and are trusted by ACS. The claims-based authentication modules in the Orders application pipeline validate the token returned from ACS that the customer sends to the Orders application (**2** in the diagram). The modules extract the user identifier claim, and the Orders application can then look up the corresponding customer details in the Customers database table.

Employees at Trey Research's partners authenticate using ACS (**3** in the diagram), but—because ACS is configured to trust the partner's directory federation service—these employees can sign into the Orders application (**4** in the diagram) using credentials stored in their own corporate directory.

Finally, Trey Research's employees are authenticated by Trey Research's own Active Directory over the corporate network when they sign into their computer (**5** in the diagram) and can then access Trey Research's internal applications (**6** in the diagram). When these employees access the Orders application, ACS authenticates their current credentials against Active Directory through ADFS (**7** in the diagram), and they are issued with a suitable token containing claims that the claims-based authentication modules in the Orders application pipeline can use to identify them (**8** in the diagram).

The following sections of this guide explore in more detail how Trey Research implemented this architecture in the Orders application.

Access Control Service Configuration

Trey Research uses a setup program (in the **TreyResearch.Setup** project of the example) to configure ACS without needing to use the Windows Azure web portal. The setup program uses the classes in the **ACS.ServiceManagementWrapper** project to access and configure ACS.

> For more information about the **ACS.ServiceManagementWrapper**, see "Access Control Service Samples and Documentation."

The following table shows how Trey Research configured ACS for authenticating visitors to the Orders application.

ACS artifact	Setting
Windows Live ID identity provider	Default rule group that contains rules to pass the visitor's ID through as both the **NameIdentifier** and the **UserName** (Windows Live ID does not reveal visitor's email addresses).
Google identity provider	Default rule group that contains a rule to pass the visitor's ID through as the **NameIdentifier** and a rule to pass the visitor's email address through as the **UserName**.
Yahoo! Identity provider	Default rule group that contains a rule to pass the visitor's ID through as the **NameIdentifier** and a rule to pass the visitor's email address through as the **UserName**.
Trey Research ADFS Identity provider	Default rule group that contains a rule to pass the visitor's user ID through as the **NameIdentifier**, a rule to pass the visitor's email address through as the **UserName**, and a rule to pass the visitor's account groups as the **Roles** claim.
Partner ADFS identity providers	Default rule group that contains a rule to pass the visitor's user ID through as the **NameIdentifier**, a rule to pass the visitor's email address through as the **UserName**, and a rule to pass the visitor's account groups as the **Roles** claim.

Handling Multiple User IDs

Typically, each registered user of the Orders application has a single unique user identifier, which links that customer to the corresponding customer details stored in the database. This works well when there is only a single authentication source, such as when using ASP.NET Forms authentication. However, when authentication takes places in multiple realms (such as different identity providers, each of which authenticates users within a specific realm such as live.com or google.com) the same customer may have more than one unique identifier.

This situation also arises when using more than one instance of ACS. At the moment, Trey Research uses only one ACS namespace configured in the US North datacenter. However, it may decide in the future to configure additional ACS namespace instances in other datacenters as it expands the deployment of the Orders application to other national or international datacenters. Each instance of ACS may, to minimize user privacy concerns, return a *different* unique user identifier from each instance for the same customer.

This means that applications must be designed in such a way that it is possible to link more than one user identifier claim value (received from an identity provider or STS) to a single registered customer. Trey Research accomplishes this by using two tables in the database. The main Customer table contains a row for each of the registered customers, with a unique customer identifier set by the accounts team at Trey Research as the key. The ACSIdentity child table contains a row for each unique identifier claim value for each customer, linked to the relevant customer through the customer key value.

An important advantage of this approach is that it makes it possible to migrate users from one authentication mechanism, to another. For example, when Trey Research changed from using ASP.NET Forms authentication to claims-based authentication, the developers could have provided a transition mechanism for existing customers. When a customer first authenticates using a social identity provider or ADFS (when the unique user identifier in the claims is not already in the ACSIdentity table), Trey Research could provide a page that allowed the user to also authenticate through ASP.NET Forms authentication, and then link the new claims-based identity to their existing customer details.

The example application available with this guide does not fully implement the authentication architecture described here. This is done to remove the requirement to configure a suitable ADFS and Windows Active Directory instance. The example application is configured to use only one instance of ACS and three social identity providers, and does not use a separate ACSIdentity table. In addition, it does not use roles to authorize user actions. For information about using role claims to authorize visitors, see "Federated Identity for Web Applications" in "A Guide to Claims-Based Identity and Access Control (2nd Edition)."

Authentication Implementation

Figure 3 shows a high-level view of the services and classes used for authentication and authorizatio in the Orders application. You'll see details of the classes identified in the schematic in the following sections of this chapter.

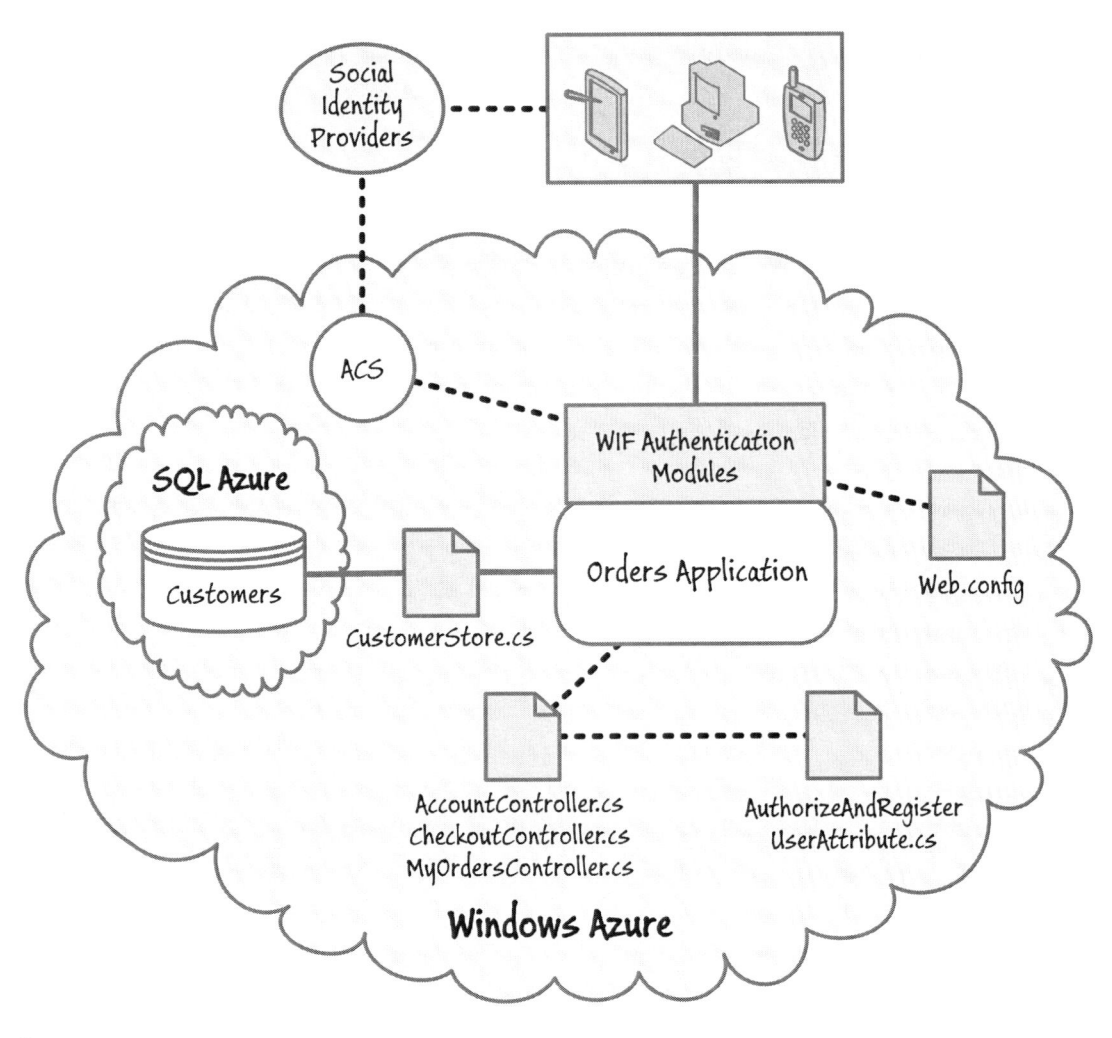

Figure 3
Overview of visitor authentication and authorization in the hybrid Orders application

Authentication with Windows Identity Foundation

Trey Research uses Windows Identity Foundation (WIF) to check for the presence of a valid token that contains claims when each visitor accesses the website. The following extracts from the Web.config file in the **Orders.Website** project show the relevant settings.

```xml
<configSections>
  <section name="microsoft.identityModel" type="..." />
  ...
</configSections>
...
<system.web>
  ...
  <authorization>
    <allow users="*" />
  </authorization>
  <authentication mode="Forms">
    <forms loginUrl="~/Account/LogOn" timeout="2880" />
  </authentication>
  ...
  <httpModules>
    <add name="WSFederationAuthenticationModule"
         type="..." />
    <add name="SessionAuthenticationModule"
         type="..." />
  </httpModules>
</system.web>
...
<microsoft.identityModel>
  <service>
    <audienceUris>
      <add value="https://127.0.0.1" />
    </audienceUris>
    <federatedAuthentication>
      <wsFederation passiveRedirectEnabled="true"
          issuer="https://treyresearch.accesscontrol.
                  windows.net/v2/wsfederation"
          realm="https://127.0.0.1" requireHttps="true" />
      <cookieHandler requireSsl="true" />
    </federatedAuthentication>
    <serviceCertificate>
      <certificateReference x509FindType="FindByThumbprint"
                            findValue="..." />
    </serviceCertificate>
    <applicationService>
      <claimTypeRequired>
        <claimType
            type="http://schemas.xmlsoap.org/ws/2005/05/
                  identity/claims/name" optional="true" />
      </claimTypeRequired>
```

```xml
    </applicationService>
    <issuerNameRegistry
       type="Microsoft.IdentityModel.Tokens.
            ConfigurationBasedIssuerNameRegistry, ...">
      <trustedIssuers>
        <add thumbprint="..."
             name="https://treyresearch.accesscontrol.
                   windows.net/" />
      </trustedIssuers>
    </issuerNameRegistry>
    <certificateValidation
       certificateValidationMode="None" />
  </service>
</microsoft.identityModel>
```

These settings insert the two WIF modules **WSFederation-AuthenticationModule** and **SessionAuthenticationModule** into the HTTP pipeline so that they are executed for each request. The settings in the **microsoft.identityModel** section specify that the modules will redirect requests that require authentication to https://treyresearch.accesscontrol.windows.net, which is the namespace Trey Research configured in ACS. The version of the Web.config file you see here is used when the application is running in the local compute emulator, so the audience URI and realm is the local computer.

The **applicationService** section shows that the Orders website accepts a **Name** claim in the token presented to the WIF modules, though this is optional. The **trustedIssuers** section specifies that the application trusts ACS, and specifies the thumbprint of the certificate that the WIF modules can use to validate a token sent by the visitor.

When the WIF modules detect a request from a visitor that must be authenticated they first check for a valid token from the trusted issuer in the request. If one is not present, the modules redirect the request to ACS. If a valid token is present, the modules extract the claims and make them available to the application code so that it can use the claims to authorize user actions.

The class named **IdentityExtensions** in the **Orders.Website.Helpers** project contains two methods that Trey Research uses to extract the values of claims. The **GetFederatedUsername** method extracts the value of the **IdentityProvider** claim (the name of the original claim issuer for this visitor, such as Windows Live ID or Google) and the **Name** claim, and then concatenates them to create a federated user name. The **GetOriginalIssuer** method simply returns the name of the original claim issuer for this visitor.

> When using WIF to authenticate visitors you must edit the values in the Web.config file for the audience URI and relying party realm if you change the URL of your application, such as when deploying it to the cloud after testing in the Local Compute environment.

```csharp
// C#
private const string ClaimType =
    "http://schemas.microsoft.com/accesscontrolservice"
    + "/2010/07/claims/identityprovider";

public static string GetFederatedUsername(this
                    IClaimsIdentity identity)
{
  var originalIssuer = identity.Claims.Single(
              c => c.ClaimType == ClaimType).Value;
  var userName = string.Format(
        CultureInfo.InvariantCulture, "{0}-{1}",
        originalIssuer, identity.Name);
  return userName;
}

public static string GetOriginalIssuer(this
                    IClaimsIdentity identity)
{
  return identity.Claims.Single(
              c => c.ClaimType == ClaimType).Value;
}
```

ASP.NET Request Validation

By default, ASP.NET checks for dangerous content in all values submitted with requests. This includes HTML and XML elements. Trey Research uses a custom class that allows requests to contain security tokens in XML format, but still protects the site from other dangerous input. This is a better alternative to turning off ASP.NET request validation altogether.

The custom class is named **WsFederationRequestValidator** and is defined in the **Orders.Website** project. The code, shown below, checks if the request is a form post containing a **SignInResponseMessage** result from a WS Federation request. If it is, the code allows the request to be processed by returning **true**. If not, the code allows the standard ASP.NET request validation handler to validate the request content.

> Turning off the default ASP.NET request validation mechanism is not a good idea. Instead, use a custom request validator that allows requests containing XML documents (which hold the user claims) to be processed.

```C#
public class WsFederationRequestValidator
            : RequestValidator
{
  protected override bool IsValidRequestString(
        HttpContext context, string value,
        RequestValidationSource requestValidationSource,
        string collectionKey,
        out int validationFailureIndex)
  {
    validationFailureIndex = 0;
    if (requestValidationSource
      == RequestValidationSource.Form &&
         collectionKey.Equals(
           WSFederationConstants.Parameters.Result,
           StringComparison.Ordinal))
    {
      if (WSFederationMessage.CreateFromFormPost(
        context.Request) as SignInResponseMessage != null)
      {
        return true;
      }
    }
    return base.IsValidRequestString(context, value,
         requestValidationSource, collectionKey,
         out validationFailureIndex);
  }
}
```

Trey Research inserted the custom request validator into the HTTP pipeline by adding it to the Web.config file for the Orders website, as shown here.

```
<system.web>
  <httpRuntime requestValidationType="Orders.Website.
      WsFederationRequestValidator, Orders.Website" />
  ...
</system.web>
```

Visitor Authentication and Authorization

Some pages in the Orders application, such as the Home page and the list of products for sale, do not require visitors to be authenticated; visitors can browse these pages anonymously. However, the pages related to placing an order, managing a visitor's account, and viewing existing orders require the visitor to be authenticated.

Trey Research allows visitors to log on and log off using links at the top of every page. These are defined in the layout page that acts as a master page for the entire site. The layout page, named **_Layout.cshtml** and located in the **Views/Shared** folder of the Orders website, contains the following code and markup.

```cshtml
@if (User.Identity.IsAuthenticated)
{
  <li>@Html.ActionLink("My Orders", "Index",
                       "MyOrders")</li>
  <li>@Html.ActionLink("Log Out", "LogOff",
                       "Account")</li>
}
else
{
  <li>@Html.ActionLink("Log On", "LogOn", "Account")</li>
}
```

The WIF modules automatically populate an instance of a class that implements the **IClaimsIdentity** interface with the claims in the token presented by a visitor, and assign this to the current context's **User.Identity** property. The code can check if a user is authenticated simply by reading the **IsAuthenticated** property, as shown in the code above. You can also use all of the other methods and properties of the **User.Identity**, such as testing if the visitor is in a specific role.

The Custom Logon Page

The **ActionLink** in the **_Layout.cshtml** page that generates the "Log On" hyperlink loads the **AccountController** class (in the Controllers folder of the **Orders.Website** project) and then calls the **LogOn** action method. This behavior is also defined in the Web.config file you saw earlier, specifying the page to display when the user accesses a secured page when not authenticated.

```xml
<authentication mode="Forms">
  <forms loginUrl="~/Account/LogOn" timeout="2880" />
</authentication>
```

The **Logon** method, shown below, creates suitable values for the realm and the return URL so that the visitor will go back to the correct page after logging on. It creates the return URL only if the visitor was redirected here from another page. It uses another method named **GetRealm**, which is defined within the **AccountController** class, to get the current realm in the form **http://[*host*].[*path*]/**. Finally, the **LogOn** method returns the login page view with the customized Trey Research look and feel.

```csharp
[HttpGet]
public ActionResult LogOn()
{
  ViewData["realm"] = this.GetRealm();
  ViewData["returnUrl"] = Request.UrlReferrer != null
    ? Request.UrlReferrer.AbsoluteUri : this.GetRealm();
  ViewData["acsNamespace"] = CloudConfiguration
     .GetConfigurationSetting("acsNamespace", null);
  return View();
}
```

The **Logon** view itself, located in the Views\Account folder of the **Orders.Website** project, generates a custom logon page containing buttons for each of the configured identity providers (such as Windows Live ID, Google, and Yahoo!). Visitors can choose an identity provider to use, and then log in using this provider. They are then returned to the appropriate page of the Orders application.

The custom logon page is created by modifying the standard page that ACS creates. The custom page uses considerable amounts of client-side JavaScript to generate the visible content and to handle user interaction.

Using a Custom Authorization Attribute

Some of the pages in the website, such as the "Checkout" and "My Orders" pages, require visitors to have already been authenticated. To implement this behavior, Trey Research uses a custom attribute named **AuthorizeAndRegisterUser**. For example, the **MyOrdersController** class has this attribute applied to the entire class (rather than to a single method). The following code shows an outline of the **MyOrdersController** class.

```csharp
C#
[AuthorizeAndRegisterUser]
public class MyOrdersController : Controller
{
  ...
}
```

The custom **AuthorizeAndRegisterUserAttribute** class, shown below, is defined in the CustomAttributes folder of the **Orders.Website** project). It extends the standard **AuthorizeAttribute** class that is used to restrict access by callers to an action method.

```csharp
C#
public class AuthorizeAndRegisterUserAttribute
          : AuthorizeAttribute
{
  private readonly ICustomerStore customerStore;

  public AuthorizeAndRegisterUserAttribute()
          : this(new CustomerStore())
  { }

  public AuthorizeAndRegisterUserAttribute(
                  ICustomerStore customerStore)
  {
    this.customerStore = customerStore;
  }

  public override void OnAuthorization(
      System.Web.Mvc.AuthorizationContext filterContext)
  {
```

> Using a custom authentication attribute is a good way to perform other tasks related to authentication, such as obtaining a user's details from a database, without needing to include this code in every class. It also makes it easy to specify which pages or resources require users to be authenticated.

```
...
if (!filterContext.HttpContext.User.
                Identity.IsAuthenticated)
{
  base.OnAuthorization(filterContext);
  return;
}
var federatedUsername =
    ((IClaimsIdentity)filterContext.
    HttpContext.User.Identity).GetFederatedUsername();
var customer = this.customerStore.FindOne(
                                federatedUsername);
if (customer == null)
{
  // Redirect to registration page.
  var redirectInfo = new RouteValueDictionary();
  redirectInfo.Add("controller", "Account");
  redirectInfo.Add("action", "Register");
  redirectInfo.Add("returnUrl",
                  filterContext.HttpContext.Request
                      .Url.AbsolutePath);
  filterContext.Result
      = new RedirectToRouteResult(redirectInfo);
  }
 }
}
```

The **AuthorizeAndRegisterUserAttribute** attribute creates an instance of the **CustomerStore** repository class that is used to access the data for this customer in the SQL Azure **Customers** table (as shown in Figure 3 earlier in this chapter). If the visitor is not authenticated, the attribute passes the request to the base **AuthorizeAttribute** instance which forces the user, though the WIF modules, to authenticate with ACS.

If the visitor is authenticated, the attribute uses the **GetFederatedUsername** method you saw earlier to get the user name from the authentication token claims, and the **FindOne** method of the **CustomerStore** class to retrieve this visitor's details from the **Customers** table. If this visitor has not yet registered, there will be no existing details in the **Customers** table and so the code redirects the visitor to the Register page.

Customer Details Storage and Retrieval

Using claims-based identity with social identity providers usually means that your application must maintain information about each registered user of the application because the token that authenticated users present to the application contains only a minimal set of claims (typically just a user identifier or name). Trey Research maintains this information in the **Customers** database table in SQL Azure. This table contains the details customer provide when they register, and additional information that is maintained for each customer by the head-office administration team (such as the credit limit).

To connect to the database, Trey Research used the Microsoft Entity Framework and created a class named **CustomerStore**, located in the Orders.Website.DataStores folder of the website project. This class exposes several methods for storing and retrieving customer details, including the **FindOne** method used in the previous code extract. This is the code of the **FindOne** method.

```C#
public Customer FindOne(string userName)
{
  ...
  using (var database
         = TreyResearchModelFactory.CreateContext())
  {
    var customer
      = this.sqlCommandRetryPolicy.ExecuteAction(
            () => database.Customers.SingleOrDefault(
                    c => c.UserName == userName));
    if (customer == null)
    {
      return null;
    }
    return new Customer
    {
      CustomerId = customer.CustomerId,
      UserName = customer.UserName,
      FirstName = customer.FirstName,
      LastName = customer.LastName,
      Address = customer.Address,
      City = customer.City,
      State = customer.State,
      PostalCode = customer.PostalCode,
      Country = customer.Country,
      Email = customer.Email,
      Phone = customer.Phone
    };
  }
}
```

This method searches for an entity using the federated user name. If it finds a match, it creates a new **Customer** instance and populates it with the data from the database. If not, it returns **null** to indicate that this visitor is not registered.

You can see that the **FindOne** method shown in the previous code first obtains a reference to the database by calling the **CreateContext** method of the **TreyResearchModelFactory** class (located in the DataStores folder of the **Orders.Website** project). This method uses another class named **Cloud-Configuration** to read the connection string for the database from the ServiceConfiguration.cscfg file, or return **null** if the setting is not found.

```csharp
// C#
public static TreyResearchDataModelContainer
        CreateContext()
{
  return new TreyResearchDataModelContainer(
          CloudConfiguration.GetConfigurationSetting(
              "TreyResearchDataModelContainer", null));
}
```

After obtaining a reference to the database, the code uses the Enterprise Library Transient Fault Handling Application Block to execute the method that connects to the data store and retrieves the details for the specified customer. The Transient Fault Handling Application Block provides a ready-built mechanism for attempting an action a specified number of times with a specified delay between each attempt. It exposes events that you can use to monitor the execution and collect information about execution failures.

> *The Transient Fault Handling Application Block can be used to manage connectivity to SQL Azure databases, Service Bus queues, Service Bus topics, and the Windows Azure Cache. The block is part of the Enterprise Library Integration Pack for Windows Azure. For more information, see "Microsoft Enterprise Library" at http://msdn.microsoft.com/entlib/.*

Although they could have used code to generate the policy dynamically at runtime, the developers at Trey Research chose to define the retry policy that the block uses in the application's Web.config file. The following extract from the file shows the definition of the policy.

```xml
<RetryPolicyConfiguration
    defaultRetryStrategy="Fixed Interval Retry Strategy"
    defaultAzureStorageRetryStrategy
      ="Fixed Interval Retry Strategy"
    defaultSqlCommandRetryStrategy="Backoff Retry Strategy"
  >
  <incremental name="Incremental Retry Strategy"
               retryIncrement="00:00:01"
               initialInterval="00:00:01"
               maxRetryCount="10" />
  <fixedInterval name="Fixed Interval Retry Strategy"
               retryInterval="00:00:05"
               maxRetryCount="6"
               firstFastRetry="true" />
  <exponentialBackoff name="Backoff Retry Strategy"
               minBackoff="00:00:05"
               maxBackoff="00:00:45"
               deltaBackoff="00:00:04"
               maxRetryCount="10" />
</RetryPolicyConfiguration>
```

You can see that it contains a default strategy that retries the action every second for a maximum of ten attempts. There is also a strategy for accessing Windows Azure Storage (tables, queues, and blobs) that retries the action every five seconds for a maximum of six attempts, with the first retry occurring immediately after a failure. Finally, there is a strategy for SQL database connections that retries the action after 5 seconds, and then adds 4 seconds to the delay between each attempt up to a maximum of 45 seconds delay and ten attempts.

The **CustomerStore** class that contains the **FindOne** method initializes the retry policy in its class constructor using the static **GetDefaultSqlCommandRetryPolicy** method of the **RetryPolicyFactory** class (which is part of the Transient Fault Handling Application Block and is exposed when you add the block to your application). The constructor also adds a handler for the **Retrying** method of the block. This handler writes information to Windows Azure diagnostics each time the block detects a connection failure.

```csharp
C#
this.sqlCommandRetryPolicy
  = RetryPolicyFactory.GetDefaultSqlCommandRetryPolicy();

this.sqlCommandRetryPolicy.Retrying += (sender, args)
    => TraceHelper.TraceInformation("Retry - Count:{0},"
       + "Delay:{1}, Exception:{2}",
       args.CurrentRetryCount, args.Delay,
       args.LastException);
```

If the Transient Fault Handling Application Block cannot complete the specified action within the specified number of attempts, it will throw an exception that you can handle in your code.

With the retry policy defined and everything initialized, the code simply needs to call the **ExecuteAction** method, specifying as a parameter the action to execute. In this case, the action to execute is the **SingleOrDefault** method on the **Customers** collection exposed by the **TreyResearchDataModelContainer** class.

```csharp
C#
    var customer
      = this.sqlCommandRetryPolicy.ExecuteAction(
          () => database.Customers.SingleOrDefault(
                c => c.UserName == userName));
```

*This code uses the simplest of the **ExecuteAction** overloads available in the Transient Fault Handling Application Block. Other overloads allow you to return a result from the method, and perform the execution asynchronously. For more details of all the overloads, see "ExecuteAction Method" on MSDN.*

Authenticating Access to Service Bus Queues and Topics

Trey Research uses Service Bus queues and topics to pass messages between the Windows Azure hosted Orders application, transport partners, and the on-premises audit logging service. The section "Securing Message Queues, Topics, and Subscriptions" in Chapter 4, "Implementing Reliable Messaging and Communications with the Cloud," describes the security configuration Trey Research used to protect the Service Bus queues and topics by authenticating users and authorizing the actions each user can carry out.

Unlike the typical ACS authentication process described so far in this chapter, Service Bus "users" are most likely to be services and other components, rather than visitors using a web browser. However, the overall principles are the same: each request to access a Service Bus queue or topic must be authenticated by a trusted identity provider, and the STS must issue a token that indicates successful authentication. For Service Bus authentication, the token must contain claims that can be used to authorize actions such as sending messages to the queue or topic.

For example, applications and users that want to access the service that exposes reporting data (described in the section "How Trey Research Uses the SQL Azure Reporting Service" in Chapter 2, "Deploying the Orders Application and Data in the Cloud") must authenticate with ACS and present a token containing a suitable set of claims to Service Bus.

ACS can be configured to act as both the identity provider and the token issuer (STS) for Service Bus endpoints. Figure 4 shows an overview of the process. Clients obtain a token from ACS that contains the relevant claims (such as "Listen", "Send", or "Manage") and present this token to Service Bus when accessing a queue, topic, or Service Bus Relay endpoint.

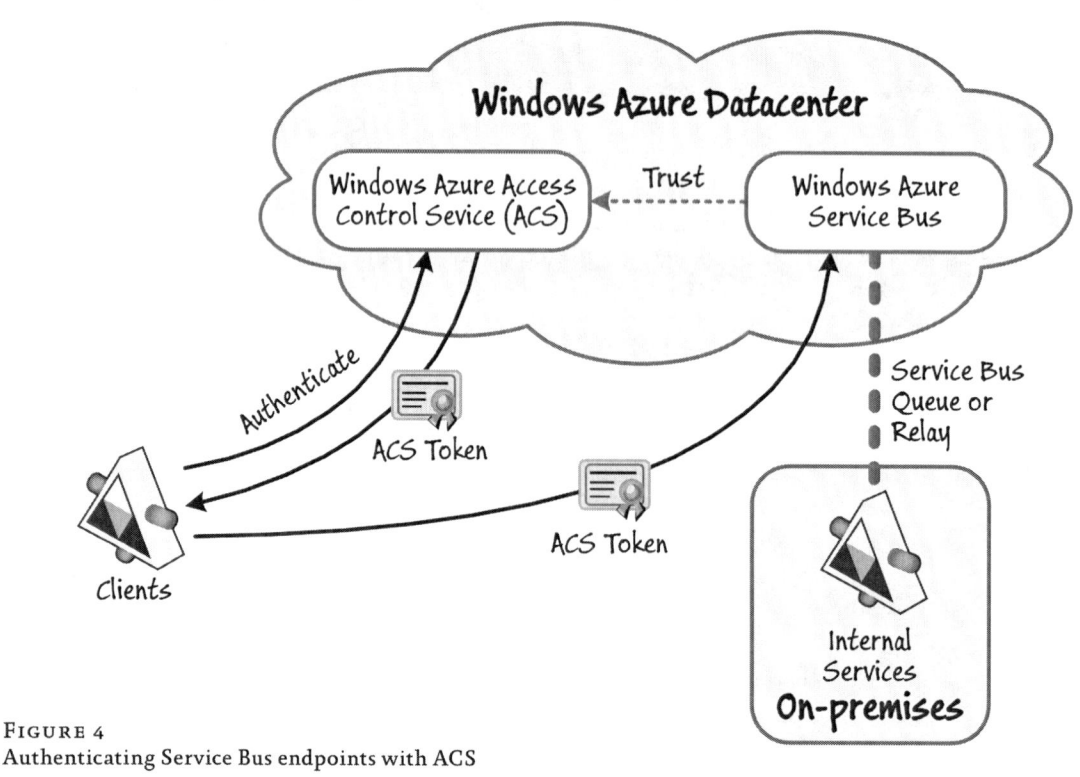

Figure 4
Authenticating Service Bus endpoints with ACS

For more information about how ACS can be used as both the identity provider and token issuer for authenticating and authorizing requests to Service Bus endpoints, see the section "Windows Azure Service Bus Authentication and Authorization" in "Appendix B - Authenticating Users and Authorizing Requests."

Summary

This chapter described how Trey Research tackled the challenge of authenticating visitors to the Orders application, and authenticating access to Service Bus queues and topics when using it to communicate between cloud-hosted and on-premises applications and services. The chapter focused on the use of claims-based identity and authentication. This is the most common and fastest-growing technique for Windows Azure hosted applications because it can take advantage of the frameworks and services specifically designed to make this approach easier.

Claims-based authentication also offers advantages in that it makes it simpler for administrators to manage lists of users and permissions, and it supports single sign-on and a choice of identity providers to make authentication easier for users. It is also a vital technique for securing services such as Windows Azure Service Bus topics and queues.

The chapter also described how the fictional organization named Trey Research implemented authentication and authorization in its hybrid Orders application.

More Information

All links in this book are accessible from the book's online bibliography available at: *http://msdn.microsoft.com/en-us/library/hh968447.aspx*.

- "Access Control Service Samples and Documentation" at *http://acs.codeplex.com/releases/view/57595*.
- "Access Control Service 2.0" at *http://msdn.microsoft.com/en-us/library/windowsazure/gg429786.aspx*.
- "How To: Implement Token Transformation Logic Using Rules" at *http://msdn.microsoft.com/en-us/library/gg185955.aspx*.
- "Securing Services" in the MSDN WCF documentation at *http://msdn.microsoft.com/en-us/library/ms734769.aspx*.
- "Service Bus Authentication and Authorization with the Access Control Service" at *http://msdn.microsoft.com/en-us/library/hh403962.aspx*.
- "Securing and Authenticating a Service Bus Connection" at *http://msdn.microsoft.com/en-us/library/dd582773.aspx*.
- "ACS How Tos" on MSDN at *http://msdn.microsoft.com/en-us/library/gg185939.aspx* provides a comprehensive list of tutorials on using ACS, including configuring identities and identity providers.
- The Identity Management home page at *http://msdn.microsoft.com/en-us/security/aa570351.aspx* contains more information about Windows Identity Foundation.
- A detailed exploration of claims-based authentication, authorization, and Windows Azure Access Control Service can be found at "A Guide to Claims-Based Identity and Access Control (2nd Edition)" at *http://msdn.microsoft.com/en-us/library/ff423674.aspx*.

- "Federated Identity for Web Applications" in "A Guide to Claims-Based Identity and Access Control (2nd Edition)" at *http://msdn.microsoft.com/en-us/library/ff359110.aspx*.
- Transient Fault Handling Application Block at *http://msdn.microsoft.com/en-us/library/hh680934(v=pandp.50).aspx*.
- "ExecuteAction Method" at *http://msdn.microsoft.com/en-us/library/microsoft.practices.transientfaulthandling.retrypolicy.executeaction.aspx*.

4 Implementing Reliable Messaging and Communications with the Cloud

After Trey Research moved the web application that enables customers to place orders to the Windows Azure™ technology platform, migrated the various databases used by the application, and secured access to the application so that only authenticated customers can place orders, the next step was to consider how the details of orders could be passed to the various transport partners for shipping, and recorded for audit and compliance purposes. This aspect of the system is critical as it supports the core business function of Trey Research fulfilling customers' orders; the order handling process requires a reliable mechanism for transmitting orders securely and reliably.

In this chapter, you will see how Trey Research addressed the various challenges associated with implementing the messaging and communications layer capable of handling orders in the cloud by using Windows Azure Service Bus and Windows Azure Connect.

> *Chapter 5, "Processing Orders in the Trey Research Solution," describes how Trey Research use this communications and messaging layer as a foundation supporting the business logic for actually processing orders placed by customers.*

> Reliable communication between the Orders application and the transport partners is essential. If the communication mechanism is prone to failure, messages could be lost, orders might not be fulfilled, and customers may go elsewhere.

Scenario and Context

In the original implementation of the Orders system, the elements of the Orders application ran on-premises, and the order processing workflow was performed in an environment that was completely controlled by Trey Research. Figure 1 illustrates the original application, with the components that handle the order processing highlighted.

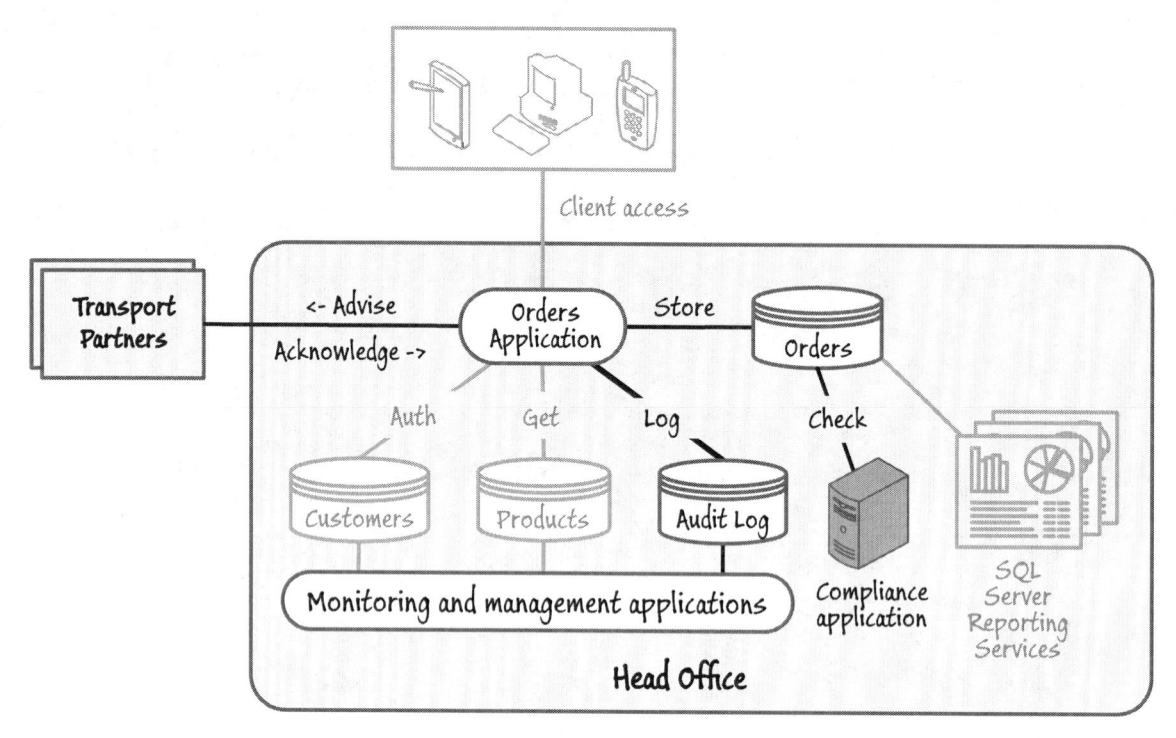

FIGURE 1
The order processing components in the on-premises application

In the on-premises solution, when a customer places an order the application stores the order details in the Orders table in the on-premises database. The Audit Log table in the on-premises database holds a range of information including runtime and diagnostic information, together with details of unusual orders such as those over a specific total value. The Orders application then sends a message to the appropriate transport partner. This message indicates the anticipated delivery date and packaging information for the order (such as the weight and number of packages). The transport partner sends a message back to the Orders application after the delivery is completed so that the Orders database table can be updated.

Due to the nature of the products Trey Research manufactures, it must also ensure that it meets legal requirements for the distribution of certain items, particularly for export to other countries and regions. These requirements include keeping detailed records of the sales of certain electronic components that may be part of Trey Research' products, and hardware items that could be used in ways not originally intended. Analyzing the contents of orders is a complex and strictly controlled process accomplished by a legal compliance application from a third party supplier, and it runs on a separate specially configured server.

When the Orders application moved to the cloud, Trey Research had to consider how to implement this business logic using the new architecture. As you will recall from Chapter 2, "Deploying the Orders Application and Data in the Cloud," Trey Research deployed the data to the SQL Azure™ technology platform. The Orders database was replicated in each datacenter, and the Orders application was modified to access the database co-located in the same datacenter that the user is connected to. Additionally, with the expectation that the volume of orders was likely to increase exponentially, the compliance application was relocated to the cloud to take advantage of the inherent scalability of Windows Azure; the compliance application is multi-threaded and can take full advantage of the power of the platform on which it runs, so it was considered appropriate only to deploy it to a single datacenter. However, for regulatory reasons, it was necessary to retain the audit log on-premises. Figure 2 shows the structure of the resulting hybrid solution, again with the order processing elements highlighted.

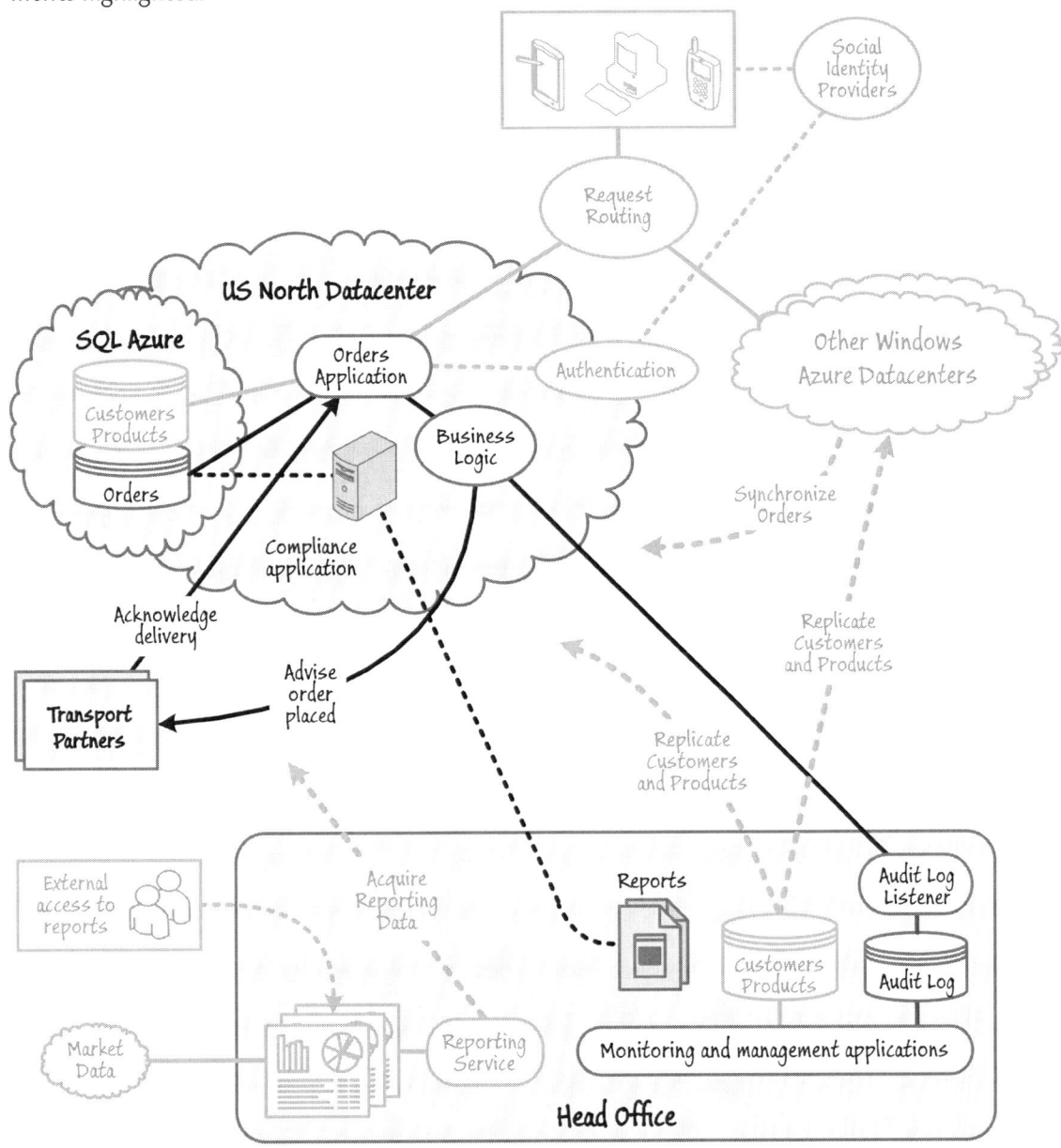

Figure 2
The hybrid version of the Trey Research solution

As far as a customer is concerned, the Orders application works in similar way to the original solution, but the logic that processes an order is now implemented as follows:
- When a customer places an order, the Orders application:
  - Stores the order details in the Orders table of the database in the local SQL Azure datacenter. All orders are synchronized across all Windows Azure datacenters so that the order status information is available to visitors irrespective of the datacenter to which they are routed.
  - Sends an advice message to the appropriate transport partner. The transport company chosen depends on the delivery location.
  - Sends any required audit information, such as orders with a value over $10,000, to the Audit Log table of the database located in the head office datacenter. The on-premises management and monitoring applications can examine this information.
- The third-party compliance application running in the cloud continually validates the orders in the Orders table for conformance with legal restrictions and sets a flag in the database table on those that require attention by managers. It also generates a daily report that it stores in a secure location in the head office datacenter.
- When transport partners deliver an order to the customer they send a message to the Orders application (running in the datacenter that originally sent the order advice message) so that it can update the Orders table in the database.

Keep in mind that, for simplicity, some of the features and processes described here are not fully implemented in the example we provide for this guide, or may work in a slightly different way. This is done to make it easier for you to install and configure the example, without requiring you to obtain and configure Windows Azure accounts in multiple datacenters, and for services such as Windows Azure Data Sync and SQL Server Reporting Services.

In its simplest terms, the high-level structure of the cloud-based elements of the hybrid solution is reasonably straightforward. The Orders application running in the cloud naturally maps to a Windows Azure web role, as described in Chapter 1, "The Trey Research Scenario," while the business logic that actually processes orders can be implemented as a Windows Azure worker role. The operation of the order processing logic must be scalable to handle the expected growth in demand as Trey Research expand their customer base, and it must be reliable because orders must not be mislaid or duplicated.

The order processing business logic divides itself naturally into three specific areas; how to communicate with transport partners, how to determine which orders to audit, and how to pass orders to the compliance application to ensure that they do not violate regulatory requirements. Trey Research considered the implementation options for each of these areas in turn.

Communicating with Transport Partners

A key part of the order processing mechanism concerns the communication with the transport partners. The worker role must examine each order and direct the order details to the most appropriate transport partner. The definition of "most appropriate" is application-specific and may change over time, but currently it is an economic decision based on the proximity of the customer to the Trey Research manufacturing plant from where the orders are shipped. Orders for local customers (customers based in the same or a neighboring state as the Trey Research manufacturing plant) use a local transport partner, while orders for more remote customers require a distance transport partner capable of shipping goods by rail or air if necessary.

> *After much negotiation and evaluation, Trey Research decided to use Contoso, Inc. to provide the local transportation services, while Fabrikam, Inc. was selected as the distance transport partner.*

During the design phase, the development team at Trey Research insisted that all communications with the transport partners had to fulfill a number of criteria:

- The solution must be responsive. All communication must operate in a timely manner that does not block the operations in the Orders application or adversely affect the experience of customers placing orders.
- The communication mechanism must be robust and reliable. Once a customer places an order, and that order is confirmed, it must be fulfilled. The system must not be able to lose orders as messages are passed from the Trey Research solution to the appropriate transport partner, even if the connection with the transport partner fails or the transport partner's system suffers a failure.
- The solution must be scalable. It must be possible to easily add further transport partners to the solution without needing to rewrite the order processing logic. Furthermore, it must be possible to host the Orders application at multiple sites in the cloud, again without requiring that this code is rewritten to handle multiple instances of the application.
- The solution must be flexible. The actual list of transport partners may vary over time, and the rationale for selecting which partner to use for a specific order may also change as delivery costs fluctuate.
- All data should be considered sensitive and must be protected appropriately. All reasonable security measures should be taken to prevent an unauthorized third party from intercepting the details of orders placed by a customer.

> The way in which messages are passed between distributed components in a hybrid application must be reliable, robust, responsive, scalable, and secure.

Choosing a Communications Mechanism

In the original on-premises application, communication between Trey Research and the transport partners was achieved through a set of web service operations. Some of the calls to these operations originated from within Trey Research's network, but others required Trey Research to expose an on-premises web service that partners called to update the delivery status. After moving to the cloud, given the challenges listed above, Trey Research considered several options for implementing the connectivity between the equivalent business logic hosted in the worker role and the transport partners. The following sections summarize some of these options, together with their perceived advantages and limitations.

Electronic Data Interchange (EDI)

The worker role could connect to the transport partner over an interface, protocol, and format that the worker role and the transport partner both understand, such as EDIFACT, RosettaNet, cXML, and BASDA. These are commonly accepted and well understood standards that many larger organizations employ to exchange data with other businesses. Furthermore, most modern EDI protocols are asynchronous as the corresponding business processes are expected to be long-lived; this can help to ensure that the worker role remains responsive during times of high demand.

However, the worker role may require additional software and infrastructure to connect to an EDI interface. Microsoft BizTalk® Server provides adapters for many well-known protocols and formats, but this solution requires passing all orders through BizTalk Server running on-premises. For more information, see *"Hybrid Reference Implementation Using BizTalk Server, Windows Azure, Service Bus and SQL Azure."* Each transport partner may expose a different EDI interface, making it difficult to easily extend the Trey Research business logic to additional partners, although it may be possible to abstract these differences into a connector layer within the worker role.

This approach would still require implementing a connector for each partner. Trey Research also had to consider that not all transport partners would necessarily provide an EDI interface to their systems. Finally, in this approach security and message protection is governed by the transport partner rather than by Trey Research.

Web Services (Push Model)

If the transport partner exposes a web services interface into its delivery system, the worker role could utilize this interface and invoke the appropriate operations to push the details of orders across to the transport partner. Web services are a common, well-understood, and maturing technology. Additionally, it is usually a straightforward task to invoke web service operations from a worker role. From a security perspective, requests can be easily encrypted, although the degree of security and message protection available is managed by the transport partner providing the web service.

There are some possible issues with this approach. Primarily, if the transport partner does not provide a web service interface, then this approach cannot be used. If a transport partner does provide such an interface, Trey Research also had to assess the possible complexity arising in the highly likely scenario that different transport partners may implement a different set of web service operations, and expect request messages in different formats. These differences could be abstracted by building a connector layer within the worker role and constructing custom connectors for each transport partner, but it adds complexity to the Trey research solution.

Another issue is that the web service may not provide an appropriate level of reliability. The worker role may be unaware if the transport partner's system raises an error that causes the details of the order to be lost; the transport partner may not know to call the operation that provides the details of the order again.

Connectivity is also an issue; if the web service at the transport partner is temporarily unavailable or a connection cannot be established, then Trey Research will not be able to send the details of any orders.

Web Services (Pull Model)

Web services can provide a secure, scalable, and reliable communication mechanism if they are implemented correctly. For this reason, Trey Research considered turning the previous option around by implementing a web service as part of the worker role, and exposing operations that transport partners invoke to retrieve (or pull) information about orders from the Trey Research application. This approach would give Trey Research full control over the degree of message protection and security available, the web service can take advantage of the inherent scalability of the worker role hosted by Windows Azure, and it affords better reliability; if the transport partner's system fails while processing an order, it can reconnect to the web service and retrieve the details of the order again when it restarts.

The worker role can expose the same web services interface to all transport partners. New transport partners can be easily integrated without modifying the worker role. The worker role does not have to wait while the transport partner retrieves the details of orders so the system remains responsive to customers. Furthermore, the worker role can take advantage of Service Bus Relay to build location-independence and security into the solution; the transport partner can connect to a well-known endpoint advertised through Service Bus Relay which can authenticate the transport partner through the Windows Azure Access Control Service (ACS), and then transparently route messages to the web service endpoints published by the worker role.

However, transport partners would be expected to develop their own software to connect to the web service; they may be unwilling or unable to develop and deploy custom software specifically for integrating with Trey Research. If the transport partner is willing to connect to the web service, it is the responsibility of the transport partner to query whether there are any orders to be shipped. If the transport partner does not perform this query often enough, then orders may not be dispatched in a timely manner, leading to customer complaints. If the transport partner's system fails to query the web service successfully, orders will not be shipped. Scalability within the transport partners systems may also be an issue. As the volume of orders increases, transport partners may not query the web service sufficiently often, causing a backlog of orders to build up.

Windows Azure Storage Queues

The Orders application runs in a Windows Azure role, so Trey Research considered posting the details of orders as messages to a Windows Azure storage queue; transport partners could connect to this queue to retrieve the orders to be shipped. Delivery acknowledgement messages could be posted back to the worker role through another queue. This mechanism is relatively simple, reliable, scalable, and secure; Windows Azure storage queues are managed and maintained within a datacenter, and Trey Research has full control over which transport partners would have access rights to connect to the queue to retrieve and post messages. Additionally, the semantics of the retrieve operation can be implemented in a reliable manner; if the transport partner fails with an error after receiving a message but before processing it, the message can be transparently returned to the queue from where it can be received when the transport partner restarts.

Windows Azure provides a REST API for accessing Windows Azure storage queues, so the transport partner can implement their system by using any technology that can connect to the web and transmit REST requests.

Again, there are some issues surrounding this approach. As with some of the preceding options, each transport partner must be willing to implement software that connects to the Windows Azure storage queue and integrate it into their own solution. Also, to prevent a transport partner from receiving an order intended for a different transport partner, Trey Research must create a separate queue for each partner. This approach may complicate the logic inside the worker role, and also makes it more difficult to add or remove transport partners. Finally, security is controlled by using storage account keys rather than ACS, so each transport partner would have to be granted access to the entire storage account rather than an individual queue. To ensure that each transport partner only has access to the relevant queue, each queue must be created within a different storage account with its own key.

Windows Azure Service Bus Queues

To counter some of the complexity issues of using Windows Azure storage queues, a similar but more advantageous option is to post the details of orders as messages to a Windows Azure Service Bus queue; transport partners can connect to this queue to retrieve the orders to be shipped. Delivery acknowledgement messages can be posted back to the worker role through another queue. This approach is highly scalable; the worker role can post messages to the Service Bus queue as quickly as orders are placed. The Service Bus infrastructure can buffer messages until they have been retrieved by the transport partner. It also offers improved reliability; after the worker role has posted a message to a Service Bus queue, it will not be lost. It will either remain on the queue until it expires (the expiration period for a message is configurable) or a transport partner retrieves it. Like a Windows Azure storage queue, Service Bus queues support reliable retrieve operations so if the transport partner fails after receiving a message, the message can be transparently returned to the queue. Security is highly configurable and flexible, especially when compared to that available for Windows Azure storage queues. It is managed through the Windows Azure Access Control Service (ACS).

However, as before, each transport partner must be prepared to connect to the appropriate Service Bus queues to retrieve messages and send delivery acknowledgement messages. If the transport partner is not amenable to this approach, and instead insists that orders are passed across using its own system-defined interfaces (such as a set of web services), then Trey Research may need to build a custom component to retrieve messages from the queue and convert them into the appropriate format for the transport partner, and transmit them using an agreed protocol. The same component can include the logic for waiting for a delivery acknowledgement from the transport partner and posting a message to the worker role. Again, as before, to prevent a transport partner from receiving an order intended for a different transport partner, Trey Research may need to create a separate queue for each partner. This approach may complicate the logic inside the worker role, and also makes it more difficult to add or remove transport partners.

> *For information comparing the features and possible uses of Windows Azure storage queues and Windows Azure Service Bus queues, see the article "Windows Azure Queues and Windows Azure Service Bus Queues - Compared and Contrasted" on MSDN.*

Windows Azure Service Bus Topics and Subscriptions

Service Bus queues provide an attractive and scalable alternative, except for the need to create and manage a separate queue for each partner. Therefore, the final option assessed by Trey Research was to post the details of orders as messages to a Windows Azure Service Bus topic; transport partners subscribe to this topic to retrieve the orders to be shipped. Messages acknowledging receipt of the order details and messages indicating that delivery was completed are posted back to the worker role through a Service Bus queue.

Like Service Bus queues, Service Bus topics and subscriptions are highly scalable and reliable, with configurable security. However, they are more flexible than using Service Bus queues to transmit messages to a transport partner; the worker role can add metadata to messages that indicate which transport partner should process them, and then post these messages to a Service Bus topic. Each transport partner can connect to the Service Bus topic through its own subscription, which can filter the messages based on this metadata so that each transport partner receives only the orders that it should process. Topics also enable messages to be routed to multiple destinations, so orders with a value over $10,000 can additionally be directed to the Audit Log Listener.

The only real drawback to this approach, in common with most of the options described previously, is that each transport partner must be prepared to connect to the appropriate Service Bus topic to retrieve messages. Alternatively, Trey Research can build custom connectivity components to integrate with the transport partners' systems. There are also some limitations imposed on topics; for example, a topic can currently have a maximum of 2000 subscriptions and can support up to 100 concurrent connections (the limit of 100 concurrent connections also applies to queues). However, Trey Research considered that the Orders system was unlikely to hit either of these two limits.

How Trey Research Communicates with Transport Partners

In the end, Trey Research decided to send orders from the worker role to the transport partners by using a Service Bus topic. Each transport partner receives messages by using a subscription that filters the orders. In this way, each transport partner receives only the orders that it should ship. For more information about using Service Bus queues, topics, and subscriptions, see *"Queues, Topics, and Subscriptions"* on MSDN.

> Service Bus queues provide a reliable and scalable mechanism for communicating between services running in the cloud and applications running on-premises.

To bridge the potential technology gap between the systems implemented by the transport partners and the Service Bus, Trey Research constructed a set of connectivity components to translate messages retrieved from the Service Bus and convert them into format expected by the transport partner. The location of these connectivity components depends on the relationship that Trey Research has with the transport partner:

- In the case of the local transport partner, Contoso, Trey Research was able to convince the partner to install a connector and integrate it into their own proprietary system. Trey Research provided credentials necessary to enable the connector to listen to the appropriate Service Bus subscription. The transport partner's own system uses this connector to retrieves the details of orders from the subscription. Additionally, the connector exposes an interface that the transport partner's system uses to post acknowledgment messages back to the Service Bus queue that the Orders application listens on.

   Implementing a connector as part of the transport partner's system does not force the transport partner to incorporate .NET Framework code into their solution. The features of the Windows Azure Service Bus are exposed through a series of HTTP REST APIs (the Windows Azure SDK simply provides a .NET Framework wrapper around these APIs), so the transport partner can use any familiar technology that can generate REST requests and consume REST responses, including the Java programming language.

- The distance transport partner, Fabrikam, is a multinational organization, and the operations staff were not willing to allow Trey Research to install software on their own servers, preferring Trey Research to connect using the interfaces that they provide to their systems. To accommodate this requirement, Trey Research implemented an adapter for posting orders to Fabrikam, and this adapter is hosted within the worker role. This mechanism enables the logic that posts messages to transport partners to remain the same, regardless of whether the partner is the local or a distance partner. If Trey research decides to add a new distance transport partner in the future, Trey Research simply needs to create and install an appropriate adapter.

> Implementing adapters and connectors enables applications to remain independent of the communication mechanism. If necessary, the Service Bus topic used by Trey Research could be switched to a different means of transferring information, and only the adapters and connectors would need to change.

Figure 3 summarizes the technologies that Trey Research decided to use to implement messaging between the Orders application and the transport partners. The following sections describe the approach that Trey Research used to build their messaging solution based on these technologies.

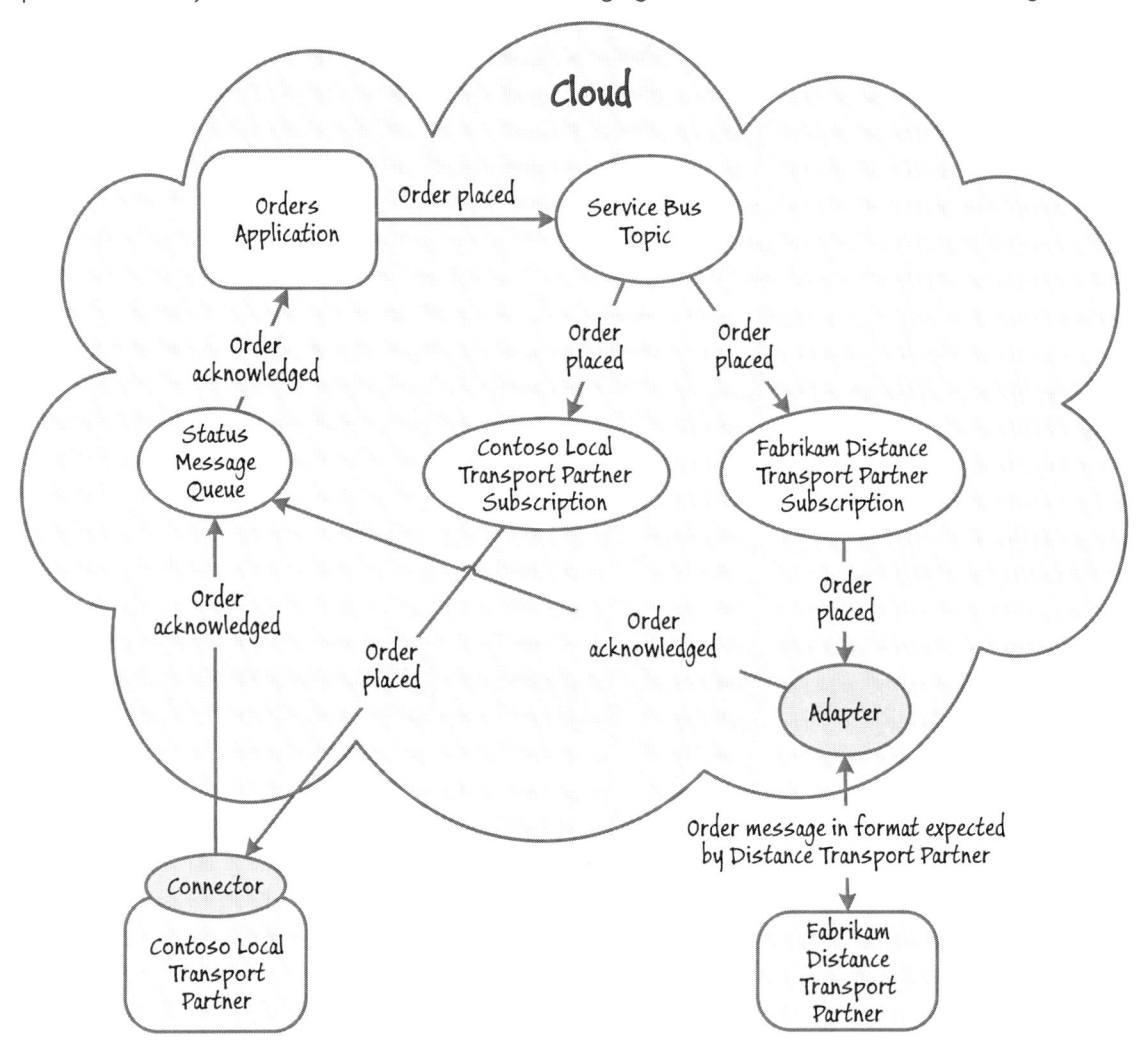

FIGURE 3
Messaging technologies used by Trey Research to communicate with transport partners

The sample Trey Research application that you can download for this guide implements many of the technologies and techniques described here. However, to simplify installation and setup, and reduce the prerequisites and the requirements for users to establish extensive Windows Azure accounts, the feature set and some of the implementation details differ from the text of this guide.

Using the Windows Azure SDK, you can implement applications that send and receive messages by using the **MessageSender** and **MessageReceiver** classes in the **Microsoft.ServiceBus.Messaging** namespace. However, these operations are synchronous. For example, the **Send** method of the **MessageSender** class waits for the send operation to complete before continuing, and similarly the **Receive** method of the **MessageReceiver** class either waits for a message to be available or until a specified timeout period has expired. These methods are really just façades in front of a series of HTTP REST requests, and the Service Bus queues and topics are remote services being accessed over the Internet. Therefore, your applications should assume that:

- Send and receive operations may take an arbitrarily long time to complete, and your application should not block waiting for these operations to finish.
- A sender can post messages at any time, and a receiver may need to listen for messages on more than one queue.
- Send and receive operations could fail for a variety of reasons, such as a failure in connectivity between your application and the Service Bus in the cloud, a security violation caused by a change in the security implemented by the Service Bus queue or topic (an administrator might decide to revoke or modify the rights of an identity for some reason), the queue being full (they have a finite size), and so on. Some of these failures might the result of transient errors, while others may be more permanent.

Trey Research decided to implement a library that added wrappers around the Service Bus queue and topic functionality available in the **Microsoft.ServiceBus.Messaging** namespace. This library is provided with the sample solution, in the **Orders.Shared** project. The classes located in the Communication folder of this project encapsulate the existing **MessageSender**, **MessageReceiver**, and **BrokeredMessage** classes (amongst others). The purpose of the new classes is to abstract the send and receive functionality so that all send and receive operations are performed asynchronously. This library also incorporates elements of the security model implemented by Trey Research; for more information, see the section "Securing Messages, Queues, Topics, and Subscriptions" later in this chapter.

> *For additional information and guidelines on optimizing performance when using Windows Azure Service Bus messaging, see the topic "Best Practices for Performance Improvements Using Service Bus Brokered Messaging" on MSDN.*

The following sections describe the structure of this library, the classes that it provides, and how these classes extend the functionality provided by Service Bus queues, topics, and subscriptions.

Sending Messages to a Service Bus Queue Asynchronously
Trey Research uses a Service Bus queue to enable transport partners to communicate with the Orders application. To send a message to a Service Bus queue by using the Orders.Shared library, an application performs the following steps:

1. Create a **BrokeredMessage** object and populate it with the required information. The **BrokeredMessage** class is the type provided by Microsoft in the **Microsoft.ServiceBus.Messaging** namespace.

2. Create a **ServiceBusQueueDescription** object and specify the Service Bus namespace, the queue name, and a set of valid credentials in the form of an access key and the name of the associated identity. The **ServiceBusQueueDescription** class is a member of the **Orders.Shared** project.

3. Create a **ServiceBusQueue** object using the **ServiceBusQueueDescription** object. The **ServiceBusQueue** type encapsulates asynchronous functionality for sending messages. Creating an instance of the **ServiceBusQueue** type connects to the underlying Service Bus queue in **PeekLock** mode.

4. Call the **Send** method of the **ServiceBusQueue** object. The parameter to the **Send** method must be a **BrokeredMessageAdapter** object that wraps the **BrokeredMessage** object created earlier. The **ServiceBusQueue** class contains an instance of the **MessageSenderAdapter** class (defined in the Communication\Adapters folder in the **Orders.Shared** project) which implements the **IMessageSenderAdapter** interface. The **Send** method uses this **MessageSenderAdapter** object to send the message.

*The **MessageSenderAdapter** class is actually just a wrapper class that was created to simplify unit testing with mock objects.*

*For an example of using the **ServiceBusQueue** type to send messages, see the **SendToUpdateStatusQueue** method in the **OrderProcessor** class in the **TransportPartner** project.*

> The **ServiceBusTopic** and **ServiceBusSubscription** classes in the **Orders.Shared** project implement a similar approach to **ServiceBusQueue**, encapsulating asynchronous functionality based on the **MessageSender** and **MessageReceiver** classes respectively.

The *MessageSenderAdapter*, *MessageReceiverAdapter*, and *BrokeredMessageAdapter* classes enable the unit tests (in the *Orders.Shared.Tests* project) to construct mock senders, receivers, and brokered messages.

The following code fragment shows the implementation of the **Send** method in the **ServiceBusQueue** class, together with the relevant members used by the **Send** method:

> The **Guard** method that is used by methods in the **ServiceBusQueue** class and elsewhere checks that the named parameter has been initialized; it should not be null or an empty string.

```C#
public class ServiceBusQueue
{
  private readonly ServiceBusQueueDescription description;
  ...
  private readonly IMessageSenderAdapter senderAdapter;
  ...

  public ServiceBusQueue(
    ServiceBusQueueDescription description)
  {
    Guard.CheckArgumentNull(description, "description");
    this.description = description;
    ...
    var sender = messagingFactory.CreateMessageSender(
      this.description.QueueName.ToLowerInvariant());
    this.senderAdapter = new MessageSenderAdapter(sender);
    ...
  }
  ...

  public void Send(IBrokeredMessageAdapter message)
  {
    Guard.CheckArgumentNull(message, "message");

    this.Send(message, this.senderAdapter);
  }

  public void Send(IBrokeredMessageAdapter message,
                   IMessageSenderAdapter sender)
  {
    Guard.CheckArgumentNull(message, "message");
    Guard.CheckArgumentNull(sender, "sender");

    Task.Factory
      .FromAsync(sender.BeginSend, sender.EndSend, message,
            null,TaskCreationOptions.AttachedToParent)
      .ContinueWith(
```

```
    taskResult =>
    {
      try
      {
        if (taskResult.Exception != null)
        {
          TraceHelper.TraceError(
            taskResult.Exception.ToString());
        }
      }
      finally
      {
        message.Dispose();
      }
    });
  }
  ...
}
```

In the **ServiceBusQueue** class, the processing performed by the **Send** method requires attaching the processing as a child task by using the **TaskCreationOptions.AttachedToParent** option. In this way, a failure in the child task while sending the message can be detected and handled by the parent, enabling the parent to abandon the **Receive** operation more easily. In this example, any exceptions are simply logged by using the static **TraceError** message of the **TraceHelper** class. The **TraceHelper** class is defined in the Helpers folder in the **Orders.Shared** project. This class simply acts as a wrapper around the trace event handlers provided by the **System.Diagnostics** library and is described in more detail in Chapter 7, "Monitoring and Managing the Orders Application."

Receiving Messages from a Service Bus Queue and Processing Them Asynchronously

The **ServiceBusQueue** class creates and exposes a **MessageReceiver** object that you can use to receive messages, through the **GetReceiver** method. This is an ordinary message receiver object with no additional functionality, and calling the **Receive** method on this object performs a synchronous receive operation. In its simplest form, a receiver using this technique may spend a lengthy period of time being blocked while waiting for messages to appear. Additionally, when a message arrives, it may require significant effort to perform the required processing, during which time more messages may arrive. These messages will not be processed until the receiver finishes its current work and retrieves the next message. If a message is urgent, this response may not be acceptable.

> Make sure your code correctly disposes of a **BrokeredMessage** instance you create after use to ensure that all of the resources it uses are released.

> Notice that the **ServiceBusQueue** class does not utilize the Transient Fault Handling Exception Block. This is because using the Transient Fault Handling Application Block to start asynchronous processes does not provide the same flexibility as using a **Task** object. When considering using the Transient Fault Handling Application Block, you should weigh up the advantages of the declarative neatness of the way in which critical code can be executed and retried, against the fine control that you may require when running this code as a background task.

The **MessageReceiver** class also supports asynchronous operations through the **BeginReceive** and **EndReceive** operations. The **ServiceBusReceiverHandler** type, also in the **Orders.Shared** project, extends this functionality to provide a class that can receive and process messages asynchronously while decoupling the business logic and exception-handling process from the code that connects to the queue.

The **ServiceBusReceiverHandler** class provides a method called **ProcessMessages** that an application can use to asynchronously wait for messages arriving on a Service Bus queue and process them (the application specifies the queue to listen on as a parameter to the constructor of this class.) The following code sample shows the constructor and the implementation of the **ProcessMessages** method.

```C#
public class ServiceBusReceiverHandler<T>
{
  private readonly IMessageReceiverAdapter receiver;
  private Func<T, ServiceBusQueueDescription, string, Task>
    messageProcessingTask;

  public ServiceBusReceiverHandler(
    IMessageReceiverAdapter receiver)
  {
    ...
    this.receiver = receiver;
  }

  ...

  // The Func parameter (that returns the Task) allows the
  // caller more control on the task result and the
  // exception handling
  public void ProcessMessages(Func<T,
      ServiceBusQueueDescription, string, Task>
    taskForProcessingMessage,
    CancellationToken cancellationToken)
  {
    ...
    this.messageProcessingTask = taskForProcessingMessage;

    this.ReceiveNextMessage(cancellationToken);
  }

  ...
}
```

The **ProcessMessages** method expects a delegate as its first parameter. This delegate should reference a method that will be run each time a message is received. The purpose of this delegated method is to perform whatever business logic the application requires on receipt of each message (for a detailed example of this logic, see the section "Receiving and Processing an Order in a Transport Partner" in Chapter 5, "Processing Orders in the Trey Research Solution"). The **ProcessMessages** method stores this delegate locally and then calls the local **ReceiveNextMessage** method, as shown in the following code sample.

```C#
...
public TimeSpan? MessagePollingInterval { get; set; }
...
private void ReceiveNextMessage(
  CancellationToken cancellationToken)
{
  if (this.MessagePollingInterval.HasValue)
  {
    Thread.Sleep(this.MessagePollingInterval.Value);
  }

  Task.Factory
    .FromAsync<TimeSpan,
      IBrokeredMessageAdapter>(this.receiver.BeginReceive,
                               this.receiver.EndReceive,
                               TimeSpan.FromSeconds(10),
                               null,
                               TaskCreationOptions.None)
    .ContinueWith(
      taskResult =>
      {
        // Start receiving the next message as soon as we
        // received the previous one.
        // This will not cause a stack overflow because the
        // call will be made from a new Task.

        this.ReceiveNextMessage(cancellationToken);

        if (taskResult.Exception != null)
        {
          TraceHelper.TraceError(
            taskResult.Exception.Message);
        }

        this.ProcessMessage(taskResult.Result);
      },
      cancellationToken);
}
```

The **ReceiveNextMessage** method implements a simple polling strategy; it sleeps for a configurable period of time before attempting to receive a message from the queue (the message queue is read in **PeekLock** mode). The receive operation is performed asynchronously, and if a message is available the method starts a new task to listen for any subsequent messages and then calls the **ProcessMessage** method to process the newly received message.

```C#
private void ProcessMessage(
  IBrokeredMessageAdapter message)
{
  if (message != null)
  {
    ...
    this.messageProcessingTask(message.GetBody<T>(),
                              queueDescription, token)
      .ContinueWith(
        processingTaskResult =>
        {
          if (processingTaskResult.Exception != null)
          {
            if (message.DeliveryCount <= 3 &&
               !(processingTaskResult.Exception.
                 InnerException is InvalidTokenException))
            {
              // If the abandon fails, the message will
              // become visible anyway after the lock
              // times out
              Task.Factory.FromAsync(message.BeginAbandon,
                  message.EndAbandon, message,
                  TaskCreationOptions.AttachedToParent)
                .ContinueWith(
                  taskResult =>
                  {
                    if (taskResult.Exception != null)
                    {
                      TraceHelper.TraceError(
                        "Error while message abandon: {0}",
                        taskResult.Exception.
                          InnerException.Message);
                    }

                    var msg = taskResult.AsyncState
                      as BrokeredMessage;
                    if (msg != null)
                    {
                      msg.Dispose();
```

```
            }
          });
      }
      else
      {
        Task.Factory.FromAsync(
            message.BeginDeadLetter,
            message.EndDeadLetter, message,
            TaskCreationOptions.AttachedToParent)
          .ContinueWith(
          taskResult =>
          {
            if (taskResult.Exception != null)
            {
              TraceHelper.TraceError(
                "Error while sending message to "
                + "the DeadLetter queue: {0}",
                taskResult.Exception.
                  InnerException.Message);
            }

            var msg = taskResult.AsyncState
              as BrokeredMessage;
            if (msg != null)
            {
              msg.Dispose();
            }
          });
        TraceHelper.TraceError(
          processingTaskResult.
            Exception.TraceInformation());
      }
    }
    else
    {
      Task.Factory
        .FromAsync(message.BeginComplete,
          message.EndComplete, message,
          TaskCreationOptions.AttachedToParent)
        .ContinueWith(
          taskResult =>
          {
            if (taskResult.Exception != null)
            {
              TraceHelper.TraceError(
                "Error while executing "
```

```
                                    + "message. Complete: {0}",
                                    taskResult.Exception.
                                        InnerException.Message);
                            }
                            var msg = taskResult.AsyncState
                                as BrokeredMessage;
                            if (msg != null)
                            {
                                msg.Dispose();
                            }
                        });
                }
            });
        }
    }
```

The **ProcessMessage** method invokes the delegated method provided by the receiving application, which it has stored in **messageProcessingTask** property, to process the message. The **ProcessMessage** method implements a simple but effective policy for handling exceptions raised while receiving or processing messages. For example, if a system exception occurs while receiving the message, the **ProcessMessage** method will attempt to abandon the message and release any locks; a subsequent invocation of the **ReceiveNextMessage** method may be able to read the message successfully if the error was only transient. However, if the same message fails to be received three times, or processing fails as the result of an authentication failure (if the simple web token received from the transport partner is invalid), the message is posted to the dead letter queue. If the message is processed successfully, the **ProcessMessage** method calls the asynchronous version of the **Complete** method to remove the message from the queue.

> The polling interval acts as a regulator to help prevent parts of the system becoming overloaded. The ideal value for the polling interval depends on the computing resources available at the transport partner, the expected volume of orders, and the number of worker role instances. For example, specifying a small polling interval if the transport partners have limited computing resources is probably a poor choice, especially during periods of high demand when a large number of orders are generated. In this case a lengthier interval between messages allows the transport partners' systems to function more effectively, with the topic effectively acting as a load-leveling buffer.

Figure 4 illustrates the flow of control through a **ServiceBusReceiverHandler** object when a receiving application retrieves and processes a message.

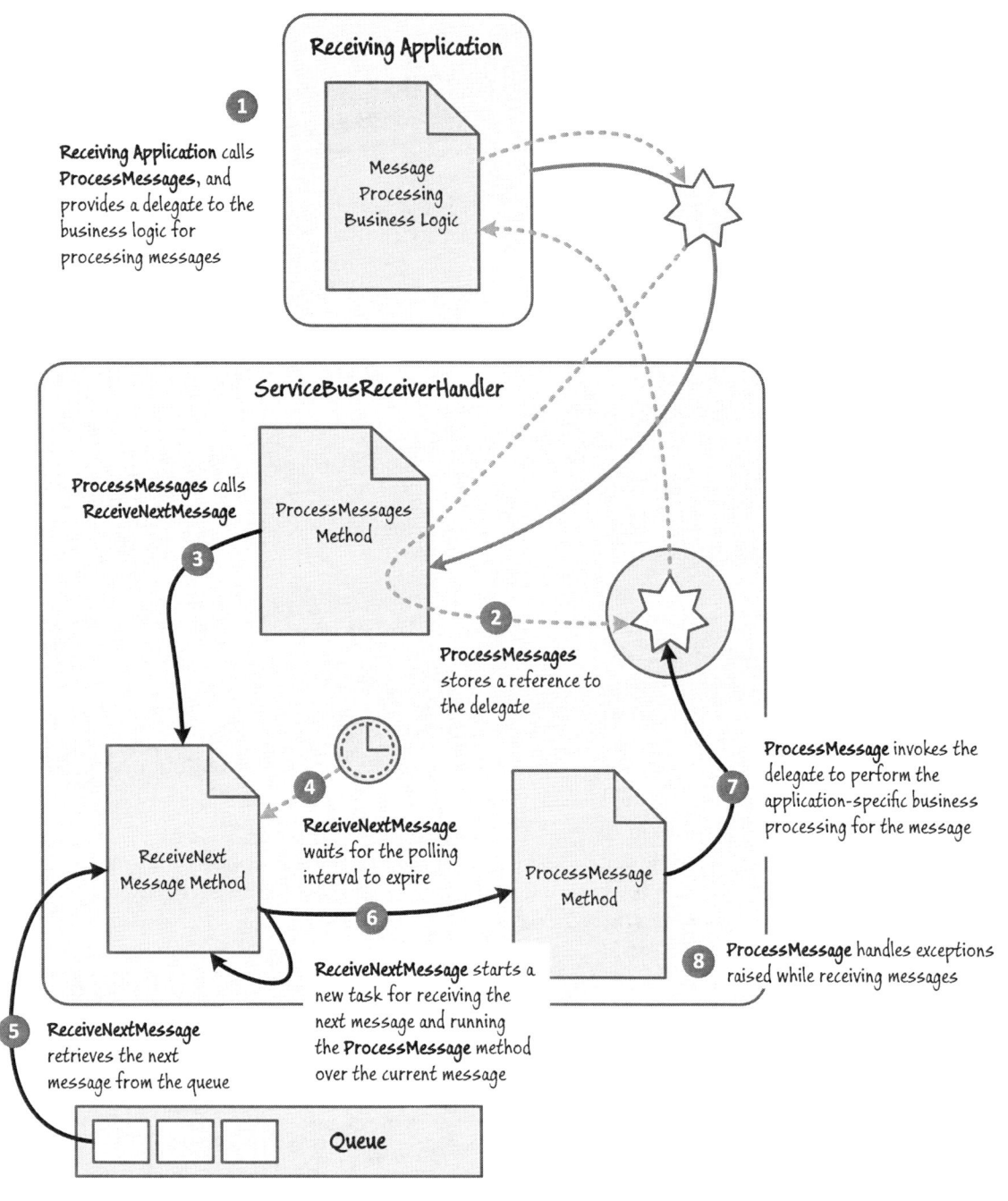

FIGURE 4
Flow of control when receiving messages through a ServiceBusReceiverHandler object

Sending Messages to a Service Bus Topic

Although transport partners send messages to the Orders application through a Service Bus queue, the worker role in the Trey Research implementation uses a Service Bus topic to send orders to each transport partner. Service Bus topics are similar to queues with one important difference; the messages posted to a topic can be filtered by using a Service Bus subscription and directed to a specific listener attached to that subscription. The filtering is based on metadata added to the message before it is sent, and only subscriptions that specify a filter that matches the value of this metadata will receive the message.

Trey Research used this mechanism to add a property called **TransportPartnerName** to each order message that specifies the transport partner that should process the message. In this way, each order message will be received only by the intended transport partner that should ship the order. Additionally, Trey Research also added a property called **OrderAmount** to each message. The auditing application subscribes to the same topic as the transport partners, but filters messages, retrieving and auditing the details of all orders with a value of more than $10,000 as specified by this property. The following code, taken from the **Execute** method of the **NewOrderJob** class in the Jobs folder of the **Orders.Workers** project, shows an example of how the Trey Research solution populates the properties of a message to direct it to a specific transport partner.

```C#
var brokeredMessage = new BrokeredMessage(msg)
{
  ...
  Properties = {
    { "TransportPartnerName", transportPartnerName },
    ...
    { "OrderAmount", orderProcess.Order.Total } },
  ...
};
```

> *The **NewOrderJob** class is described in more detail in Chapter 5, "Processing Orders in the Trey Research Solution."*

The Service Bus methods used to send messages to a topic are very similar to those used to send messages to a queue. However, to manage the minor differences in these methods, the developers at Trey Research created two custom classes specially targeted at making it easier to use Service Bus topics. These classes are **ServiceBusTopicDescription** and **ServiceBusTopic**, and are located in the Communication folder of the **Orders.Shared** project.

In the Trey Research sample solution there are only small differences between the **ServiceBusQueueDescription** class and the **ServiceBusTopicDescription** class. The **ServiceBusQueue** class (which encapsulates the functionality of a Service Bus queue) and the **ServiceBusTopic** class (which provides similar functionality, but for a Service Bus topic) also differ, but in a few more significant ways, primarily due to the way in which Trey Research uses these types rather than any underlying differences in the mechanisms that they expose for sending messages:

- Unlike the **ServiceBusQueue** class, the **ServiceBusTopic** class does not instantiate a receiver. Clients will subscribe to a topic when required by creating a suitable receiver. This removes coupling between the sender and receiver, and allows different clients to subscribe and receive messages without needing to reconfigure the topic. It also separates the business logic from the message routing concerns.
- The **ServiceBusQueue** class sends the message and only raises an exception (which it logs) if sending the message fails. In contrast, the **ServiceBusTopic** class accepts two **Action** delegates that it executes when the message has been sent, or when there is an error. This approach enables Trey Research to incorporate more extensive exception handling when sending order details to a transport partner than was deemed necessary when posting order status messages back to the Orders application.
- The **ServiceBusQueue** class uses the static **FromAsync** method of the **Task.Factory** class to send messages asynchronously. In contrast, the **ServiceBusTopic** class uses the Enterprise Library Transient Fault Handling Application Block to detect transient errors when posting a message to a topic, and transparently retries the **Send** operation when appropriate. The rationale behind this approach is similar to that described in the previous point.

The following code shows the definition of the **Send** method in the **ServiceBusTopic** class. As described above, two **Action** methods are passed to the method as parameters (one to execute after a message is sent and one to execute if sending fails), together with a function that creates the message, and a state object.

```C#
public void Send(Func<BrokeredMessage> createMessage,
                 object objectState,
                 Action<object> afterSendComplete,
                 Action<Exception, object> processError)
{
    ...
}
```

The **Send** method uses the Transient Fault Handling Block. The constructor for the **ServiceBusTopic** class initializes the block, loads the default policy for Windows Azure Service Bus connectivity, and sets up a handler for the **Retrying** event that writes information to Windows Azure diagnostics.

*Code can only read the body of a **BrokeredMessage** instance once. When you implement a method that uses a **BrokeredMessage** instance and may be executed more than once, as is the case when using the Transient Fault Handling Application Block, you must create and populate the **BrokeredMessage** instance each time you call the method. This is why the **Send** method accepts a function that creates the message, instead of accepting an existing message instance.*

```C#
this.serviceBusRetryPolicy = RetryPolicyFactory.
  GetDefaultAzureServiceBusRetryPolicy();

this.serviceBusRetryPolicy.Retrying += (sender, args) =>
   TraceHelper.TraceWarning("Retry in ServiceBusTopic - "
      + "Count:{0}, Delay:{1}, Exception:{2}",
      args.CurrentRetryCount, args.Delay,
      args.LastException);
```

The **Send** method then calls one of the asynchronous overloads of the **ExecuteAction** method of the Transient Fault Handling Block and passes in the required parameters, as shown in the following code extract. These parameters are the asynchronous start and end delegates, an action to execute after the process completes successfully, and an action to execute if it fails after retrying a number of times (the retry policy parameters are specified in the configuration file).

```C#
this.serviceBusRetryPolicy.ExecuteAction<BrokeredMessage>(
  ac =>
  {
    var message = createMessage();
    var dictionary
        = (objectState as Dictionary<string, object>);
    if (dictionary.ContainsKey("message"))
    {
      dictionary["message"] = message;
    }
    else
    {
      dictionary.Add("message", message);
    }
    this.sender.BeginSend(message, ac, objectState);
  },
  ar =>
  {
    this.sender.EndSend(ar);
    return (ar.AsyncState as Dictionary<string,
            object>)["message"] as BrokeredMessage;
  },
  (message) =>
  {
    try
    {
      afterSendComplete(objectState);
    }
    catch (Exception ex)
```

```
        {
          TraceHelper.TraceError(ex.Message);
        }
        finally
        {
          message.Dispose();
        }
      },
      e =>
      {
        processError(e, objectState);
        var message = (objectState as Dictionary<string,
                    object>)["message"] as BrokeredMessage;
        message.Dispose();
      });
```

The asynchronous start delegate (**ac**) first calls the function passed to the **Send** method as the **createMessage** parameter to create the **BrokeredMessage** instance. Next, it obtains a reference to a **Dictionary** stored in the object state (which is also passed to the **Send** method as a parameter) and adds to it a reference to the **BrokeredMessage** instance. It must hold on to a reference to the **BrokeredMessage** so that it can be disposed correctly afterwards. The code then calls the **BeginSend** method of the Service Bus **MessageSender** instance referenced by the **sender** property of the **ServiceBusTopic** class. It passes as parameters the **BrokeredMessage** instance to send, a reference to the callback provided by the Transient Fault Handling Application Block, and a **Dictionary** containing a copy of the message as the object state. A reference to this copy of the brokered message is maintained so that the code can dispose it and correctly release the resources it uses. This occurs in the actions that are executed after sending the message, regardless of whether the send operation is successful or if it fails.

> *The **Dictionary** provides a thread-safe object that holds the state information referenced by the **BeginSend** and **EndSend** methods that send a message to the queue asynchronously.*

The asynchronous end delegate (**ar**) first calls the **EndSend** method of the Service Bus **MessageSender** instance. Next, it extracts the **Dictionary** containing the message from the object state and returns it as a **BrokeredMessage** instance. This is passed to the action that is executed when the message is successfully sent.

If the process successfully posts the message to the topic, it invokes the **Action** referenced by the **afterSendComplete** parameter. If it fails to execute this action, the code uses the **TraceHelper** class to log an error message.

If the process fails to post the message to the topic, it invokes the **Action** referenced by the **processError** parameter. The code passes to the **processError** action the exception returned from the **MessageSender** class and the object state containing the message. After the **processError** action completes, the code obtains a reference to the **BrokeredMessage** instance stored in the **objectState** variable and disposes it.

Subscribing to a Service Bus Topic

One of the major advantages of using Service Bus topics to distribute messages is that it provides a level of decoupling between the sender and receivers. The sender can construct a message with additional properties that filters within the topic use to redirect the message to specific receivers. However, receivers must subscribe to receive messages, and the number of subscribers is independent of the topic. For example, Trey Research can add new transport partners and arrange to send messages to these new partners simply by editing the filter criteria in the topic. New receivers can subscribe to a topic and receive messages that match the appropriate filtering criteria. Trey Research could add additional subscribers that listen for messages and pass them to auditing or other types of services.

In the Orders application, Trey Research created a single Service Bus topic for each deployed instance of the application (in other words, there is one topic per datacenter). All of the transport partners subscribe to all of these topics, and receive messages destined for them based on the filter rules Trey Research established for the choice of transport partner.

The Service Bus subscriptions and filters themselves are created by the **SetupServiceBusTopicAndQueue** method in the setup program in the **TreyResearch.Setup** project. The following code shows the relevant parts of this method.

```csharp
C#
private static void SetupServiceBusTopicAndQueue()
{
  ...
  // Create one subscription per transport partner with
  // corresponding filter expression.
  var transportPartners = new[] {
    "Contoso", "Fabrikam" };
  for (int i = 0; i <= 1; i++)
  {
    string transportPartnerName = transportPartners[i];
    string formattedName = transportPartnerName.Replace(
      " ", string.Empty).ToLowerInvariant();
    ...

    var serviceBusTopicDescription =
      new ServiceBusSubscriptionDescription
      {
        Namespace = ServiceBusNamespace,
        TopicName = TopicName,
        SubscriptionName = string.Format(
          "{0}Subscription", formattedName),
        Issuer = Issuer,
        DefaultKey = DefaultKey
      };

    var serviceBusSubscription =
      new ServiceBusSubscription(
        serviceBusTopicDescription);
```

```csharp
    string filterExpression = string.Format(
      "TransportPartnerName = '{0}'",
        transportPartnerName);
    serviceBusSubscription.CreateIfNotExists(
     filterExpression);
    ...
  }
}
```

Trey Research implemented two custom classes (**ServiceBusSubscriptionDescription** and **ServiceBusSubscription**, located in the Communication folder of the **Orders.Shared** project) for connecting to subscriptions. The **ServiceBusSubscriptionDescription** class specifies the properties for a subscription, indicating which topic in which Service Bus namespace to connect to and the name of the subscription to use. The following code example shows the definition of this class. Note that in the Trey Research example, the code populates the **SubscriptionName** property with the name of the transport partner; remember that each Service Bus subscription filters messages by using this property.

C#
```csharp
public class ServiceBusSubscriptionDescription
{
  public string Namespace { get; set; }
  public string TopicName { get; set; }
  public string SubscriptionName { get; set; }
  ...
}
```

The constructor of the **ServiceBusSubscription** class accepts a populated instance of the **ServiceBusSubscriptionDescription** class and connects to the specified topic and subscription, as shown in the following code. This method also creates a **MessageReceiver** for the topic subscription.

C#
```csharp
public ServiceBusSubscription(
       ServiceBusSubscriptionDescription description)
{
  ...
  var runtimeUri
    = ServiceBusEnvironment.CreateServiceUri("sb",
                 this.description.Namespace,
                 string.Empty);
  var messagingFactory
    = MessagingFactory.Create(runtimeUri, ...);

  this.receiver
    = messagingFactory.CreateMessageReceiver(
          this.description.TopicName.ToLowerInvariant()
          + "/subscriptions/" +
```

```
                        this.description.SubscriptionName
                                .ToLowerInvariant(),
                        ReceiveMode.PeekLock);
}
```

Receiving Messages from a Topic and Processing Them Asynchronously

To receive a message from a topic, an application can use a **ServiceBusReceiverHandler** object initialized with the receiver encapsulated within a **ServiceBusSubscription** object. For more information about the **ServiceBusReceiverHandler** class, see the section "Receiving Messages from a Service Bus Queue and Processing Them Asynchronously" earlier in this chapter. The following code example shows how an application can create a **ServiceBusReceiverHandler** object to receive **NewOrderMessage** messages from the subscription (the **NewOrderMessage** class is described in Chapter 5, "Processing Orders in the Trey Research Solution.")

> The message polling interval you specify for receiving messages from a queue or topic must take into account variables specific to your environment (such as CPU processing power) and the expected volume of work (such as the number of orders to process and number of worker role instances).

```C#
var serviceBusSubscription = new ServiceBusSubscription(
          ...);
var receiverHandler
  = new ServiceBusReceiverHandler<NewOrderMessage>
          (serviceBusSubscription.GetReceiver())
{
  MessagePollingInterval = TimeSpan.FromSeconds(2)
};
...
```

An application can then call the **ProcessMessages** method of the **ServiceBusReceiverHandler** instance it just retrieved, and pass it a delegate specifying the code to be executed as each message is received. Again, this process was described in the section "Receiving Messages from a Service Bus Queue and Processing Them Asynchronously" earlier in this chapter. The following code shows an example:

```C#
...
receiverHandler.ProcessMessages(
  (message, queueDescription, token) =>
  {
      return Task.Factory.StartNew(
        () => this.ProcessMessage(message,
                                  queueDescription),
        this.tokenSource.Token,
        TaskCreationOptions.None,
```

```
    context);
},
this.tokenSource.Token);
```

Implementing Adapters and Connectors for Translating and Reformatting Messages
As described in the section "Selected Option for Communicating with Transport Partners" earlier in this chapter, Trey Research uses connectors and adapters for retrieving messages from the Service Bus subscription for each transport partner, and then translates these messages into a format that the transport partner understands before handing the message off for processing.

> *In the solution code provided with this guide, mock versions of the local and distance transport partners are both implemented by means of a Windows Forms application. For the local transport partner, Contoso, the connector is integrated into the Windows Forms application. For the distance transport partner, Fabrikam, the connector is implemented in a similar manner as part of the Windows Forms code for this partner. However, this is for simplicity and demonstration purposes only; in the real implementation Trey Research incorporates the adapter for the distance partner into the worker role as described earlier.*

The Trey Research solution includes two sample transport partners; one that handles deliveries to local customers that reside in the same state or neighboring states as Trey Research and another that delivers goods to more distant customers. These transport partners are defined in the **ContosoTransportPartner** and **FabrikamTransportPartner** Windows Forms classes in the **TransportPartner** project. Both transport partners implement their own systems for tracking and delivering packages.

Contoso, the local transport partner runs a connector on its own infrastructure that connects directly to the Windows Azure Service Bus to retrieve and send messages. This functionality is implemented in the **Connector** class in the Connectivity folder. Fabrikam, the distance transport partner exposes a service interface, and an adapter running as part of the Trey Research solution interacts with the Service Bus and reformats messages into service calls; responses from the service are reformatted as messages and posted back to the Service Bus. The adapter is implemented in the **Adapter** class, also located in the Connectivity folder.

When the transport partner receives a request to deliver an order, the connector or adapter (depending on the transport partner) posts an acknowledgement message to a Service Bus queue. This queue constitutes a well-known but secure endpoint, available to all transport partners. The **Connector** and **Adapter** classes are both descendants of the **OrderProcessor** class (defined in the Connectivity folder in the **TransportPartner** project), and this class actually handles the connectivity between the transport partner and the Service Bus. In the **FabrikamTransportPartner** Windows Forms class, the flow of control is:

- The **OnLoad** method instantiates the **Adapter** object and invokes its **Run** method. The **Run** method of the **Adapter** class is inherited from the **OrderProcessor** class.

- The **Run** method in the **OrderProcessor** class creates a **ServiceBusReceiverHandler** object to connect to the Service Bus subscription on which it expects to receive orders, and calls the **ProcessMessages** method of this object.

- The first parameter to the **ProcessMessages** method in the **ServiceBusReceiverHandler** class is a delegated function (specified as a lambda expression in the sample code) that provides the business logic to be performed when an order is received from the topic.

- The **ServiceBusReceiverHandler** object invokes this function after it has received each message. This strategy enables you to decouple the mechanics for receiving a message from a queue or topic (as implemented by the **ServiceBusReceiverHandler** class) from the logic for converting this message into the format expected by the transport partner and sending this request to the internal system implemented by the partner.

The following example, taken from the OrderProcessor.cs file, shows how this code is structured.

```C#
public void Run()
{
    ...
    foreach (...)
    {
        ...
        var receiverHandler = new
          ServiceBusReceiverHandler<...>(...);

        receiverHandler.ProcessMessages(
          (message, ..., ...) =>
          {
             return Task.Factory.StartNew(
                // Message conversion logic goes here.
                // The message parameter contains the body of
                // the message received from the topic.
                () => this.ProcessMessage(
                       message, ...),
                ...);
          }, ...);
    }
}
```

In the **OrderProcessor** class, the lambda expression invokes the local **ProcessMessage** method (not to be confused with **ServiceBusReceiverHandler.ProcessMessages**) to pass the message to the local partner's internal system and wait for a response by calling the **ProcessOrder** method (this method provides logic that is specific to the transport partner and is implemented in the **Adapter** class.) Because the **ProcessMessage** method runs by using a separate task, it can wait synchronously for the **ProcessOrder** method to complete without affecting the responsiveness of the application. The following code example shows part of the implementation of the **ProcessMessage** method in the **OrderProcessor** class.

> Many of the details of the **ProcessMessage** method, such as the purpose of the **trackingId** variable, and the operations performed by the **ProcessOrder** method in the **Connector** and **Adapter** classes provided in the sample solution are explained in detail in Chapter 5, "Processing Orders in the Trey Research Solution".

```C#
protected virtual void ProcessMessage(
  NewOrderMessage message,
  ServiceBusQueueDescription queueDescription)
{
  var trackingId = this.ProcessOrder(
    message, queueDescription);

  if (trackingId != Guid.Empty)
  {
    ...
    this.SendOrderReceived(message,
      queueDescription, statusMessage, trackingId, token);
  }
}
```

When the order has been processed by the transport partner, the **ProcessMessage** method invokes the local **SendOrderReceived** method of the **OrderProcessor** object to send an appropriate response message back to the Order application through the Service Bus queue specified as the second parameter to the **ProcessMessage** method.

> *The details of the **SendOrderReceived** method are also described in Chapter 5, "Processing Orders in the Trey Research Solution."*

Correlating Messages and Replies

Unlike web service operations, messaging implemented by using Service Bus Message queues and topics is an inherently one-way mechanism. Although sometimes viewed as a limitation, this is actually what makes this form of messaging extremely responsive; a sender does not have to wait for a response from a distant, possible unreliable receiver whenever it posts a message. However, there will inevitably be cases when a sender expects some form of reply, even if it is only an acknowledgement that the receiver has actually received the message. This is precisely the situation in the Trey Research scenario. When the Orders application posts the details of an order to a topic, the application expects to receive a response that indicates the order has been received.

However, there may be a significant time between these two events, and the Orders application must not block waiting for the response to arrive. To address this situation, Trey Research implements two-way messaging by using a combination of Service Bus topics and queues. The Orders application posts order messages to a Service Bus topic, and expects the responses from the various transport partners to appear on a separate Service Bus queue. The key question is how does the Orders application know which response belongs to which order message? The answer lies in using message correlation.

When the worker role for the Orders application sends an order message to a transport partner, it populates the **MessageId** property with the identifier for the order (this identifier is generated when the order is created), and it also specifies the name of the queue on which the Orders application expects a response in the **ReplyTo** property, as shown in the following code sample taken from the **Execute** method in the **NewOrderJob** class.

```C#
var brokeredMessage = new BrokeredMessage(msg)
{
  MessageId = msg.OrderId.ToString(),
  ...
  ReplyTo = this.replyQueueName
};
```

The transport partner constructs an **OrderStatusUpdateMessage** object as a reply and then posts this message to the queue specified by the **ReplyTo** property of the original order message. In the Trey Research example, this logic occurs in the **SendToUpdateStatusQueue** method (invoked by the **SendOrderReceived** method) in the **OrderProcessor** class. Chapter 5, "Processing Orders in the Trey Research Solution" describes the message flow of through the transport partners in more detail.

The worker role receives the response on the specified Service Bus queue. When a response message is received it is used to update the status of the order in the Orders database. This functionality is implemented in the **StatusUpdateJob** class in the worker role, which is also described in detail in Chapter 5, "Processing Orders in the Trey Research Solution."

Securing Message Queues, Topics, and Subscriptions

A key requirement of the messaging solution is that all messages should be protected from unauthorized access. As described in Chapter 3, "Authenticating Users in the Orders Application," the Windows Azure Service Bus uses the ACS to protect Service Bus queues, topics, and subscriptions. To connect to a queue, topic, or subscription, an application must present a valid authentication token.

To secure the communication channels Trey Research defined rules in ACS to allow the local and distance transport partners to connect to the Service Bus subscription on which the Orders application posts order messages; the transport partners are granted the "Listen" claim to the subscription to enable them to receive messages only. The worker role in the Orders application is granted the "Send" claim over the topic to enable it to post messages only.

> *For more information about these claims and how to configure ACS to authenticate clients and authorize access to Service Bus artifacts, see "Service Bus Authentication and Authorization with the Access Control Service" on MSDN.*

For the Service Bus queue that the worker role listens to for response messages, the privileges are reversed; the transport partners are granted the "Send" claim to the queue while the worker role has the "Listen" claim.

> *The various ACS rules and rule groups and identities used by the Orders application and the transport partners are created by the setup program in the **TreyResearch.Setup** project.*

> Passing the address to which a receiving application should post a response by using the **ReplyTo** property of a message decouples the receiving application from using a specific hard-coded queue.

For completeness, the following table summarizes how Trey Research configured ACS to enable Service Bus authentication for applications and services connecting to the various Service Bus queues, topics, and subscriptions in the Orders application.

Service Bus artifact	Setting
Service identities.	AuditLogListener, Fabrikam, HeadOffice, Contoso, NewOrderJob, NewOrdersTopic, owner, StatusUpdateJob.
Default Service Bus (relying party).	Name: ServiceBus. Realm: http://treyresearch.servicebus.windows.net/ Claim issuer: ACS. Token type: SWT. Rule groups: • Default rule group containing: If name identifier="owner" emit "Manage", "Send", and "Listen" action claims.
Service Bus topic (relying party) for sending new order details to transport partners and the on-premises audit log.	Name: NewOrdersTopic Realm: http://treyresearch.servicebus.windows.net/neworderstopic Claim issuer: ACS. Token type: SWT. Rule groups: • Default rule group for ServiceBus • Rule group containing: If name identifier="NewOrderJob" emit "Send" action claim. Subscriptions: • Local ("Contoso") and distance ("Fabrikam") shipping partners. • Audit log service.
Service Bus queue (relying party) that transport partners use to send messages to the Orders application that: • Acknowledge receipt of new order details. • Indicate that the order has been delivered.	Name: OrderStatusUpdateQueue Realm: http://treyresearch.servicebus.windows.net/orderstatusupdatequeue Claim issuer: ACS. Token type: SWT. Rule groups: • Default rule group for Service Bus. • Rule group containing: If name identifier="Contoso" emit "Send" action claim. • Rule group containing: If name identifier="Fabrikam" emit "Send" action claim. • Rule group containing: If name identifier="StatusUpdateJob" emit "Listen" action claim.
Transport partner (relying party) for local deliveries (Contoso, Inc.)	Name: Contoso Realm: http://treyresearch.servicebus.windows.net/neworderstopic/subscriptions/contososubscription Claim issuer: ACS. Token type: SWT. Rule groups: • Default rule group for ServiceBus. • Rule group containing: If name identifier="Contoso" emit "Listen" action claim.
Transport partner (relying party) for distance deliveries (Fabrikam Inc.)	Name: Fabrikam Realm: http://treyresearch.servicebus.windows.net/neworderstopic/subscriptions/fabrikamsubscription Claim issuer: ACS. Token type: SWT. Rule groups: • Default rule group for ServiceBus. • Rule group containing: If name identifier="Fabrikam" emit "Listen" action claim.
On-premises management and monitoring application (relying party). Subscribes to Topic to collect audit log messages.	Name: AuditLogListener Realm: http://treyresearch.servicebus.windows.net/neworderstopic/subscriptions/auditloglistenersubscription Claim issuer: ACS. Token type: SWT. Rule groups: • Default rule group for ServiceBus. • Rule group containing: If name identifier="AuditLogListener" emit "Listen" action claim.

The worker role and transport partners are each configured with the appropriate storage account key, and they present this information when they connect to the Service Bus queue, topic, or subscription in the form of a simple web token. For example, the **Run** method in the **NewOrderJob** class in the worker role uses the following code to extract the key and issuer details from the application configuration and store them in a **ServiceBusTopicDescription** object, which is in turn used to create a **ServiceBusTopic** object.

```C#
public void Run()
{
  ...
  this.serviceBusNamespace = CloudConfiguration.
    GetConfigurationSetting("serviceBusNamespace",
      string.Empty);
  this.acsNamespace = CloudConfiguration.
    GetConfigurationSetting("acsNamespace", string.Empty);
  var topicName = CloudConfiguration.
    GetConfigurationSetting("topicName", string.Empty);
  var issuer = CloudConfiguration.
    GetConfigurationSetting("newOrdersTopicIssuer",
      string.Empty);
  var defaultKey = CloudConfiguration.
    GetConfigurationSetting("newOrdersTopicKey",
      string.Empty);
  ...

  var serviceBusTopicDescription =
    new ServiceBusTopicDescription
    {
      Namespace = this.serviceBusNamespace,
      TopicName = topicName,
      Issuer = issuer,
      DefaultKey = defaultKey
    };

  this.newOrderMessageSender =
      new ServiceBusTopic(serviceBusTopicDescription);
  ...
}
```

The constructor in the **ServiceBusTopic** class uses this information to create a token provider for a **MessageFactory** object. The **MessageFactory** object is used to construct the **MessageSender** object that the **ServiceBusTopic** object utilizes to actually post messages to the underlying Service Bus topic.

```C#
...
private readonly ServiceBusTopicDescription description;
private readonly TokenProvider tokenProvider;
private readonly MessageSender sender;
...

public ServiceBusTopic(
  ServiceBusTopicDescription description)
{
  ...

  this.description = description;
  this.tokenProvider = TokenProvider.
    CreateSharedSecretTokenProvider(
      this.description.Issuer,
      this.description.DefaultKey);

  var runtimeUri = ServiceBusEnvironment.
    CreateServiceUri("sb", this.description.Namespace,
                     string.Empty);
  var messagingFactory = MessagingFactory.Create(
    runtimeUri, this.tokenProvider);
  this.sender = messagingFactory.CreateMessageSender(
    this.description.TopicName.ToLowerInvariant());
  ...
}
```

The constructors of the **ServiceBusQueue** and **ServiceBusSubscription** classes follow a similar pattern.

Securing Messages

To help prevent spoofing, Trey Research also implements a mechanism to verify the identity of a transport partner posting messages to the Service Bus queue on which the Orders application listens. This helps to ensure that a rogue third party is not somehow impersonating a valid transport partner and sending fake messages. To accomplish this, each time a transport partner sends a message, it adds a simple web token in the header of the message that indicates the identity of the sender, and the receiver in the Orders application validates the token when each message arrives.

Adding tokens to the header and validating them cannot be achieved just by configuring the Service Bus artifacts and ACS. Instead, Trey Research uses the following code to obtain a token from ACS. This code is taken from the **OrderProcessor** class; this is the base class from which the **Adapter** and **Connector** classes used by the transport partners descend.

```csharp
private string GetToken(ServiceBusQueueDescription
                            queueDescription)
{

  var realm = string.Format("urn:{0}/{1}",
        queueDescription.QueueName,
        HttpUtility.UrlEncode(this.acsServiceIdentity));

  var token = GetTokenFromAcs(string.Format(
          "https://{0}.accesscontrol.windows.net/",
          queueDescription.SwtAcsNamespace),
       this.acsServiceIdentity, this.acsPassword, realm);

  return token;
}

private string GetTokenFromAcs(string acsNamespace,
             string serviceIdentity, string password,
             string relyingPartyRealm)
{
  // request a token from ACS
  var client = new WebClient();
  client.BaseAddress = acsNamespace;
  var values = new NameValueCollection();
  values.Add("wrap_name", serviceIdentity);
  values.Add("wrap_password", password);
  values.Add("wrap_scope", relyingPartyRealm);
  byte[] responseBytes = client.UploadValues(
         "WRAPv0.9/", "POST", values);
  string response = Encoding.UTF8.GetString(responseBytes);
  return HttpUtility.UrlDecode(
      response
        .Split('&')
        .Single(value =>
          value.StartsWith("wrap_access_token=",
            StringComparison.OrdinalIgnoreCase))
        .Split('=')[1]);
}
```

The **GetTokenFromAcs** method (also shown in the previous code example) sends a request for a token to ACS. The **GetToken** method passes to it values extracted from the application's configuration file for the service identity name, password, and realm to create the appropriate token for this sender's identity.

After obtaining a suitable token, the transport partner can add it to the message that it posts to the Service Bus queue.

When a message is received, the receiver can extract the token and use it to verify the identity of the sender. For example, the worker role uses the **IsValidToken** method in the **StatusUpdateJob** class shown in the following code example to establish whether the token extracted from a message is valid.

```C#
private bool IsValidToken(Guid orderId, string token)
{
  string transportPartner;
  ...

  string acsServiceNamespace = CloudConfiguration.
    GetConfigurationSetting("acsNamespace", null);
  string acsUsername = CloudConfiguration.
    GetConfigurationSetting("acsUsername", null);
  string acsPassword = CloudConfiguration.
    GetConfigurationSetting("acsUserKey", null);

  var acsWrapper = new ServiceManagementWrapper(
    acsServiceNamespace, acsUsername, acsPassword);
  var relyingParty = acsWrapper.
    RetrieveRelyingParties().
    SingleOrDefault(
      rp => rp.Name.Contains(transportPartner));

  var keyValue = string.Empty;

  if (relyingParty != null)
  {
    var key = relyingParty.
      RelyingPartyKeys.
      FirstOrDefault();
      ...
    keyValue = Convert.ToBase64String(key.Value);
  }

  // Values for trustedAudience:
  //   urn:[queue-name]/[partner-name]
  var trustedAudience = string.Format("urn:{0}/{1}",
    CloudConfiguration.GetConfigurationSetting(
      "orderStatusUpdateQueue", string.Empty),
    HttpUtility.UrlEncode(transportPartner));

  var validator = new TokenValidator(
    "accesscontrol.windows.net",
    RoleEnvironment.GetConfigurationSettingValue(
```

```
        "acsNamespace"),
    trustedAudience, keyValue);

    return validator.Validate(token);
}
```

The **IsValidToken** method uses the classes in the **ACS.ServiceManagementWrapper** project to retrieve information about the various relying parties configured in ACS. For more information about the **ACS.ServiceManagementWrapper**, see "Access Control Service Samples and Documentation" at *http://acs.codeplex.com/releases/view/57595*.

The **IsValidToken** method also uses a separate class named **TokenValidator** in the **Orders.Workers** project to actually validate the token given the ACS hostname, the service namespace, the audience value, and the signing key.

Sending Orders to the Audit Log

Currently, all orders with a value over $10,000 must be audited, and the audit log is held on-premises. The order processing logic must be able to quickly determine the total cost of an order and direct the details to the audit log. This processing must happen quickly without impinging on the responsiveness to customers, it must be scalable as the volume of orders increases, and it must be flexible enough to allow the auditing criteria to be quickly changed, again without extensively rewriting the code for the worker role. As with orders sent to transport partners, all audit information is extremely sensitive and must be protected against unauthorized access, especially as it traverses the network.

Choosing a Mechanism for Sending Orders to the Audit Log

Once Trey Research had settled on the use of Service Bus topics as the mechanism for communicating with transport partners, they decided to use the same approach for auditing messages. When a customer places an order, the total value of the order is calculated and added as a property called **OrderAmount** to the message. If the value is more than $10,000, it is picked up by the audit subscription and sent to the on-premises auditing application at Trey Research. Figure 5 highlights how this technology fits into the Trey Research solution. Notice that the audit log uses the same Service Bus topic as the transport partners, but with a subscription that applies a different filter.

> *Remember that if a message posted to a topic satisfies the filter associated with more than one subscription, a copy of the message will be routed to all matching subscriptions.*

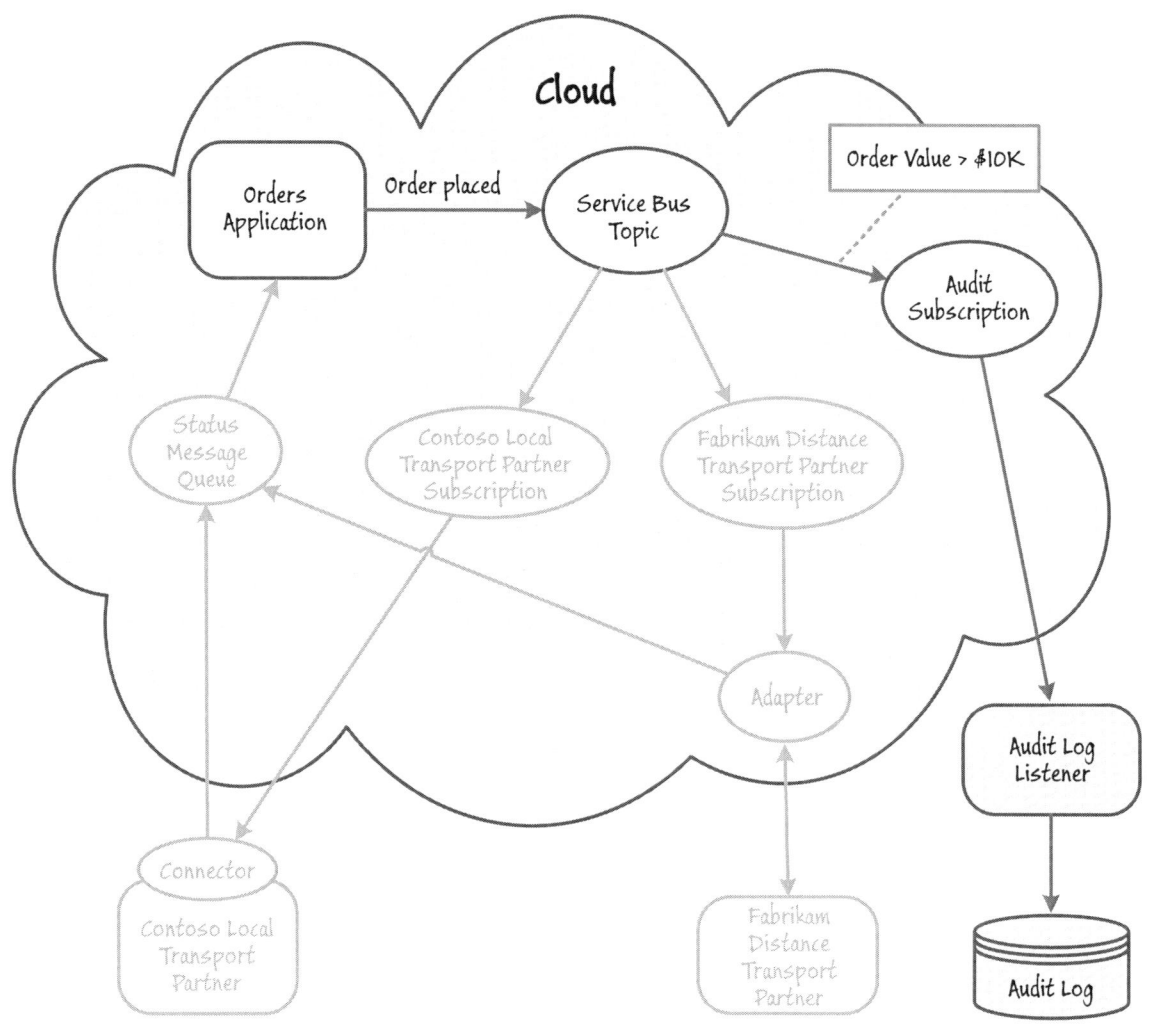

FIGURE 5
Messaging technology used by Trey Research to route orders to the audit log

How Trey Research Sends Orders to the Audit Log

The total value of the order is added as the **OrderAmount** property to every order message posted by the worker role to the Service Bus topic. The Trey Research application identifies all orders that require auditing by creating a Service Bus subscription with an appropriate filter. The code that creates this Service Bus subscription is located in the **SetupAuditLogListener** method in the setup program in the **TreyResearch.Setup** project. The following code example shows the parts of this method that configure the filter.

```C#
private static void SetupAuditLogListener()
{
  var formattedName = AuditLogListener.Replace(" ",
    string.Empty).ToLowerInvariant();
  ...
  var serviceBusTopicDescription =
    new ServiceBusSubscriptionDescription
    {
      Namespace = ServiceBusNamespace,
      TopicName = TopicName,
      SubscriptionName = string.Format(
        "{0}Subscription", formattedName),
      Issuer = Issuer,
      DefaultKey = DefaultKey
    };

  var serviceBusSubscription =
    new ServiceBusSubscription(serviceBusTopicDescription);
  const int AuditAmount = 10000;
  var filterExpression = string.Format(
    "OrderAmount > {0}", AuditAmount);
  serviceBusSubscription.CreateIfNotExists(
    filterExpression);

  ...
}
```

The Trey Research Head Office web application in the **HeadOffice** project includes the **AuditController** class, which connects to this subscription and retrieves orders to be audited. The **DownloadLogs** method in this class contains the code that actually retrieves the details of the orders to be audited. Note that this method connects to the Service Bus topic in each datacenter in which the Orders application runs; each instance of the Orders application posts messages to the topic in its local datacenter. The name of the subscription to use, the name of the topic, and the security keys are stored in the configuration file with the application.

```C#
public ActionResult DownloadLogs()
{
  ...
  var serviceBusNamespaces = WebConfigurationManager.
    AppSettings["AuditServiceBusList"].Split(',').ToList();
  ...

  foreach (var serviceBusNamespace in serviceBusNamespaces)
```

```csharp
{
    // Connect to servicebus, download messages from the
    // Audit log subscription, save to database.
    var serviceBusTopicDescription =
      new ServiceBusSubscriptionDescription
      {
        Namespace = serviceBusNamespace,
        TopicName = WebConfigurationManager.
          AppSettings["topicName"],
        SubscriptionName = WebConfigurationManager.
          AppSettings["subscriptionName"],
        Issuer = WebConfigurationManager.
          AppSettings["issuer"],
        DefaultKey = WebConfigurationManager.
          AppSettings["defaultKey"]
      };

    var serviceBusSubscription = new
      ServiceBusSubscription(serviceBusTopicDescription);

    // MessagePollingInterval should be configured taking
    // into consideration variables such as CPU
    // processing power, expected volume of orders to
    // process and number of worker role instances
    var receiverHandler =
      new ServiceBusReceiverHandler<NewOrderMessage>(
        serviceBusSubscription.GetReceiver()) {
          MessagePollingInterval =
            TimeSpan.FromSeconds(2) };

    receiverHandler.ProcessMessages(
      (message, queueDescription, token) =>
      {
        return Task.Factory.StartNew(
          () => this.ProcessMessage(
            message, queueDescription),
          ...);
      },
      ...);
}

return RedirectToAction("Index");
}
```

The **ProcessMessage** method (called by the **ProcessMessages** method of the **ServiceBusReceiverHandler** object) simply saves the order message data to a local SQL Server database.

```C#
public void ProcessMessage(NewOrderMessage message,
  ServiceBusQueueDescription queueDescription)
{
  // Save the AuditLog to the database
  var auditLog = new AuditLog
  {
    OrderId = message.OrderId,
    OrderDate = message.OrderDate,
    Amount = Convert.ToDecimal(message.Amount),
    CustomerName = message.CustomerName
  };

  this.auditLogStore.Save(auditLog);
}
```

Verifying Orders to Ensure Regulatory Compliance

The final challenge concerns integration with the compliance application. This application examines orders for compliance with export restrictions and government regulations for technical products. The compliance application communicates with the Orders database using a standard SQL Server connection string, and executes queries to determine compliance on a pre-determined schedule. Additionally, the compliance application generates reports that are stored in a secure on-premises location within the Trey Research Head Office.

When the application was deployed on-premises, it accessed the Orders database and the secure reporting location that were also located on-premises. Now that the Orders database is located in the cloud, the compliance application must connect to a SQL Azure instance. This is a simple configuration issue that can be easily resolved. However, the volume of traffic between the compliance application and the Orders database is considerable as the compliance application executes its many data queries and searches. These factors led Trey Research to consider in more depth how the application itself should be deployed.

The source code for this application is confidential and not available; a government department specifies the processes it must follow and certifies the operation. This makes it difficult if not impossible to refactor the application as a worker role. In addition, the reporting functionality requires authenticated connectivity to the appropriate server, and all data transmitted over this connection must be secure.

Choosing Where to Host the Compliance Application

For hosting the compliance application, Trey Research decided to install and configure the application using the Windows Azure Virtual Machine (VM) role. This solution balances the need to configure, deploy, and maintain a VM role in the cloud, close to the orders data being examined, against the alternative of retaining the compliance application on-premises and either connecting to the orders data in the cloud or transferring the data from the cloud to an on-premises database.

The compliance application needs to access the secure location where it stores the various reports that it generates. This location is on an on-premises server, and Trey Research decided to use Windows Azure Connect to provide an authenticated, secure virtual network connection between the VM role and this server.

Trey Research chose to deploy the VM role to the US North Data Center as it is the closest datacenter to the Head Office, hopefully minimizing any network latency that may result from connecting across the on-premises/cloud boundary.

> When deciding whether to deploy an application to a VM role, you need to consider the benefits of reducing the network overhead of a chatty application such as the compliance application connecting to a database in the cloud against the cost of maintaining and managing the VM role.

How Trey Research Hosted the Compliance Application

The VM role that hosts the compliance application examines data held in the SQL Azure Orders database. The VM role is deployed to the US North Data Center, but the compliance application generates reports that are stored in a secure on-premises location within the Trey Research head office infrastructure. The compliance application also sends data to the monitoring application, which is also located on-premises; this application exposes a series of Distributed Component Object Model (DCOM) interfaces to which the compliance application connects for this purpose.

Trey Research implemented a separate small domain with its own Domain Name System (DNS) service in the on-premises infrastructure specifically for hosting the Windows Azure Connect endpoint software, the reporting data, and the monitoring application. Reports are stored in a share protected by using an Access Control List (ACL). Access is granted to an account defined within the domain. The compliance application, which is joined to the domain, provides these credentials when writing reports to this share. The same approach is used to protect the DCOM interface exposed by the monitoring application.

This section is provided for information only, showing how a solution could be implemented. The Trey Research example application does not actually include the compliance application or the corresponding VM role.

This domain has a trust relationship with the primary domain within Trey Research, and the management application running in the primary domain can periodically retrieve the reporting data and analyze the information logged by the monitoring application. Figure 6 shows the structure of the compliance system.

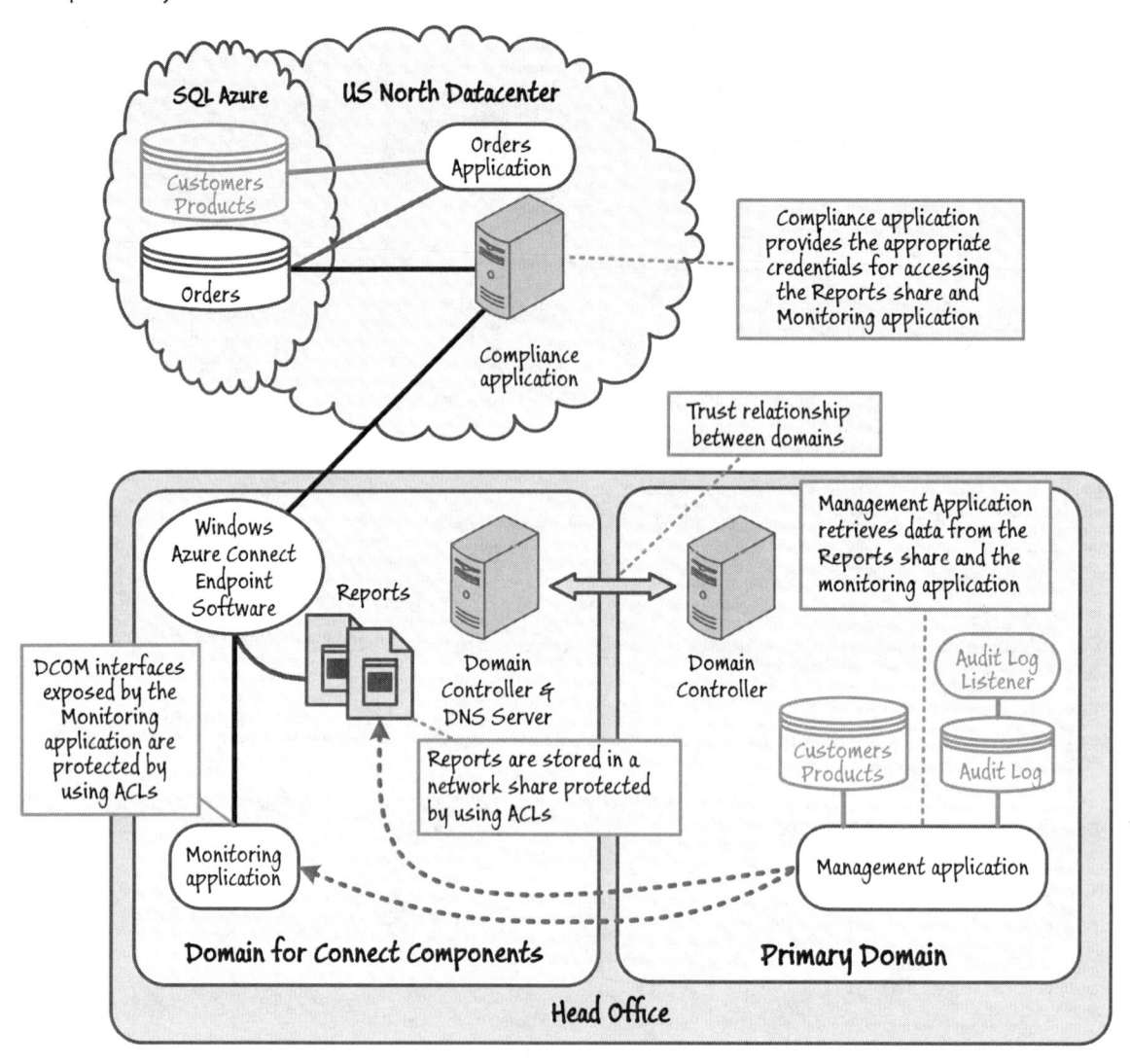

Figure 6
Structure of the compliance system

Summary

This chapter has looked at how Trey Research used two important Windows Azure technologies to implement a reliable cross-boundary communication layer based on Service Bus topics, subscriptions, and queues. These technologies provide a foundation that you can use to construct elegant hybrid solutions comprising components that need to communicate across the cloud/on-premises divide.

Service Bus queues enable you to implement asynchronous messaging that helps to remove the temporal dependency between the client application posting a request and the service receiving the request. Message-oriented applications are highly suited to use in cloud environments as they can more easily handle the variable volumes and peak loading that is typical of many commercial systems, and can easily be made robust enough to handle network and communications failure. Using Service Bus queues, you can implement a number of common messaging patterns and adapt them as appropriate to the requirements of your system.

Service Bus topics and subscriptions enable you to intelligently route messages to services. An application can post messages to a topic and include metadata that a filter can use to determine which subscriptions to route the message through. Services listening on these subscriptions then receive all matching messages. This simple but powerful mechanism enables you to address a variety of scenarios, and easily construct elegant solutions for these scenarios.

Finally, Windows Azure Connect enables you to establish a virtual network connection between a role hosted in the cloud and your on-premises infrastructure, and is suitable for situations where you need a direct connection between components rather than a message-oriented interface. You can share data across this network connection in a similar manner to accessing resources shared between computers running on-premises.

More Information

All links in this book are accessible from the book's online bibliography available at: *http://msdn.microsoft.com/en-us/library/hh968447.aspx*.

- "Hybrid Reference Implementation Using BizTalk Server, Windows Azure, Service Bus and SQL Azure" at *http://msdn.microsoft.com/en-us/windowsazure/hh547113(v=VS.103).aspx*.
- "An Introduction to Service Bus Queues" at *http://blogs.msdn.com/b/appfabric/archive/2011/05/17/an-introduction-to-service-bus-queues.aspx*.
- "Windows Azure Queues and Windows Azure Service Bus Queues - Compared and Contrasted" at *http://msdn.microsoft.com/en-us/library/windowsazure/hh767287(v=vs.103).aspx*.
- "Building loosely-coupled apps with Windows Azure Service Bus Topics and Queues" at *http://channel9.msdn.com/Events/BUILD/BUILD2011/SAC-862T*.
- "Best Practices for Leveraging Windows Azure Service Bus Brokered Messaging API" at *http://windowsazurecat.com/2011/09/best-practices-leveraging-windows-azure-service-bus-brokered-messaging-api/*.
- "Best Practices for Performance Improvements Using Service Bus Brokered Messaging" at *http://msdn.microsoft.com/en-us/library/windowsazure/hh528527.aspx*.
- "Queues, Topics, and Subscriptions" at *http://msdn.microsoft.com/en-us/library/windowsazure/hh367516.aspx*.
- "Service Bus Authentication and Authorization with the Access Control Service" at *http://msdn.microsoft.com/en-us/library/windowsazure/hh403962.aspx*.

- "Using Service Bus Topics and Subscriptions with WCF" at *http://blogs.msdn.com/b/tomholl/archive/2011/10/09/using-service-bus-topics-and-subscriptions-with-wcf.aspx.*
- "How to Simplify & Scale Inter-Role Communication Using Windows Azure Service Bus" at *http://windowsazurecat.com/2011/08/how-to-simplify-scale-inter-role-communication-using-windows-azure-service-bus/.*
- "Service Bus REST API Reference" at *http://msdn.microsoft.com/en-us/library/windowsazure/hh780717.aspx.*
- "Overview of Windows Azure Connect" at *http://msdn.microsoft.com/en-us/library/gg432997.aspx.*

5 Processing Orders in the Trey Research Solution

Having established an asynchronous messaging layer for communicating between the Orders application and the transport partners based on Service Bus topics, subscriptions, and queues, the developers at Trey Research were able to turn their attention to implementing the business logic for processing orders. This is the primary function of the Orders application running on the Windows Azure™ technology platform, and it was important for Trey Research to establish a robust mechanism that ensures orders will not be mislaid.

In this chapter, you will see how Trey Research designed the order processing and reporting logic to take full advantage of the scalability of the messaging framework, while ensuring that customers' orders are managed correctly and reliably.

Scenario and Context

Trey Research utilizes the services of external transport partners to ship orders to customers. These transport partners may implement their own systems, and no two transport partners necessarily follow the same procedures for handling orders; they simply guarantee that once Trey Research has provided them with the details of an order and a shipping address, they will collect the goods from the Trey Research manufacturing plant and deliver the goods to the customer.

For the reasons described in Chapter 4, "Implementing Reliable Messaging and Communications with the Cloud," Trey Research decided to use Service Bus topics, subscriptions, and queues to communicate with the transport partners. Using this infrastructure, after storing the details of a new order in the local database running on the SQL Azure™ technology platform, the Orders application posts these details to a well-known Service Bus topic and the transport partners each use their own Service Bus subscription to retrieve orders and ship them (each subscription has a filter that ensures that a transport partner receives only the orders that it should ship, and it does not have access to the orders intended for other transport partners.) Each transport partner responds with one or more messages indicating the current state of the shipping process; these messages are posted to a Service Bus queue. The Orders application retrieves these messages and uses them to update the status of the orders in the local SQL Azure database. Customers can use the Orders application to examine the status of their orders.

Figure 1 illustrates the logical flow of messages and data through the Orders application and the transport partners. Note that although the Order application always sends a single order message to a transport partner for each new order placed, the message or messages returned by the transport partner depend entirely on the internal systems implemented by that transport partner, and the Orders application simply records the details of each response as a status message in the SQL Azure database. This variation is reflected in the Trey Research sample solution. The local transport partner, Contoso, sends two messages; the first message acknowledges receipt of the order and responds with a tracking ID for the order while the second message is sent when the order has actually been shipped. The distance transport partner, Fabrikam, only sends a reply acknowledging the order, again containing a tracking ID.

> *Many commercial partners provide their own web applications that customers can use to query the delivery status of an order; the customer simply has to provide the tracking ID. This functionality is outside the scope of Trey Research.*

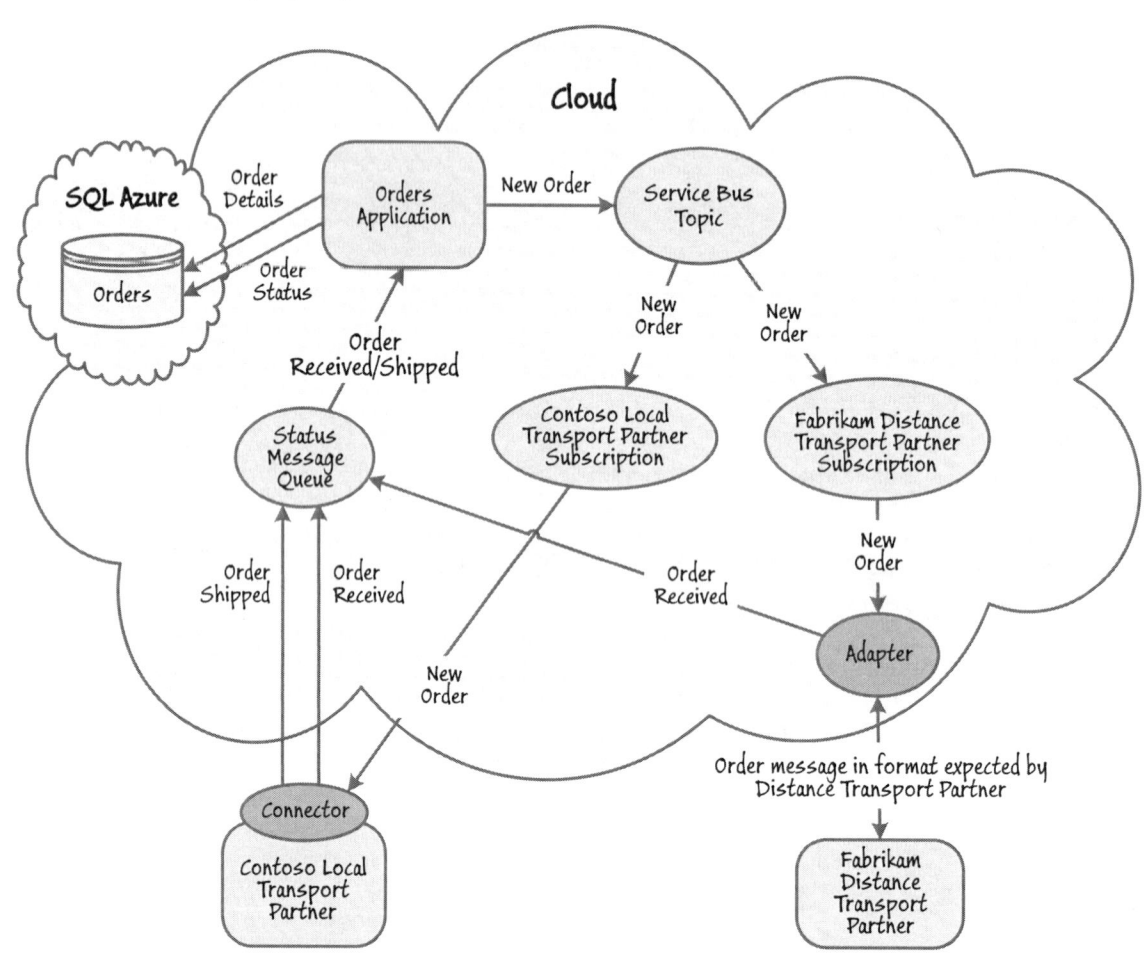

FIGURE 1
Logical flow of messages and data when a customer places an order

Orders with a value above $10,000 must be audited, and they are retrieved from the Service Bus topic by using a separate Service Bus subscription and directed to the audit log. The implementation of this part of the order process is described in the section "How Trey Research Sends Orders to the Audit Log" in Chapter 4, "Implementing Reliable Messaging and Communications with the Cloud." Additionally, each order is subjected to checking against export regulations by a separate compliance application. The section "How Trey Research Hosted the Compliance Application" (also in Chapter 4) describes how Trey Research configured the solution to support this requirement.

Neither of these two aspects of the order process is covered further in this chapter.

Processing Orders and Interacting with Transport Partners

Although simple in theory, there were a number of technical and logistical challenges, spread across two main areas of concern, which Trey Research had to address when implementing the order process.

- The process must be reliable and scalable.

    Once a customer has placed an order, the order must be fulfilled and should not be lost. Service Bus topics and queues provide a reliable mechanism for routing and receiving messages and, once posted, a message will not be lost. However, the act of posting a message involves connecting to a topic or queue across the Internet, and this is a potential source of failure. The order process must guarantee that orders are posted successfully.

    Additionally, the volume of orders may vary significantly over time, so the business logic implementing the order handling process must be scalable to enable it to post messages quickly without consuming excessive resources.

- The order process must be decoupled from the internal logic, number, and location of transport partners, as well as the location of the datacenter running the Orders application.

    The transport partners internal systems may be unable to connect directly to the Service Bus subscriptions and queues provided by Trey Research. Administrators at transport partners may be unwilling to install Trey Research's code on their systems.

    Furthermore, Trey Research may work with the services of different transport partners over time. The solution must be flexible enough to allow transport partners to be enrolled or removed from the system quickly and easily.

    Finally, the Orders application may be running in more than one datacenter, each hosting its own set of Service Bus topics, subscriptions, and queues. Each transport partner must be prepared to receive orders from a subscription in any of these datacenters, and post response messages back to the queue in the correct datacenter.

The following sections describe how Trey Research resolved these issues in their implementation.

How Trey Research Posts Messages to a Topic in a Reliable Manner

When a customer places an order the Orders application, the web role saves the order to a database. Subsequently, the worker role retrieves the order, posts the delivery details as a message to a transport partner through a Service Bus topic, and updates the status of the order to indicate that it has been sent. It is vital that the send and update operations both succeed, or that a corresponding recovery action or notification occurs if one of these operations fails; this is essentially the definition of a transaction. However, at the time of writing, Service Bus messaging only provides transactional behavior within the Service Bus framework. Windows Azure Service Bus does not currently support the use of the Microsoft Distributed Transaction Coordinator (DTC), so you cannot combine SQL Azure database operations with a Service Bus send or receive operation within the same transaction.

For this reason, Trey Research decided to implement a custom implementation that keeps track of the success or failure of each send operation that posts order messages to the Service Bus topic, and maintains the status of the order accordingly. This mechanism implements a pseudo-transaction; it is arguably a more complex process than may be considered necessary, but it can be more successful in countering transient faults (such as those caused by an intermittent connection to a Service Bus topic) and when using asynchronous messaging operations. This approach is also highly extensible; it is likely to be most useful when there are many operations to perform, or where there is a complex inter-relationship between operations.

Figure 2 shows a high-level overview of the solution Trey Research implemented to ensure that orders placed by customers are stored in the database and the delivery request is successfully sent to a transport partner and the audit log.

> It is vital to design your messaging code in such a way that it can cope with failures that may arise at any point in the entire messaging cycle, and not just when code fails to send or receive a message. Frameworks, components, and documentation are available to help you implement retry logic for Windows Azure messaging, such as *"Best Practices for Leveraging Windows Azure Service Bus Brokered Messaging API."*

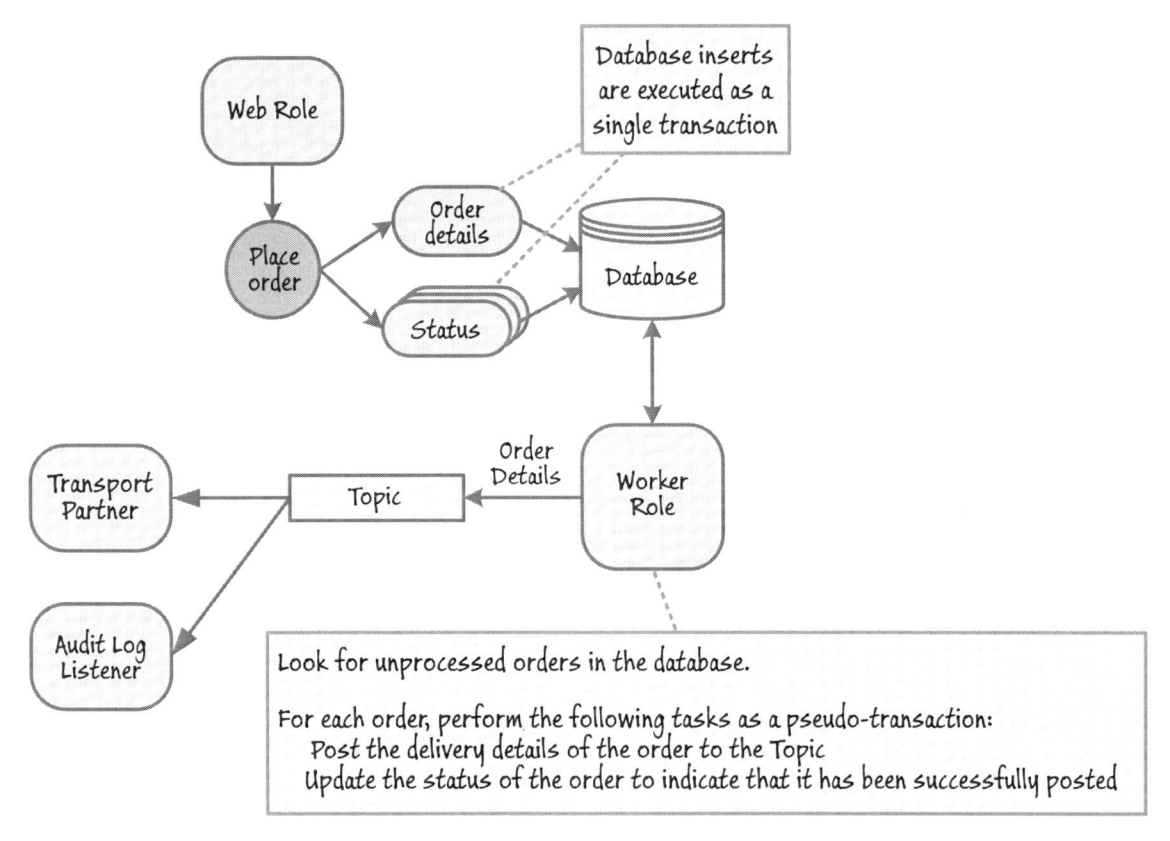

FIGURE 2
The custom transactional and retry mechanism implemented for the Trey Research Orders application

The implementation uses separate database tables that store the details of the order and the current status of each order. When a customer places an order, the web role populates these tables using a database transaction to ensure that they all succeed. If there is an error, it notifies the administrator.

This approach separates the task of saving the order from the tasks required to process the order, which are carried out by the worker role, and releases the web role to handle other requests. It also means that the web role can display information such as the order number to the customer immediately, and the customer will be able to view the current status of all of their orders without needing to wait until the order processing tasks have completed.

If you need to perform complex processing on messages before posting them to a queue, or handle multiple messages and perhaps combine them into one message, you might consider doing this in the web role and then storing the resulting message in the database. This approach can improve performance and reduce the risk of failure for worker roles, but can have a corresponding negative impact on the responsiveness of the web roles. It also splits the logic for the order processing task across the worker roles and web roles, with a resulting loss of separation of responsibility.

The worker role then carries out the tasks required to complete the process. It sends each message to a Service Bus topic that passes it to the transport partner to notify that a delivery is pending, and to the audit log when the order value exceeds a specific amount.

The worker role also listens for a response from a transport partner that indicates the message was received. This response will contain information such as the delivery tracking number that the worker role stores in the database (this part of the process is outside the scope of the custom retry mechanism and is not depicted in Figure 2).

At each stage of the overall process, the worker role updates the order status in the database to keep track of progress. For example, the status details may indicate that an order has been accepted and sent to a transport partner, but an acknowledgement has not yet been received.

Typically, the status rows will also contain a count of the number of times the send process has been attempted for each item, so the worker role can abandon the process and raise an exception, or place the message in a dead letter queue for human intervention, if a specified number of retries has been exceeded.

The following sections provide more detail on how Trey Research designed and implemented this mechanism using SQL Azure and the messaging layer defined in Chapter 4, "Implementing Reliable Messaging and Communications with the Cloud."

Recording the Details of an Order

A customer places an order by using the "Checkout" facility in the Orders web application implemented by the **Orders.Website** project. This operation invokes the **AddressAndPayment** action handler in the **CheckoutController** class. The **AddressAndPayment** method creates an **Order** object using the data entered by the customer, and uses the Entity Framework (EF) to add this order to the SQL Azure database. The following code sample shows the relevant sections of the **AddressAndPayment** method. Note that the customer and address information is added to the **Order** object by using the **TryUpdateModel** method.

The sample Trey Research application that you can download for this guide implements many of the technologies and techniques described here. However, to simplify installation and setup, and reduce the prerequisites and the requirements for users to establish extensive Windows Azure accounts, the feature set and some of the implementation details differ from the text of this guide.

```C#
public ActionResult AddressAndPayment(
  FormCollection values)
{
  var order = new Order();
  this.TryUpdateModel(order);

  var identity = User.Identity as IClaimsIdentity;
  var userName = identity.GetFederatedUsername();
```

```csharp
  var cartId = ShoppingCart.GetCartId(this.HttpContext);
  var cartItems = this.cartStore.FindCartItems(cartId);

  order.OrderDetails = cartItems.Select(
    i => new OrderDetail
      { ProductId = i.ProductId, Quantity = i.Count,
        Product = i.Product });
  order.UserName = userName;
  order.OrderDate = DateTime.Now;

  // Save the order
  this.ordersStore.Add(order);

  ...
}
```

For each order that is created, the **Add** method of the underlying entity model also creates a collection of **OrderDetail** objects and populates these with the information about each item in the order. The information that describes the order (the name, address, and contact details for the customer together with the date the order was placed, the total value of the order, and a unique order ID) is saved to the **Order** table in the SQL Azure database. The line items that comprise the order (the product and the quantity required) are saved to the **OrderDetail** table. The following code sample shows the **Add** method.

C#
```csharp
public void Add(Order order)
{
  ...
  var orderId = Guid.NewGuid();

  var orderToSave = new Entities.Order
    {
      OrderId = orderId,
      UserName = order.UserName,
      OrderDate = order.OrderDate,
      Address = order.Address,
      City = order.City,
      State = order.State,
      PostalCode = order.PostalCode,
      Country = order.Country,
      Phone = order.Phone,
      Email = order.Email,
      Total = order.OrderDetails.Sum(
             d => d.Quantity * d.Product.Price)
    };

  using (var database
```

```csharp
            = TreyResearchModelFactory.CreateContext())
{
    database.Orders.AddObject(orderToSave);

    foreach (var orderDetail in order.OrderDetails)
    {
        var detailToSave = new OrderDetail
            {
                ProductId = orderDetail.ProductId,
                OrderId = orderId,
                Quantity = orderDetail.Quantity
            };
        database.OrderDetails.AddObject(detailToSave);
    }
    ...
}
```

Processing an order requires the Orders application to keep track of the status of an order. The Orders application records information in two tables; **OrderStatus** and **OrderProcessStatus**. These rows indicate the current processing status of the order and the most recent operation carried out by the order processing code. After it has created the **Order** and **OrderDetail** entities, the **Add** method continues by creating rows for the **OrderStatus** and **OrderProcessStatus** tables.

C#
```csharp
    ...
    var status = new Entities.OrderStatus
        {
            OrderId = orderId,
            Status = "TreyResearch: Order placed",
            Timestamp = DateTime.UtcNow
        };
    database.OrderStatus.AddObject(status);

    var orderProcess = new OrderProcessStatus
        {
            OrderId = orderId,
            ProcessStatus = "pending process"
        };
    database.OrderProcessStatus.AddObject(orderProcess);
    ...
```

Finally, the **Add** method saves all of the changes to the tables by calling the **SaveChanges** method of the data model. Notice that it does so by calling the **ExecuteAction** method of the object defined in the **sqlCommandRetryPolicy** property of the class.

```C#
...
this.sqlCommandRetryPolicy.ExecuteAction(
    () => database.SaveChanges());
order.OrderId = orderId;
...
```

The sqlCommandRetryPolicy property implements an example of a Transient Fault Handling Application Block policy for accessing a SQL Azure database. For more information and further examples, see the section "Customer Details Storage and Retrieval" in Chapter 3, "Authenticating Users in the Orders Application." Additional information is also available online; see the topic "The Transient Fault Handling Application Block" on MSDN.

At this point, the order is ready to be sent for processing.

Sending an Order to a Service Bus Topic from the Orders Application

The order processing logic is initiated by the worker role in the **Orders.Workers** project. The worker role uses two classes that encapsulate the logic for sending orders to transport partners, and for receiving order status and acknowledgement messages from transport partners. These two classes, **NewOrderJob** and **StatusUpdateJob**, are referred to as "job processors." The constructor for the **WorkerRole** class (defined in the **Orders.Workers** project) executes the **CreateJobProcessors** method to instantiate these objects.

```C#
public class WorkerRole : RoleEntryPoint
{
  private readonly IEnumerable<IJob> jobs;
  ...

  public WorkerRole()
  {
    ...
    this.jobs = this.CreateJobProcessors();
  }

  ...

  private IEnumerable<IJob> CreateJobProcessors()
  {
    return new IJob[]
    {
      new NewOrderJob(),
      new StatusUpdateJob(),
    };
  }
}
```

The job processors both implement the **IJob** interface (defined in the Jobs folder of the **Orders.Workers** project). This interface defines methods for starting and stopping long-running jobs:

```C#
public interface IJob
{
  void Run();
  void Stop();
}
```

Having created the job processors, the **Run** method in the worker role starts each job processor executing by using a separate **Task**. The worker role keeps track of the state of each task, polling every 30 seconds, and if necessary restarting any job processors that have failed. The following code example shows the **Run** method of the worker role.

```C#
public class WorkerRole : RoleEntryPoint
{
  private readonly IEnumerable<IJob> jobs;
  private readonly List<Task> tasks;
  private bool keepRunning;

  ...

  public override void Run()
  {
    this.keepRunning = true;

    // Start the jobs
    foreach (var job in this.jobs)
    {
      var t = Task.Factory.StartNew(job.Run);
      this.tasks.Add(t);
    }

    // Control and restart a faulted job
    while (this.keepRunning)
    {
      for (int i = 0; i < this.tasks.Count; i++)
      {
        var task = this.tasks[i];
```

```
      if (task.IsFaulted)
      {
        // Observe unhandled exception
        if (task.Exception != null)
        {
          TraceHelper.TraceError(
            "Job threw an exception: " +
            task.Exception.InnerException.Message);
        }
        else
        {
          TraceHelper.TraceError(
            "Job Failed and no exception thrown.");
        }

        var jobToRestart = this.jobs.ElementAt(i);
        this.tasks[i] =
          Task.Factory.StartNew(jobToRestart.Run);
      }
    }

    Thread.Sleep(TimeSpan.FromSeconds(30));
  }
 }
}
```

The following section describes how the **NewOrderJob** class sends orders to transport partners. The **UpdateStatusJob** class is described in the section "Receiving Acknowledgement and Status Messages in the Orders Application," later in this chapter.

The NewOrderJob Class

The worker role uses the **NewOrderJob** job processor to query the database for new orders and dispatch them to the appropriate transport provider.

The **NewOrderJob** class implements a reliable mechanism that sends messages to the Service Bus topic. It tracks whether the send operation was successful, and if necessary can retry failed send operations later. If (and only if) the message was sent successfully is the status of the order updated in the database.

> The mechanism implemented by the **NewOrderJob** class is a simplified adaptation of the Scheduler-Agent-Supervisor pattern documented by Clemens Vasters. For more information, see the article "Cloud Architecture - The Scheduler-Agent-Supervisor Pattern."

Figure 3 shows a high-level view of the tasks that the **NewOrderJob** accomplishes and the ancillary classes it uses. The **ServiceBusTopic** and **ServiceBusTopicDescription** classes are described in Chapter 4, "Implementing Reliable Messaging and Communications with the Cloud," but the remaining classes are described later in this section.

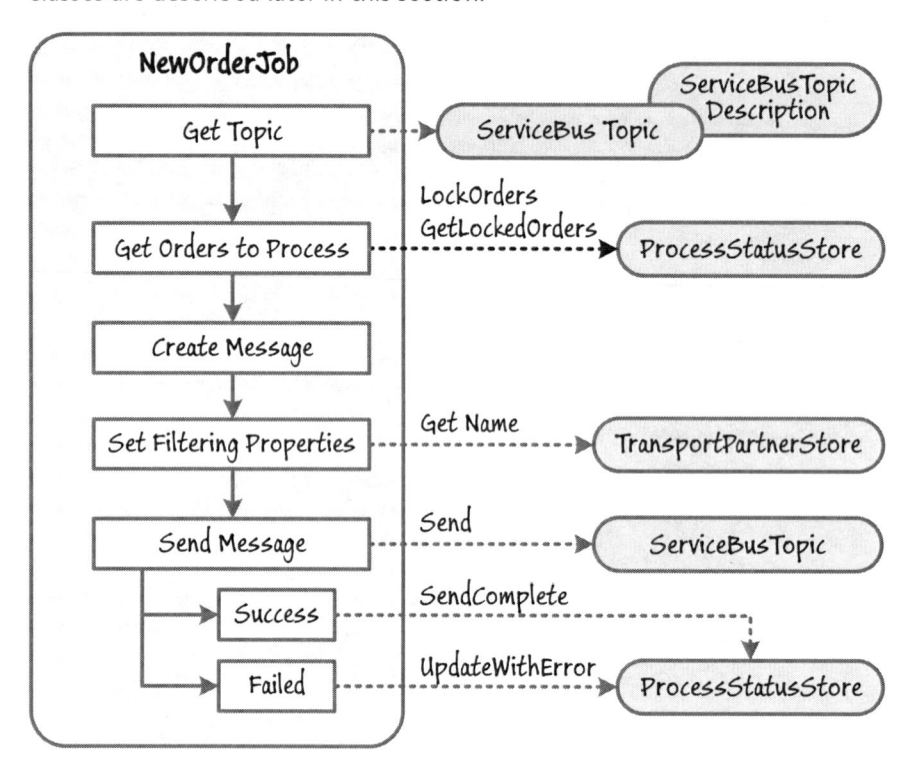

FIGURE 3
Message flow for the order processing system

Recall that the worker role starts the **NewOrderJob** job processor by calling the **Run** method. This method creates an instance of the **ServiceBusTopicDescription** class and populates it with the required values for the target topic, then uses this to create an instance of the **ServiceBusTopic** class named **newOrderMessageSender**.

```C#
var serviceBusTopicDescription
    = new ServiceBusTopicDescription
{
  Namespace = this.serviceBusNamespace,
  TopicName = topicName,
  Issuer = issuer,
  DefaultKey = defaultKey
};
this.newOrderMessageSender
    = new ServiceBusTopic(serviceBusTopicDescription);
```

The **NewOrderJob** job processor can then use the **ServiceBusTopic** instance to send the message, which it does by calling its **Execute** method.

```C#
while (this.keepRunning)
{
  this.Execute();
  Thread.Sleep(TimeSpan.FromSeconds(10));
}
```

> The **Execute** method processes orders in batches, up to 32 at a time. Each order in a batch is posted asynchronously by a separate task created by the **Send** method of a **ServiceBusTopic** object. The **NewOrderJob** job processor sleeps for 10 seconds between invocations of the **Execute** method to prevent resource starvation. This value was selected based the expected volume of orders and profiling the performance of the application, but it may be changed if the number of orders increases significantly.

Locking Orders for Processing

The details of orders are held in the local SQL Azure database (in tables named **Order** and **OrderDetails**), but to help keep track of the status of orders and ensure that they are processed reliably Trey Research defined two further tables in the same database.

The first of these, the **OrderStatus** table, contains the status rows for each order. Each row indicates the most recent publicly visible status of the order as displayed in the website page when customers view the list of their current orders. New rows are added with a timestamp; to maintain history information existing rows are not updated. The following table describes the columns in the **OrderStatus** table.

Column	Description
OrderID	The foreign key that links the row to the related row in the Order table.
Status	A value that indicates the most recent publicly visible status of the order. Possible values are **Order placed**, **Order sent to transport partner**, **Order received** (by transport partner), and **Order shipped** (when the goods have been delivered).
Timestamp	The UCT date and time when this change to the status of the order occurred.

The other table, **OrderProcessStatus**, contains the data that the retry mechanism uses to determine whether an order message has been successfully posted to the Service Bus topic. The following table describes the columns in the **OrderProcessStatus** table.

Column	Description
OrderID	The foreign key that links the row to the related row in the **Order** table.
ProcessStatus	A value used internally by the worker role that indicates the status of the process. Possible values are **pending process**, **processed**, **error**, and **critical error**. These values are defined in the **ProcessStatus** class located in the Stores folder of the **Orders.Workers** project.
LockedBy	The ID of the worker role that is processing the order (obtained from **RoleEnvironment.CurrentRoleInstance.Id**) or **NULL** if the order is not currently being processed.
LockedUntil	The UCT date and time when the current processing of the order will timeout, or **NULL** if the order is not currently being processed.
Version	A value used to support optimistic locking by EF for rows in this table. It has the **ConcurrencyMode** property set to **Fixed** and the **StoreGeneratedPatterns** property set to **Computed**. It is automatically updated by EF when rows are inserted or updated.
RetryCount	The number of times that processing of the order has been attempted so far. It is incremented each time a worker role attempts to process the order. After a number of failed attempts, the worker role sets the process status to **critical error** and raises an exception to advise administrators that there is a problem.
BatchId	A GUID that identifies this batch of orders being processed. Each time a worker role locks a batch of orders that it will start processing it assigns the same **BatchId** to all of these so that it can extract them after the lock is applied in the database.

As an optimization mechanism, the **Execute** method retrieves orders from the SQL Azure database in batches, but it needs to ensure that the same orders are not going to be retrieved by another concurrent invocation of the **Execute** method. Therefore, the **Execute** method locks a batch of rows in the Order table that it will process, and then retrieves this batch of rows as a collection of **OrderProcessStatus** objects by calling the **LockOrders** and **GetLockedOrders** methods in the **Process-StatusStore** class (located in the Stores folder of the **Orders.Workers** project).

```csharp
var batchId = this.processStatusStore.LockOrders(
        RoleEnvironment.CurrentRoleInstance.Id);
var ordersToProcess
    = this.processStatusStore.GetLockedOrders(
        RoleEnvironment.CurrentRoleInstance.Id, batchId);
```

The **LockOrders** method in the **ProcessStatusStore** class locks the orders by executing a SQL statement that sets the values in the **OrderProcessStatus** table rows. It assigns the current worker role instance ID and the batch ID to each one that has not already been processed, has not resulted in a critical error and been abandoned, and is not already locked by this or another instance of the **NewOrderJob** job processor. Also, notice that the **LockOrders** method only locks the first 32 orders available, and only for a specified period of time. This approach prevents the **NewOrderJob** job processor from causing a bottleneck by attempting to process too many orders at a time, and also prevents a failed instance of the **NewOrderJob** job processor causing an order to be locked indefinitely.

```C#
public Guid LockOrders(string roleInstanceId)
{
    using (var database =
           TreyResearchModelFactory.CreateContext())
    {
        var batchId = Guid.NewGuid();
        var commandText = "UPDATE TOP(32) OrderProcessStatus "
            + "SET LockedBy = {0}, LockedUntil = {1}, "
            + "BatchId = {2} "
            + "WHERE ProcessStatus != {3} "
            + "AND ProcessStatus != {4} "
            + "AND (LockedUntil < {5} OR LockedBy IS NULL)";
        this.sqlCommandRetryPolicy.ExecuteAction(
            () => database.ExecuteStoreCommand(
                commandText, roleInstanceId,
                DateTime.UtcNow.AddSeconds(320), batchId,
                ProcessStatus.Processed,
                ProcessStatus.CriticalError,
                DateTime.UtcNow));
        return batchId;
    }
}
```

The **NewOrderJob** job processor locks orders, processes them and finally sends "New Order" messages asynchronously, which means that after sending the message, the execution continues. Under some possible but unlikely circumstances, the **Execute** method of an instance of the **NewOrderJob** class can be called when there are still some orders being processed by that same instance. This will occur if processing the full batch of 32 orders takes more than 10 seconds (see the earlier note regarding the sleep interval between invocations of the **Execute** method).

To prevent the job processor from attempting to handle the same order more than once, the developers added the **BatchId** to each batch of orders as they are locked. This is a random value generated for each batch of orders. Subsequently, when the **Execute** method processes each order, it only fetches the orders that have the appropriate value for the **BatchId**.

The **GetLockedOrders** method in the **ProcessStatusStore** class queries the **OrderProcessStatus** table to retrieve the rows that were successfully locked by the **LockOrders** method.

```csharp
public IEnumerable<Models.OrderProcessStatus>
      GetLockedOrders(string roleInstanceId, Guid batchId)
{
  using (var database =
        TreyResearchModelFactory.CreateContext())
  {
    return
      this.sqlCommandRetryPolicy.ExecuteAction(
        () =>
        database.OrderProcessStatus.Where(
          o =>
          o.LockedBy.Equals(roleInstanceId,
                StringComparison.OrdinalIgnoreCase)
          && o.BatchId == batchId).Select(
            op =>
            new Models.OrderProcessStatus
              {
                LockedBy = op.LockedBy,
                LockedUntil = op.LockedUntil,
                OrderId = op.OrderId,
                ProcessStatus = op.ProcessStatus,
                Order =
                  new Models.Order
                    {
                      OrderId = op.Order.OrderId,
                      UserName = op.Order.UserName,
                      OrderDate = op.Order.OrderDate,
                      Address = op.Order.Address,
                      City = op.Order.City,
                      State = op.Order.State,
                      PostalCode = op.Order.PostalCode,
                      Country = op.Order.Country,
                      Phone = op.Order.Phone,
                      Email = op.Order.Email,
                      Total = op.Order.Total
                    }
              }).ToList());
  }
}
```

Handling Orders with Expired Locks

The **Execute** method then iterates over the collection of locked orders and sends a suitable message for each one to the Service Bus topic. However, there is a small possibility, especially during initial configuration and profiling of the application, that the batch size selected for the **NewOrderJob** job processor may be too large, or the interval between processing batches is too small, and the next iteration commences before the current batch of orders has been dispatched but after the **LockedUntil** time for one or more orders in the batch has passed. In this case these locks are considered to have expired and the next iteration of the **NewOrderJob** job processor may have relocked these orders and be in the process of sending them. So, to avoid the same order being posted twice, the **Execute** method examines the **LockedUntil** property of each order in the batch, and if the value of this property is before the current time then the order is skipped and handled in a subsequent iteration, if necessary.

The following code shows how Trey Research iterates over the orders to process and checks for expired messages.

```C#
foreach (var orderProcess in ordersToProcess)
{
  if (orderProcess.LockedUntil < DateTime.UtcNow)
  {
    // If this orderProcess expired, ignore it and let
    // another Worker Role process it.
    continue;
  }

  ...
  // Code here to create the message, add the required
  // Service Bus topic filter properties, and then
  // send the message.
  ...
}
```

Posting Orders to the Service Bus Topic

The Service Bus topic that Trey Research uses in the Orders application is configured to filter messages based on the delivery location and the total value of the order. A summary of every order must be sent to the appropriate transport partner to advise the requirement for delivery. Additionally, the details of all orders with a total value over $10,000 are sent to an on-premises service to be stored in the audit log database. The **Execute** method therefore adds the required properties to the message so that the Service Bus topic can filter them and post them to the appropriate subscribers. The following code shows how the **Execute** method uses a separate class named **TransportPartnerStore** to obtain the appropriate transport partner name.

> The use of filters in a Service Bus topic allows you to implement rudimentary business logic, and even modify the properties of messages as they pass through the topic. It also allows you to decouple senders and receivers by allowing a varying number of subscribers to listen to a topic, each receiving only messages that are targeted to them by a filter in the topic. However, topics and filters do not provide a general purpose workflow mechanism, and you should not try to implement complex logic using Service Bus topics.

*The **TransportPartnerStore** class is used to determine which transport partner a message should be sent to; the choice is based on the location of the order recipient. It is defined in the Transport-PartnerStore.cs file in Stores folder of the **Orders.Workers** project.*

C#
```
var transportPartnerName = this.transportPartnerStore
        .GetTransportPartnerName(orderProcess.Order.State);
```

Creating a New Order Message

The **Execute** method can now create the order message to send. To handle failures when sending messages, the custom retry mechanism might need to attempt to send the same message more than once. The **Execute** method cannot just create the message and pass it to the **Send** method of the **ServiceBusTopic** class because the serialization mechanism used by Service Bus messages means that the body of a message can only be read once. Therefore, the **Execute** method defines a function that creates the message dynamically and then passes this function to the **Send** method of the **ServiceBusTopic** class. The **Send** method can then invoke this method to construct a fresh copy of the message each time it is sent.

The following code shows the section of the **Execute** method that defines this function. Notice that it does so by first creating an instance of the **NewOrderMessage** class (defined in the Communication\Messages folder of the **Orders.Shared** project), which represents the specific message that Trey Research needs to send, and then builds a Service Bus **BrokeredMessage** instance from this. Finally, it sets the following properties of the **BrokeredMessage** instance:

- The **TransportPartnerName** property, previously retrieved from the **TransportPartnerStore**, is used by a Service Bus filter to direct the message to the appropriate transport partner.

- The **ServiceBusNamespace** property indicates the Service Bus namespace that contains the Service Bus queue specified in the **ReplyTo** property. The Orders application may be deployed to more than one datacenter, and the transport partner needs to know which instance of Service Bus it should use when posting a response.

- The **AcsNamespace** property allows the receiver to tell which instance of ACS was used by the sender to authenticate (the application may be configured to use ACS in more than one datacenter for authenticating partners for Service Bus access). This is required so that the receiver of the message can validate the sender, as described in the section "Securing Messages" in Chapter 4 "Implementing Reliable Messaging and Communications with the Cloud."

- The **OrderAmount** property is used by a Service Bus filter to detect orders over the specified total value and send a copy of these messages to the on-premises audit log.

- The **ReplyTo** property specifies the Service Bus queue on which the receiver should send any response messages. For example, the transport partner uses this queue to post the message that acknowledges receipt of this message, as described in the section "Correlating Messages and Replies" in Chapter 4, "Implementing Reliable Messaging and Communications with the Cloud."

```C#
Func<BrokeredMessage> brokeredMessageFunc = () =>
{
  // Send new order message
  var msg = new NewOrderMessage
  {
    OrderId = orderProcess.Order.OrderId,
    OrderDate = orderProcess.Order.OrderDate,
    ShippingAddress = orderProcess.Order.Address,
    Amount =
      Convert.ToDouble(orderProcess.Order.Total),
    CustomerName = orderProcess.Order.UserName
  };

  var brokeredMessage = new BrokeredMessage(msg)
  {
    MessageId = msg.OrderId.ToString(),
    CorrelationId = msg.OrderId.ToString(),
    Properties = { { "TransportPartnerName",
                     transportPartnerName },
                   { "ServiceBusNamespace",
                     this.serviceBusNamespace },
                   { "AcsNamespace",
                     this.acsNamespace },
                   { "OrderAmount",
                     orderProcess.Order.Total } },
    ReplyTo = this.replyQueueName
  };

  return brokeredMessage;
};
```

The message will be sent asynchronously, so the **NewOrderJob** class must also assemble an object that can be passed as the state in the asynchronous method calls, as shown in the following code. This object must contain the order ID (so that the retry mechanism can update the correct order status rows in the **OrderStatus** table) and the name of the transport partner (so that it can be displayed when customers view their existing orders).

```csharp
var objectState = new Dictionary<string, object>
{
  { "orderId", orderProcess.OrderId },
  { "transportPartner", transportPartnerName }
}
```

Sending the New Order Messages
The **Execute** method can now send the order to the Service Bus topic. It does this by calling the **Send** method of the **ServiceBusTopic** instance it created and saved in the variable named **newOrderMessageSender**. As described in the section "Sending Messages to a Service Bus Topic" in Chapter 4, "Implementing Reliable Messaging and Communications with the Cloud," the **Send** method of the **ServiceBusTopic** class takes four parameters; the function that creates the **BrokeredMessage** instance to send, the asynchronous state object, and two **Action** methods (one to execute after a message is sent and one to execute if sending fails). The **ProcessStatusStore** class defines lambda statements that are passed as these two actions, as shown in the following code.

```csharp
  this.newOrderMessageSender
    .Send(
      brokeredMessageFunc,
      objectState,
      (obj) =>
      {
        var objState = (IDictionary<string, object>)obj;
        var orderId = (Guid)objState["orderId"];
        var transportPartner
            = (string)objState["transportPartner"];
        this.processStatusStore.SendComplete(orderId,
                         transportPartner);
      },
      (exception, obj) =>
      {
        var objState = (IDictionary<string, object>)obj;
        var orderId = (Guid)objState["orderId"];
        this.processStatusStore.UpdateWithError(
                 exception, orderId);
      });
}
```

The first action, which is executed after the message has been sent successfully, extracts the order ID and transport partner name from the asynchronous state object, and passes these to the **SendComplete** method of the **ProcessStatusStore** class instance that the **NewOrderJob** class is using. The second action extracts only the order ID from the asynchronous state object, and passes it and the current **Exception** instance to the **UpdateWithError** method instance.

Completing the Reliable Send Process

The **SendComplete** method of the **ProcessStatusStore** class, called by the **Execute** method if the message is posted successfully, performs the following tasks:

- It updates the matching row in the **OrderProcessStatus** table to modify the value in the **ProcessStatus** column to "processed" (this prevents another iteration of the **NewOrderJob** job processor from attempting to send the order again).
- It adds a new row to the **OrderStatus** table to show the current status of the order ("**Order sent to transport partner**"), with the correct timestamp.
- It updates the **Order** table with the name of the transport partner that will deliver the order.

C#
```
public void SendComplete(Guid orderId,
                         string transportPartner)
{
  using (var database =
         TreyResearchModelFactory.CreateContext())
  {
    try
    {
      using (var t = new TransactionScope())
      {
        // Avoid the transaction being promoted.
        this.sqlConnectionRetryPolicy.ExecuteAction(
            () => database.Connection.Open());

        // Update the OrderProcessStatus table row
        var processStatus =
          this.sqlCommandRetryPolicy.ExecuteAction(
            () => database.OrderProcessStatus
                .SingleOrDefault(
                    o => o.OrderId == orderId));
        processStatus.ProcessStatus
            = ProcessStatus.Processed;
        processStatus.LockedBy = null;
        processStatus.LockedUntil = null;
        this.sqlCommandRetryPolicy.ExecuteAction(
            () => database.SaveChanges());

        // Add a new row to the OrderStatus table
        var status = new OrderStatus { OrderId = orderId,
          Status =
            "TreyResearch: Order sent to transport partner",
                Timestamp = DateTime.UtcNow };
        database.OrderStatus.AddObject(status);
        this.sqlCommandRetryPolicy.ExecuteAction(
            () => database.SaveChanges());
```

```csharp
    // Update the Order table row
    var order =
      this.sqlCommandRetryPolicy.ExecuteAction(
        () => database.Order.SingleOrDefault(
                    o => o.OrderId == orderId));
    order.TransportPartner = transportPartner;
    this.sqlCommandRetryPolicy.ExecuteAction(
        () => database.SaveChanges());

    t.Complete();
   }
  }
  catch (UpdateException ex)
  {
   ...
  }
 }
}
```

The **UpdateWithError** method, called by the **Execute** method if the messages was not posted successfully, generates suitable warning messages using the custom **TraceHelper** class, updates the matching **OrderProcessStatus** table row with the value "**error**", and sets the values of the **LockedBy** and **LockedUntil** columns to null to unlock the message and make it available for processing again.

The **UpdateWithError** method also checks whether the number of retry attempts has exceeded the specified maximum (defined in the ServiceConfiguration.cscfg file). If it has, it updates the **OrderProcessStatus** table row with the value "**critical error**" and generates a message that indicates the administrator should investigate the failed order process.

C#
```csharp
public void UpdateWithError(Exception exception,
                    Guid orderId)
{
  TraceHelper.TraceWarning("NewOrderJob: The Order '{0}' "
    + "couldn't be processed. Error details: {1}",
    orderId.ToString(), exception.ToString());

  using (var database
    = TreyResearchModelFactory.CreateContext())
  {
    var processStatus =
        this.sqlCommandRetryPolicy.ExecuteAction(
            () => database.OrderProcessStatus
                .SingleOrDefault(
                    o => o.OrderId == orderId));
    processStatus.ProcessStatus = ProcessStatus.Error;
    processStatus.LockedBy = null;
```

```
    processStatus.LockedUntil = null;
    processStatus.RetryCount
      = processStatus.RetryCount + 1;

    var newOrderJobRetryCountCheck = int.Parse(
      CloudConfiguration.GetConfigurationSetting(
        "NewOrderJobRetryCountCheck", "3"));

    if (processStatus.RetryCount
        > newOrderJobRetryCountCheck)
    {
      processStatus.ProcessStatus
          = ProcessStatus.CriticalError;
      TraceHelper.TraceError("NewOrderJob: The Order '{0}' "
        + "has reached {1} retries. This order requires "
        + "manual intervention.",
        orderId.ToString(), processStatus.RetryCount);
    }

    this.sqlCommandRetryPolicy.ExecuteAction(
                () => database.SaveChanges());
  }
}
```

At this point, the message has been sent and the status recorded in the SQL Azure database is consistent with the actual status of the message.

The next task is to retrieve the order from the subscription at the appropriate transport partner.

How Trey Research Decouples the Order Process from the Transport Partners' Systems

To communicate with the delivery systems of the various transport partners, Trey Research implemented a series of connectivity components in the form of connectors and adapters. These components provide an interface between the Service Bus and the transport partner. Each component is specific to the transport partner, and provides the functionality to retrieve messages from the appropriate Service Bus subscription, translate it into a format accepted by the transport partner, and then pass it to the transport partner's internal system.

Responses from the transport partner are passed back to the component, converted into Service Bus messages, and then posted to the Service Bus queue for the Order application. Figure 1 earlier in this chapter shows where these components are deployed. The following sections provide more information on the sample implementation of these components provided with the Trey Research solution.

> *The adapter and connector included with the sample solution are provided as simple examples for enabling a transport partner to interact with the Service Bus topics and queues used by Trey Research. In the real world, the internal business logic for these connectivity components may be considerably more complex than that illustrated by these samples.*

Receiving and Processing an Order in a Transport Partner

Transport partners connect to the Service Bus topic on which the **NewOrderJob** job processor has posted the order messages. Each transport partner uses its own Service Bus subscription, configured with a filter that examines the **TransportPartnerName** property of each message. The subscriptions and filters themselves are created by the setup program in the **TreyResearch.Setup** project; for more information, see the section "Subscribing to a Service Bus Topic" in Chapter 4, "Implementing Reliable Messaging and Communications with the Cloud."

Sample transport partners are provided in the **TransportPartner** project. The solution includes two transport partners; Contoso (implemented by the **ContosoTransportPartner** class) which ships orders to customers that reside in the same or an adjacent state to the Trey Research manufacturing plant, and Fabrikam (implemented by the **FabrikamTransportPartner** class) which ships orders to customers located elsewhere.

The local transport partner, Contoso, connects to the Service Bus topic by using the **Connector** class defined in the **Connectivity** folder of the **TransportPartner** project, while the distance transport partner, Fabrikam, uses the **Adapter** class defined in the same folder.

> *The section "Implementing Adapters and Connectors for Translating and Reformatting Messages" in Chapter 4, "Implementing Reliable Messaging and Communications with the Cloud," describes the rationale behind the use of connectors and adapters.*

The **Connector** and **Adapter** classes both inherit from the **OrderProcessor** class, which provides the functionality for actually connecting to the Service Bus topic and receiving messages in the **Run** method.

```C#
public void Run()
{
  var serviceBusNamespaces = ConfigurationManager.
    AppSettings["serviceBusNamespaces"].
    Split(',').ToList();
  ...

  foreach (var serviceBusNamespace in serviceBusNamespaces)
  {
    this.serviceBusSubscriptionDescription.Namespace =
      serviceBusNamespace;
    var serviceBusSubscription = new
      ServiceBusSubscription(
        this.serviceBusSubscriptionDescription);
    var receiverHandler = new
      ServiceBusReceiverHandler<NewOrderMessage>(
        serviceBusSubscription.GetReceiver())
    {
      MessagePollingInterval = TimeSpan.FromSeconds(2)
    };

    receiverHandler.ProcessMessages(
```

```
      (message, queueDescription, token) =>
      {
        return Task.Factory.StartNew(
          () => this.ProcessMessage(
            message, queueDescription),
          this.tokenSource.Token,
          TaskCreationOptions.None,
          context);
      },
      this.tokenSource.Token);
  }
}
```

Chapter 4, "Implementing Reliable Messaging and Communications with the Cloud," describes how and why the **Run** method creates a **ServiceBusReceiverHandler** object to retrieve messages. As far as the order process is concerned, there are two important aspects of this method:

- The Trey Research web application and worker role may be deployed to multiple datacenters, and the deployment at each datacenter is configured with its own set of Service Bus topics and queues in its own Service Bus namespace. Therefore the **Run** method must connect to each datacenter and listen for messages on the topic at each one. This is the purpose of the **foreach** loop; the Service Bus namespaces to which the method must connect are defined in the configuration file, and the loop connects to the Service Bus topic in each namespace.

- The call to the **ProcessMessages** method of the **ServiceBusReceiverHandler** object invokes the **ProcessMessage** method. This method provides the logic for actually processing an order message, as shown by the following code sample:

C#
```
protected virtual void ProcessMessage(
    NewOrderMessage message,
    ServiceBusQueueDescription queueDescription)
{
  var trackingId = this.ProcessOrder(message,
                                    queueDescription);

  if (trackingId != Guid.Empty)
  {
    // Get SWT from ACS.
    var token = this.GetToken(queueDescription);

    var statusMessage =
```

```
            string.Format("{0}: Order Received",
               this.TransportPartnerDisplayName);
         this.SendOrderReceived(message, queueDescription,
            statusMessage, trackingId, token);
      }
   }
```

The **ProcessMessage** method calls the **ProcessOrder** method to perform the transport partner-specific business processing for the order. **ProcessOrder** is an abstract method with implementations in the **Adapter** and **Connector** classes. In the sample transport partners, the orders are stored as a list of **ActiveOrder** objects and each order is displayed on the Windows Form that provides the user interface (the **OnOrderProcessed** event handler performs this task).

However, there is a small possibility that the same order may be received more than once; the retry mechanism in the **NewOrderJob** class that posts orders to the Service Bus topic may cause the same order message to be repeated, depending on the reliability and performance of the network connection. Consequently, before creating the **ActiveOrder** object the **ProcessOrder** method verifies that an order with the same ID as the message just received does not already exist in the list; if there is such an order, this new one is assumed to be a duplicate and is discarded.

The following code sample shows the implementation of the **ProcessOrder** method in the **Adapter** class.

C#
```
protected override Guid ProcessOrder(
   Orders.Shared.Communication.Messages.NewOrderMessage
   message, ServiceBusQueueDescription queueDescription)
{
   var processedOrder =
      this.orderStore.GetById(message.OrderId);

   if (processedOrder != null)
   {
      // This order has been received for processing more
      // than once, and will be discarded.
      return Guid.Empty;
   }

   var activeOrder = new ActiveOrder
   {
      OrderId = message.OrderId,
      ShippingAddress = message.ShippingAddress,
      Amount = message.Amount,
      ReplyTo = queueDescription.QueueName,
      ReplyToNamespace = queueDescription.Namespace,
      Status = "received",
      SwtAcsNamespace = queueDescription.SwtAcsNamespace
   };
```

```
      this.orderStore.Add(activeOrder);

      // Call the transport partner service and
      // retrieve a tracking id.
      var trackingId = this.transportServiceWrapper.
        RequestShipment(activeOrder);

      if (this.OnOrderProcessed != null)
      {
        this.OnOrderProcessed(this,
          new OrderProcessedEventArgs
          { ActiveOrder = activeOrder });
      }

      // if tracking id received, delivery request is
      // acknowledged, it is safe to update the status queue
      // with the "Order Received" status.
      return trackingId;
}
```

Note that the sample assumes that the transport partners have internal systems that react in different ways when they receive an order message. This is to provide a more real-world experience in the application.

- The local transport partner, Contoso, responds immediately with an acknowledgement message indicating that the order has been received. Later, when the order is shipped, the local transport partner sends another message.
- The distance transport partner, Fabrikam, accepts the order and generates a tracking ID. It sends a response message back to Trey Research containing this tracking ID. Many distance transport partners provide their own web applications that enable customers to log in and query the progress of the order by providing this tracking ID (these web applications are the responsibility of the transport partner; an example is not provided as part of the sample solution.)

Acknowledging an Order or Indicating that it has Shipped in a Transport Partner

After receiving an order, the local transport partner, Contoso, should acknowledge successful receipt of the message. Later on, the local transport partner sends another message when the order is dispatched. As described in the previous section, the distance transport partner, Fabrikam, only sends a single message when the order is received. In both cases, the **ProcessMessage** method calls the **SendOrderReceived** method to construct and send an appropriate message to the Orders application.

C#
```
protected virtual void ProcessMessage(
    NewOrderMessage message,
    ServiceBusQueueDescription queueDescription)
{
  var trackingId = this.ProcessOrder(message,
                                     queueDescription);

  if (trackingId != Guid.Empty)
  {
    // Get SWT from ACS.
    var token = this.GetToken(queueDescription);

    var statusMessage =
      string.Format("{0}: Order Received",
        this.TransportPartnerDisplayName);
    this.SendOrderReceived(message, queueDescription,
      statusMessage, trackingId, token);
  }
}

protected void SendOrderReceived(
  NewOrderMessage message,
  ServiceBusQueueDescription queueDescription,
  string statusMessage, Guid trackingId, string swt)
{
  this.SendToUpdateStatusQueue(message.OrderId, trackingId,
    statusMessage, queueDescription, swt);
}
```

The **SendOrderReceived** method calls the **SendToUpdateStatusQueue** method, which contains the logic for composing an **OrderStatusUpdateMessage** that it posts to a Service Bus queue. The name of the queue to use and the Service Bus namespace in which it resides were specified in the **ReplyTo** and **ServiceBusNamespace** properties of the original order message. They were used to create a **ServiceBusQueueDescription** object when the message was received by the **ProcessMessage** method of the **ServiceBusReceiverHandler** class (for more information about the **ServiceBusReceiverHandler** class, see the section "Receiving Messages from a Service Bus Queue and Processing Them Asynchronously" in Chapter 4, "Implementing Reliable Messaging and Communications with the Cloud").

When the transport partner sends a reply, it should include a security token to enable the Orders application to authenticate the response. This mechanism is also described in Chapter 4, "Implementing Reliable Messaging and Communications with the Cloud," in the section "Securing Messages." The ACS namespace in which this token is defined is provided in **AcsNamespace** property in the original order, and this value is also added to the **ServiceBusQueueDescription** object. When the transport partner sends the reply, the security token is retrieved from the specified ACS namespace and added to the response message.

```C#
public class ServiceBusReceiverHandler<T>
{
    ...
    private void ProcessMessage(
      IBrokeredMessageAdapter message)
    {
      if (message != null)
      {
        ...
        var queueDescription = new ServiceBusQueueDescription
        {
          QueueName = message.ReplyTo,
        };

        if (message.Properties.ContainsKey(
          "ServiceBusNamespace"))
        {
          queueDescription.Namespace = message.Properties[
            "ServiceBusNamespace"].ToString();
        }

        if (message.Properties.ContainsKey("AcsNamespace"))
        {
          queueDescription.SwtAcsNamespace =
            message.Properties["AcsNamespace"].ToString();
        }
        ...
      }
      ...
    }
    ...
}
```

The **ServiceBusQueueDesciption** object permeates down to the **SendToUpdateStatusQueue** method in the **queueDescription** parameter. The **SendToUpdateStatusQueue** method creates an **OrderStatusUpdateMessage** object, populating it with the acknowledgement details including the "Order Received" status message and the tracking ID generated by the transport partner. This **OrderStatusObjectMessage** object is packaged up inside a **BrokeredMessage** object and posted to the queue specified by the **queueDescription** parameter. Notice that the **CorrelationId** property of this response message is set to the order ID of the original request so that the Orders application can correlate this response with the request when it is received.

The following code sample shows the **SendToUpdateStatusQueue** method. Note that, as an optimization mechanism, this method caches a copy of the **ServiceBusQueue** object that it creates in a **Dictionary** object called **statusUpdateQueueDictionary**. When the **SendToUpdateStatusQueue** method is called again, it should find this **ServiceBusQueue** in the **Dictionary** and should not need to create it again.

```C#
private void SendToUpdateStatusQueue(Guid orderId,
  Guid trackingId, string orderStatus,
  ServiceBusQueueDescription queueDescription, string swt)
{
  var updateStatusMessage =
    new BrokeredMessage(
      new OrderStatusUpdateMessage
      {
        OrderId = orderId,
        Status = orderStatus,
        TrackingId = trackingId,
        TransportPartnerName =
          this.TransportPartnerDisplayName,
      })
      { CorrelationId = orderId.ToString() };

  updateStatusMessage.Properties.Add(
    "SimpleWebToken", swt);

  ServiceBusQueue replyQueue;
  if (this.statusUpdateQueueDictionary.
      ContainsKey(queueDescription.Namespace))
  {
    replyQueue = this.statusUpdateQueueDictionary[
      queueDescription.Namespace];
  }
  else
  {
    var description = new ServiceBusQueueDescription
    {
      Namespace = queueDescription.Namespace,
      QueueName = queueDescription.QueueName,
      DefaultKey =
        this.serviceBusQueueDescription.DefaultKey,
      Issuer = this.serviceBusQueueDescription.Issuer
    };

    replyQueue = new ServiceBusQueue(description);
    this.statusUpdateQueueDictionary.Add(
      queueDescription.Namespace, replyQueue);
```

```
    }
    var brokeredMessageAdapter =
      new BrokeredMessageAdapter(updateStatusMessage);
    replyQueue.Send(brokeredMessageAdapter);
}
```

For the local transport partner, Contoso, when an order is delivered the application calls the **SendOrderShipped** method in the **OrderProcessor** class. The **SendOrderShipped** method operates in much the same way as the **SendOrderReceived** method, calling the **SendToUpdateStatusQueue** method to create and post an **OrderStatusUpdate** message to the Service Bus queue.

Receiving Acknowledgement and Status Messages in the Orders Application

As well as starting a **NewOrderJob** object to post new orders to transport partners, each worker role in the Trey Research solution creates a **StatusUpdateJob** object to listen for status messages received from the transport partners. This class is located in the StatusUpdateJob.cs file in the Jobs folder in the Orders.Workers project. In common with the **OrderProcessor** class, the **Run** method in the **StatusUpdateJob** class employs a **ServiceBusReceiverHandler** object to actually connect to the queue and retrieve messages.

The **StatusUpdateJob** object also provides the business logic that is run for each status message as it is received, as a lambda expression, when it calls the **ProcessMessages** method of the **ServiceBusReceiverHandler** object. This lambda expression performs the following tasks:

- It checks the authentication token in the message and throws an **InvalidTokenException** exception if it is not recognized (the message may be from a rogue third party).
- It creates an order status record with the status information provided by the transport partner.
- It discards the message if it is a duplicate of an existing order (retry logic in the transport partner may cause it to send duplicate status messages if it detects a transient error).
- It updates the order with the name of the transport partner that will ship it to the customer.
- It adds the tracking ID provided by the partner to the order status record.
- It stores the order status record in the TreyResearch database (other parts of the application can query this status; for example, if the customer wishes to know whether an order has been shipped).

The following code sample shows how the **StatusUpdateJob** class defines this logic.

```csharp
C#
public void Run()
{
  ...
  receiverHandler.ProcessMessages(
    (message, replyTo, token) =>
    {
      return Task.Factory.StartNew(
        () =>
```

```csharp
{
  ...
  if (!this.IsValidToken(message.OrderId, token))
  {
    // Throw exception, to be caught by handler.
    // Will send it to the DeadLetter queue.
    throw new InvalidTokenException();
  }

  var orderStatus = new OrderStatus {
    OrderId = message.OrderId,
    Status = message.Status
  };

  using (var db =
    TreyResearchModelFactory.CreateContext())
  {
    // Checking for duplicate entries in the order
    // status table.  If a duplicate message
    // arrives, it is discarded.
    var existingStatus =
      this.sqlCommandRetryPolicy.ExecuteAction(
        () => db.OrderStatus.SingleOrDefault(
          os => os.OrderId == message.OrderId &&
          os.Status == message.Status));
    if (existingStatus != null)
    {
      return;
    }

    var order = this.sqlCommandRetryPolicy.
      ExecuteAction(
        () => db.Order.Single(o =>
          o.OrderId == message.OrderId));

    order.TransportPartner =
      message.TransportPartnerName;

    if (message.TrackingId != Guid.Empty)
    {
      order.TrackingId = message.TrackingId;
    }

    db.OrderStatus.AddObject(
      new OrderStatus {
        OrderId = orderStatus.OrderId,
```

```
                Status = orderStatus.Status,
                Timestamp = DateTime.UtcNow
            });

            this.sqlCommandRetryPolicy.ExecuteAction(
                () => db.SaveChanges());
        }
    });
},
...);
}
```

Summary

This chapter has examined how Trey Research implemented the business logic for processing orders in their Orders application. The business logic is based on Service Bus topics, subscriptions, and queues, and uses the software infrastructure described in the Chapter 4, "Implementing Reliable Messaging and Communications with the Cloud," to provide an extensible layer for reliable, asynchronous messaging.

This chapter has also described how Trey Research provided a reliable mechanism for posting messages to a Service Bus topic, detecting and handling failures, and transparently retrying to send messages when they occur.

Finally, this chapter described how the internal business logic for the transport partners was decoupled from the messaging infrastructure utilized by Trey Research; the connectors and adapters defined in the sample solution call upon the existing business services provided by the transport partners without requiring that they modify or disrupt their internal systems. These connectors and adapters also provided location independence by tracking the source of incoming order requests and routing any response messages back to the appropriate destination.

More Information

All links in this book are accessible from the book's online bibliography available at: *http://msdn.microsoft.com/en-us/library/hh968447.aspx*.
- "Best Practices for Leveraging Windows Azure Service Bus Brokered Messaging API" at *http://windowsazurecat.com/2011/09/best-practices-leveraging-windows-azure-service-bus-brokered-messaging-api/*.
- "Cloud Architecture: The Scheduler-Agent-Supervisor Pattern" at *http://vasters.com/clemensv/CommentView,guid,83f937f7-b838-43d0-ad61-74605eceafa2.aspx*.
- "Achieving Transactional Behavior with Messaging" at *http://vasters.com/clemensv/2011/10/06/Achieving+Transactional+Behavior+With+Messaging.aspx*.
- "Queues, Topics, and Subscriptions" at *http://msdn.microsoft.com/en-us/library/windowsazure/hh367516.aspx*.
- "The Transient Fault Handling Application Block" at *http://msdn.microsoft.com/en-us/library/hh680934(v=pandp.50).aspx*.

6 Maximizing Scalability, Availability, and Performance in the Orders Application

A primary reason for Trey Research migrating and reconfiguring the Orders application to run on the Windows Azure™ technology platform was to take advantage of the improved scalability and availability that this environment provides. As Trey Research expand their operations, they expect to attract an ever-increasing number of customers, so they designed their solution to be able to handle a large volume of orders from clients located anywhere in the world. This was a challenging proposition as the Orders application had to remain responsive but cost-effective, regardless of the number of customers requesting service at any given point in time. Trey Research analyzed the operations of their system, and identified three principal requirements. The solution should:

- Automatically scale up as demand increases (to process orders in a timely manner), but then scale back when demand drops to keep running costs down.
- Reduce the network latency associated with a customer accessing the Orders web application and its resources, hosted in a location remote across a public and uncontrollable (from Trey Research's perspective) network such as the Internet .
- Optimize the response time and throughput of the web application to maintain a satisfactory user experience.

This chapter describes how Trey Research met these requirements in their solution.

Scenario and Context

Trey Research implemented their solution as a web application primarily targeting customers located in the United States, so they initially deployed it to two Windows Azure datacenters; US North and US South. However, Trey Research plans to expand their operations and can foresee a time when the application will have to be deployed to other datacenters worldwide to satisfy the demands of overseas customers. For this reason, when Trey Research constructed the Orders application, they designed it to allow customers to connect to any instance, and to provide functionality that was consistent across all instances. In this way, Trey Research can start and stop instances of the Orders application and customers will always be able to view products, place orders, and query the status of their orders regardless of which instance they are connected to at any point in time. This approach also enables Trey Research to balance the load evenly across the available instances of the Orders application and maintain throughput. Additionally, if an instance becomes unavailable or a connection to an instance fails, customers can be directed to other working instances of the application.

CHAPTER SIX

In practice, this solution depends on a variety of components to provide the necessary infrastructure to determine the optimal instance of the Orders application to which a customer should connect, transparently route customer requests (and reroute requests in the event a connectivity failure with an instance occurs), and maintain consistent data across all datacenters. Additionally, the resources that the Orders application uses involve making further requests across the network; for example to retrieve the Products catalog, Customer details, or Order information. Access to these resources must be accomplished in a scalable and timely manner to provide customers with a responsive user experience.

Controlling Elasticity in the Orders Application

Trey Research noticed that the Orders application experienced peaks and troughs in demand throughout the working day. Some of these patterns were predictable, such as low usage during the early hours of the morning and high usage during the latter part of the working day, while others were more unexpected; sometimes demand increased due to a specific product being reviewed in a technical journal, but occasionally the volume of use changed for no foreseeable reason.

Trey Research needed a solution that would enable them to scale the Orders application to enable a highly variable number of customers to log in, browse products, and place orders without experiencing extended response times, while at the same time remaining cost-efficient.

Choosing How to Manage Elasticity in the Orders Application

Trey Research considered a number of options for determining how best to scale the Orders application. These options, together with their advantages and limitations, are described in the following sections.

Do Not Scale the Application

This is the simplest option. The Orders application has been designed and implemented to take advantage of concurrent web and worker role instances, and utilizes asynchronous messaging to send and receive messages while minimizing the response time to users. In this case, why not simply deploy the application to number of web and worker role instances in each possible datacenter, and allocate each role the largest possible virtual machine size, with the maximum number of CPU cores and the largest available volume of memory?

> Windows Azure provides elasticity and scale by allowing you to start and stop instances of roles on demand. However, unless you actually do manage your role instance count proactively, you are missing out on some of the major benefits that cloud computing offers.

This approach is attractive because it involves the least amount of maintenance on the part of the operations staff at Trey Research. It is also very straightforward to implement. However, it could be very expensive; hosting a web or worker role using the "Extra Large" virtual machine size (as defined by the Windows Azure pricing model) is currently 24 times more expensive on an hourly rate than hosting the same role in an "Extra Small" virtual machine. If the volume of customers for much of the time does not require the processing or memory capabilities of an extra-large virtual machine, then Trey Research would be paying to host a largely idle virtual machine. If you multiply the charges by the number of instances being hosted across all datacenters, the final sum can be a significant amount of money.

There is one other question that this approach poses; how many web and worker role instances should Trey Research create? If the number selected is too large, then the issues of cost described in the previous paragraph become paramount. However, if Trey Research create too few instances, then although the company is not necessarily paying for wasted resources, customers are likely to be unhappy due to extended response times and slow service, possibly resulting in lost business.

For these reasons this approach is probably not going to be cost effective or desirable.

Implement Manual Scaling

Clearly, some kind of scale-up and scale-down solution is required. By using the Windows Azure Management Portal it is possible to start and stop instances of web and worker roles manually; or even deploy new instances of the Orders application to datacenters around the world. Decisions about when to start, stop, or deploy new instances could be made based on usage information gathered by monitoring the application; Chapter 7, "Monitoring and Managing the Orders Application" contains more information on how to perform tasks such as these. However, this is potentially a very labor intensive approach, and may require an operator to determine when best to perform these tasks.

Some of these operations can be scripted using the Windows Azure Powershell Cmdlets, but there is always the possibility that having started up a number of expensive instances to handle a current peak in demand, the operator may forget to shut them down again at the appropriate time, leaving Trey Research liable for additional costs.

Implement Automatic Scaling using a Custom Service

Starting and stopping role instance manually was considered to be too inefficient and error prone, so Trey Research turned their attention to crafting an automated solution. Theoretically, it should be able to follow the same pattern and implement the same practices as the manual approach, except in a more reliable and less labor intensive manner. To this end, Trey Research considered configuring the web and worker roles to gather key statistical information, such as the number of concurrent requests, the average response time, the activity of the CPU and disks, and the memory utilization. This information could be obtained by using the Windows Azure Diagnostics and other trace event sources available to code running in the cloud, and then periodically downloaded and analyzed by a custom application running on-premises within the Trey Research Head Office. This custom application could then determine whether to start and stop additional role instances, and in which datacenters.

> External services that can manage autoscaling are also available. These services remove the overhead of developing your own custom solution but you must provide these services with your Windows Azure management certificate so that they can access the role instances, which may not be acceptable to your organization.

To simplify installation and setup, and reduce the prerequisites and the requirements for users to establish extensive Windows Azure accounts, the Trey Research example solution provided with this guide is designed to be deployed to a single datacenter, and is not configured to support autoscaling or request rerouting. Consequently, the sections in this chapter describing how Trey Research implemented the Enterprise Library Autoscaling Application Block and Windows Azure Traffic Manager are provided for information only.

The downside of this approach is the possible complexity of the on-premises application; it could take significant effort to design, build, and test such an application thoroughly (especially the logic that determines whether to start and stop role instances). Additionally, gathering the diagnostic information and downloading it from each datacenter could impose a noticeable overhead on each role, impacting the performance.

Implement Automatic Scaling using the Enterprise Library Autoscaling Application Block

The Microsoft Enterprise Library Autoscaling Application Block provides a facility that you can integrate directly into your web and worker roles running in the cloud, and also into on-premises applications. It is part of the Microsoft Enterprise Library 5.0 Integration Pack for Windows Azure, and can automatically scale your Windows Azure application or service based on rules that you define specifically for that application or service. You can use these rules to help your application or service maintain its throughput in response to changes in its workload, while at the same time minimize and control hosting costs. The application block enables a cloud application to start and stop role instances, change configuration settings to allow the application to throttle its functionality and reduce its resource usage, and send notifications according to a defined schedule.

The key advantages of this approach include the low implementation costs and ease of use; all you need to do is to provide the configuration information that specifies the circumstances (in terms of a schedule and performance measures) under which the block will apply instance scaling or application throttling actions.

How Trey Research Controls Elasticity in the Orders Application

Trey Research decided to use the Enterprise Library Autoscaling Application Block to start and stop instances of web and worker roles as the load from users changes. Initially, Trey Research deployed the Orders application and made it available to a small but statistically significant number of users, evenly spread across an area representing the geographical location of their expected market. By gathering statistical usage information based on this pilot, Trey Research identified how the system functions at specific times in the working day, and identified periods during which performance suffered due to an increased load from customers. Specifically, Trey Research noticed that:

- Many customers tended to place their orders towards the end of the working day (between 15:30 and 20:30 Central time, allowing for the spread of users across the United States), with an especially large number placed between 17:30 and 18:30.

- On Fridays, the peak loading tended to start and finish two hours earlier (13:30 to 18:30 Central time).
- On the weekend, very few customers placed orders.

To cater for these peaks and troughs in demand, Trey Research decided to implement the Enterprise Library Autoscaling Application Block as follows:
- The developers configured constraint rules for the application to start an additional three instances of the web and worker roles at each datacenter at 15:15 Central time (it can take 10 or 15 minutes for the new instances to become available), and to shut these instances down at 20:30 between Monday and Thursday.
- At 17:15 the application was configured to start a further two instances of each role, which are shut down at 18:30.
- On Fridays, the times at which the extra instances start and stop are two hours earlier.
- To handle unexpected demand, Trey Research also configured reactive rules to monitor the number of customer requests, and start additional instances if the average CPU usage for a web role exceeds 85% for 10 or more minutes, up to a maximum of 12 instances per datacenter. When the CPU usage drops below 50%, instances are shut down, subject to a minimum of two instances per datacenter.
- On weekends, the system is constrained to allow a maximum of four instances of each role at each datacenter, and any additional instances above this number are shut down to reduce running costs.
- When the system is inactive or lightly loaded, the system returns to its baseline configuration comprising two instances of each role per datacenter.

Hosting the Autoscaling Application Block

The Autoscaling Application Block monitors the performance of one or more roles, starting and stopping roles, applying throttling changes to configuration, or sending notifications as specified by the various constraint rules and reactive rules. The Autoscaling Application Block also generates diagnostic information and captures data points indicating the work that it has performed. For more information about the information collected, see *"Autoscaling Application Block Logging"* on MSDN.

To perform this work, the Autoscaling Application Block uses an **Autoscaler** object (defined in the **Microsoft.Practices.EnterpriseLibrary.WindowsAzure.Autoscaling** namespace), and you must arrange for this object to start running when your application executes. The Trey Research solution performs this task in the **Run** method in the **WorkerRole** class (in the **Orders.Workers** project), and stops the **Autoscaler** in the **OnStop** method:

```
C#
public class WorkerRole : RoleEntryPoint
{
  private Autoscaler autoscaler;
  ...
  public override void Run()
  {
    this.autoscaler = EnterpriseLibraryContainer.Current.
      GetInstance<Autoscaler>();
```

```
      this.autoscaler.Start();
      ...
    }
    ...
    public override void OnStop()
    {
      this.autoscaler.Stop();
      ...
    }
    ...
}
```

The information about which roles to monitor, the storage account to use for storing diagnostic data, and the location of the rules defining the behavior of the **Autoscaler** object are specified in the **<serviceModel>** section of the service information store file. This file was uploaded to blob storage and stored in the blob specified by the **<serviceInformationStores>** section of the app.config file for the worker role. For more information and an example of defining the service model for the Autoscaling Application Block, see Chapter 5, *"Making Tailspin Surveys More Elastic,"* of the *Developer's Guide to the Enterprise Library 5.0 Integration Pack for Windows Azure* on MSDN.

Defining the Autoscaling Rules

The Trey Research solution implements a combination of constraint rules and reactive rules. The constraint rules specify the schedule the **Autoscaler** object should use, in addition to the maximum and minimum number of instances of roles during each scheduled period. The **Autoscaler** initiates creation of instances of the web and worker roles, or stops existing instances, when the schedule changes the boundaries and when the instance count is outside that new boundary. The reactive rules start further instances of the web role or stop them, according to the CPU loading of the web role. As an initial starting point, Trey Research defined the following set of rules:

```xml
<?xml version="1.0" encoding="utf-8" ?>
<rules xmlns="http://schemas.microsoft.com/practices/2011/
entlib/autoscaling/rules">
  <constraintRules>
    <rule name="Weekday" enabled="true" rank="10">
      <timetable startTime="00:00:00" duration="23:59:59"
        utcOffset="-06:00">
        <weekly days=
            "Monday Tuesday Wednesday Thursday Friday"/>
      </timetable>
      <actions>
        <range target="Orders.Workers"
                  min="2" max="12"/>
        <range target="Orders.Website"
                  min="2" max="12"/>
      </actions>
    </rule>
```

```xml
<rule name="Weekend" enabled="true" rank="10">
  <timetable startTime="00:00:00" duration="23:59:59"
   utcOffset="-06:00">
    <weekly days="Sunday Saturday"/>
  </timetable>
  <actions>
    <range target="Orders.Workers"
           min="2" max="4"/>
    <range target="Orders.Website"
           min="2" max="4"/>
  </actions>
</rule>
<rule name="MondayToThursday" enabled="true" rank="2">
  <timetable startTime="15:15:00" duration="05:15:00"
   utcOffset="-06:00">
    <weekly days="Monday Tuesday Wednesday Thursday"/>
  </timetable>
  <actions>
    <range target=" Orders.Workers"
               min="4" max="12"/>
    <range target=" Orders.Website"
               min="4" max="12"/>
  </actions>
</rule>
<rule name="MondayToThursdayPeak" enabled="true"
      rank="3">
  <timetable startTime="17:15:00" duration="03:15:00"
   utcOffset="-06:00">
    <weekly days="Monday Tuesday Wednesday Thursday"/>
  </timetable>
  <actions>
    <range target=" Orders.Workers"
           min="6" max="12"/>
    <range target=" Orders.Website"
           min="6" max="12"/>
  </actions>
</rule>
<rule name="Friday" enabled="true" rank="2">
  <timetable startTime="13:15:00" duration="05:15:00"
   utcOffset="-06:00">
    <weekly days="Friday"/>
  </timetable>
  <actions>
    <range target=" Orders.Workers"
           min="6" max="12"/>
    <range target=" Orders.Website"
```

CHAPTER SIX

```xml
          min="6" max="12"/>
      </actions>
    </rule>
    <rule name="FridayPeak" enabled="true" rank="3">
      <timetable startTime="15:15:00" duration="03:15:00"
        utcOffset="-06:00">
        <weekly days="Friday"/>
      </timetable>
      <actions>
        <range target=" Orders.Workers"
               min="7" max="12"/>
        <range target=" Orders.Website"
               min="7" max="12"/>
      </actions>
    </rule>
  </constraintRules>
  <reactiveRules>
    <rule name="HotCPU" enabled="true" rank="4">
      <when>
        <greater operand="CPU" than="85" />
      </when>
      <actions>
        <scale target="Orders.Website"
               by ="1"/>
      </actions>
    </rule>
    <rule name="CoolCPU" enabled="true" rank="4">
      <when>
        <less operand="CPU" than="50" />
      </when>
      <actions>
        <scale target="Orders.Website"
               by ="-1"/>
      </actions>
    </rule>
  </reactiveRules>
  <operands>
    <performanceCounter alias="CPU"
source="AccidentReporting_WebRole" performanceCounterName=
"\Processor(_Total)\% Processor Time" timespan="00:10:00"
aggregate="Average"/>
  </operands>
</rules>
```

Hosting costs for Windows Azure services are calculated on an hourly basis, with each part hour charged as a complete hour. This means that if, for example, you start a new service instance at 14.50, and shut it down at 16.10, you will be charged for 3 hours. You should keep this in mind when configuring the schedule for the Autoscaler. For more information, see the "Pricing Overview" page for Windows Azure.

The rules were uploaded to blob storage, to the blob specified by the **<rulesStores>** section of the app.config file for the worker role.

The **CPU** operand referenced by the reactive rules calculates the average processor utilization over a 30 minute period, using the **\Processor(_Total)\% Processor Time** performance counter. The **Orders.Website** web role was modified to collect this information by using the following code shown in bold in the **StartDiagnostics** method (called from the **OnStart** method) in the file **WebRole.cs**:

```C#
public class WebRole : RoleEntryPoint
{
  ...
  private static void StartDiagnostics()
  {
    var config =
      DiagnosticMonitor.GetDefaultInitialConfiguration();
    ...
    config.PerformanceCounters.ScheduledTranferPeriod =
      Timespan.FromMinutes(10);

    config.PerformanceCounters.DataSources.Add(
      new PerformanceCounterConfiguration
      {
        CounterSpecifier =
          @"\Processor(_Total)\% Processor Time",
        SampleRate = TimeSpan.FromMinutes(30)
      });

    ...
    DiagnosticMonitor.Start(
      "DiagnosticsConnectionString", config);
  }
  ...
}
```

The performance counter data is written to the **WADPerformance-CountersTable** table in Windows Azure Table storage. The web role must be running in full trust mode to successfully write to this table.

> Using automatic scaling is an iterative process. The configuration defined by the team at Trey Research is kept constantly under review, and the performance of the solution is continuously monitored as the pattern of use by customers evolves. The operators may modify the configuration and autoscaling rules in the future to change the times and circumstances under which additional instances are created and destroyed.

Managing Network Latency and Maximizing Connectivity to the Orders Application

Trey Research initially deployed the Orders application to two datacenters, US North and US South, both located in the United States. The rationale behind this decision was the US North datacenter is located just a few miles from Trey Research, affording reasonable network response times for the compliance application hosted in this datacenter (as described in Chapter 4, "Implementing Reliable Messaging and Communications with the Cloud"), while the majority of Trey Research's customers are expected to be located in the continental United States.

However, as Trey Research expands their customer base, it expects users connecting to the Orders application to be situated farther afield, perhaps on a different continent. The distance between customers and the physical location in which the Orders application is deployed can have a significant bearing on the response time of the system. Therefore Trey Research felt it necessary to adopt a strategy that minimizes this distance and reduces the associated network latency for users accessing the Orders application.

As its customers became distributed around the world, Trey Research considered hosting additional instances of the Orders application in datacenters that are similarly distributed. Customers could then connect to the closest available instance of the application. The question that Trey Research needed to address in this scenario was how to route a customer to the most local instance of the Orders application?

Choosing How to Manage Network Latency and Maximize Connectivity to the Orders Application

Trey Research investigated a number of solutions for directing customers to the most local instance of the Orders application, including deploying and configuring a number of DNS servers around the world (in conjunction with a number of network partners) based on the DNS address of the machine from which the customer's request originated. However, many of these solutions proved impractical or expensive, leaving Trey Research to consider the two options described in the following sections.

Build a Custom Service to Redirect Traffic

Trey Research examined the possibility of building a custom service through which all customers would connect, and then deploying this service to the cloud. The purpose of this service would be to examine each request and forward it on to the Orders application running in the most appropriate datacenter. This approach would enable Trey Research to filter and redirect requests based on criteria such as the IP address of each request. The custom service could also detect whether the Orders application at each datacenter was still running, and if it was currently unavailable it could transparently redirect customer requests to functioning instances of the application. Additionally, the custom service could attempt to distribute requests evenly across datacenters that are equally close (in network terms) to the customer, implementing a load-balancing mechanism to ensure that no one instance of the Orders application became unduly overloaded while others remained idle.

This type of custom service is reasonably common, and can be implemented by using the **System.ServiceModel.Routing.RoutingService** class of Windows Communication Foundation. However, a custom service such as this is non-trivial to design, build, and test; and the routing rules that determine how the service redirects messages can quickly become complex and difficult to maintain. Additionally, this service must itself be hosted somewhere with sufficient power to handle every customer request, and with good network connectivity to all customers. If the service is underpowered it will become a bottleneck, and if customers cannot connect to the service quickly then the advantages of using this service are nullified. Furthermore, this service constitutes a single point of failure; if it becomes unavailable then customers may not be able to connect to any instance of the Orders application.

Use Windows Azure Traffic Manager to Route Customers' Requests

Windows Azure Traffic Manager is a Windows Azure service that enables you to set up request routing and load balancing based on predefined policies and configurable rules. It provides a mechanism for routing requests to multiple deployments of your Windows Azure-hosted applications and services, regardless of the datacenter location. The applications or services could be deployed in one or more datacenters.

Traffic Manager is effectively a DNS resolver. When you use Traffic Manager, web browsers and services accessing your application will perform a DNS query to Traffic Manager to resolve the IP address of the endpoint to which they will connect, just as they would when connecting to any other website or resource.

Traffic Manager addresses the network latency and application availability issues by providing three mechanisms, or policies, for routing requests:

- The **Performance** policy redirects requests from users to the application in the closest data center. This may not be the application in the data center that is closest in purely geographical terms, but instead the one that provides the lowest network latency. Traffic Manager also detects failed applications and does not route to these, instead choosing the next closest working application deployment.

- The **Failover** policy allows you to configure a prioritized list of applications, and Traffic Manager will route requests to the first one in the list that it detects is responding to requests. If that application fails, Traffic Manager will route requests to the next applications in the list, and so on.

- The **Round Robin** policy routes requests to each application in turn; though it detects failed applications and does not route to these. This policy evens out the loading on each application, but may not provide users with the best possible response times as it ignores the relative locations of the user and data center.

You select which one of these policies is most appropriate to your requirements; Performance to minimize network latency, Failover to maximize availability, or Round Robin to distribute requests evenly (and possibly improve response time as a result).

Traffic Manager is managed and maintained by Microsoft, and the service is hosted in their datacenters. This means that there is no maintenance overhead.

How Trey Research Minimizes Network Latency and Maximizes Connectivity to the Orders Application

Using the Enterprise Library Autoscaling Application Block helps to ensure that sufficient instances of the Orders application web and worker roles are running to service the volume of customers connecting to a specific datacenter at a given point in time. With this in mind, the operations staff at Trey Research decided to use Traffic Manager simply to route customers' requests to the nearest responsive datacenter by implementing the Performance policy.

The operation staff configured the policy to include the DNS addresses of the Orders application deployed to the US North and US South datacenters, and monitoring the Home page of the web application to determine availability. The operations staff selected the DNS prefix **ordersapp.treyresearch**, and mapped the resulting address (**ordersapp.treyresearch.trafficmanager.net**) to the public address used by customers, **store.treyresearch.net**. In this way, a customer connecting to the URL **http://store.treyresearch.net** is transparently rerouted by Traffic Manager to the Orders application running in the US North datacenter or the US South datacenter. Figure 1 shows the structure of this configuration.

> The selection of the Performance policy was very easy; the Failover policy is not suitable for the Trey Research scenario, and the Enterprise Library Autoscaling Application Block ensures that an appropriate number of instances of the Orders application roles will be available at each datacenter to facilitate good performance so the Round Robin policy is unnecessary.

Implementing the Round Robin policy may be detrimental to customers as they might be routed to a more distant datacenter, incurring additional network latency and impacting the response time of the application. Additionally, the Round Robin policy may conceivably route two consecutive requests from the same customer to different datacenters, possibly leading to confusion if the data cached at each datacenter is not completely consistent. The Performance policy has the advantage of reducing the network latency while ensuring that requests from the same customer are much more likely to be routed to the same datacenter.

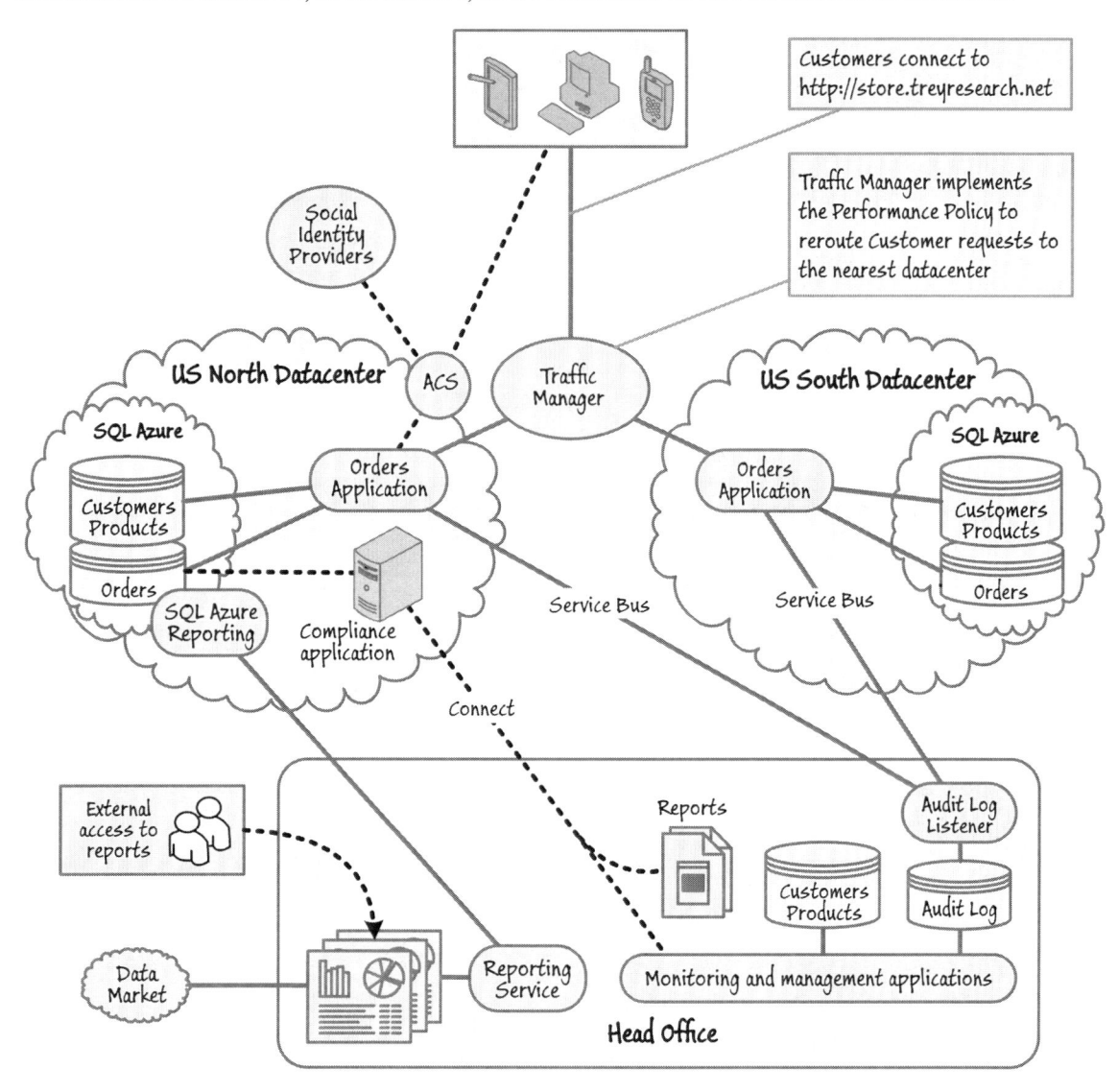

FIGURE 1
How Trey Research uses Windows Azure Traffic Manager

Notice that the Orders application in both datacenters must connect to the head office audit log listener service. Both deployments of the Orders application must also connect to all of Trey Research's transport partners; although, for simplicity, this is not shown in the diagram. Some features of the application, such as the use of the SQL Azure™ technology platform Reporting Service and the deployment of the compliance application in a Windows Azure VM Role, are not duplicated in both datacenters. The Orders data is synchronized across both datacenters and so one instance of the Reporting Service and the compliance application will provide the required results, without incurring additional hosting and service costs.

However, the designers at Trey Research realized that using a mechanism that may route users to different deployments of the application in different datacenters will have some impact. For example, data such as the user's current shopping cart is typically stored in memory or local storage (such as Windows Azure table storage or SQL Azure). When a user is re-routed to a different datacenter, this data is lost unless the application specifically synchronizes it across all datacenters.

In addition, if Trey Research configured ACS in more than one datacenter to protect against authentication issues should ACS in one datacenter be unavailable, re-routing users to another datacenter would mean they would have to sign in again.

However, Trey Research considers that both of these scenarios were unlikely to occur often enough to be an issue.

Optimizing the Response Time of the Orders Application

Windows Azure is a highly scalable platform that offers high performance for applications. However, available computing power alone does not guarantee that an application will be responsive; an application that is designed to function in a serial manner will not make best use of this platform and may spend a significant period blocked waiting for slower, dependent operations to complete. The solution is to perform these operations asynchronously, and the techniques that Trey Research adopted to implement this approach have been described in Chapter 4, "Implementing Reliable Messaging and Communications with the Cloud" and Chapter 5, "Processing Orders in the Trey Research Solution."

Aside from the design and implementation of the application logic, the key factor that governs the response time and throughput of a service is the speed with which it can access the resources and data that it needs. In the case of the Orders application, the primary data source is the SQL Azure database containing the customer, order, and product details. Chapter 2, "Deploying the Orders Application and Data in the Cloud" described how Trey Research positioned the data within each datacenter to try and minimize the network overhead associated with accessing this information. However, databases are still relatively slow when compared to other forms of data storage. So, Trey Research was left facing the question: How do you provide scalable, reliable, and fast access to the customer, order and product data as this could be key to minimizing the response time of the Orders application?

Choosing How to Optimize the Response Time of the Orders Application

Upon investigating the issues surrounding response times in more detail, Trey Research found that that there were two complimentary approaches available (both can be used, if appropriate).

Implement Windows Azure Caching

Windows Azure Caching is a service that enables you to cache data in the cloud, and provides scalable, reliable, and shared access to this data.

On profiling the Orders application, the developers at Trey Research found that it spent a significant proportion of its time querying the SQL Azure database, and the latency associated with connecting to this database, together with the inherent overhead of querying and updating data in the database, accounted for a large part of this time. By caching data with the Windows Azure Caching service, Trey Research hoped to reduce the overhead associated with repeatedly accessing remote data, eliminate the network latency associated with remote data access, and improve the response times for applications referencing this data.

The overhead associated with querying and updating data in SQL Azure are not a criticism of this database management system (DBMS). All DBMSs that support concurrent multiuser access have to ensure consistency and integrity of data, typically by serializing concurrent requests from different users and locking data. SQL Azure meets these requirements very efficiently. However, retrieving data from a cache does not have this overhead; it is simply retrieved or updated. This efficiency comes at a cost, as the application itself now has to take responsibility for ensuring the integrity and consistency of cached data.

Additionally, sooner or later any updates to cached data must be copied back to the database, otherwise the cache and the database will be inconsistent with each other or data may be lost; the cache has a finite size, and the Windows Azure Caching service may evict data if there is insufficient space available, or expire data that has remained in the cache for a lengthy period of time.

The Windows Azure Caching Service is also chargeable; it is hosted and maintained by Microsoft in their datacenters, and they offer guarantees concerning the availability of this service and the cached data, but you will be charged depending upon the size of the cache and the volume of traffic read from or written to the cache. For more information, see "*Caching, based on cache size per month*."

Configure the Content Delivery Network

The Windows Azure Content Delivery Network (CDN) is a service designed to improve the response time of web applications by caching the static output generated by hosted services, and also frequently accessed blob data, closer to the users that request them. While Windows Azure Caching is primarily useful for improving the performance of web applications and services running in the cloud, users will frequently be invoking these web applications and services from their desktop, either by using a custom application that connects to them or by using a web browser. The data returned from a web application or service may be of a considerable size, and if the user is very distant it may take a significant time for this data to arrive at the user's desktop. The CDN enables you to cache the output of web pages and frequently queried data at a variety of locations around the world. When a user makes a request, the web content and data can be served from the most optimal location based on the current volume of traffic at the various Internet nodes through which the request is routed.

> *Detailed information, samples, and exercises showing how to configure CDN are available on MSDN; see the topic "Windows Azure CDN" at* **http://msdn.microsoft.com/en-us/gg405416**. *Additionally Chapter 3, "Accessing the Surveys Application" in the guide "Developing Applications for the Cloud, 2nd Edition" provides further implementation details.*

While CDN is a useful technology, investigation by the developers at Trey Research suggested that it would not be applicable in the current version of the Orders application; CDN is ideally suited to caching web pages with static content and blob data for output or streaming to client applications, while many of the pages generated by the Orders application may be relatively dynamic, and the application does not store or emit blob data.

How Trey Research Optimizes the Response Time of the Orders Application

The Orders application uses several types of data; customer information, order details, and the product catalog. Order information is relatively dynamic, and customer details are accessed infrequently compared to other data (only when the customer logs in). Furthermore the same customer and order information tends not to be required by concurrent instances of the Order application. However the product catalog is queried by every instance of the Orders application when the user logs in. It is also reasonably static; product information is updated very infrequently. Additionally, the product catalog can comprise a large number of items. For these reasons, the developers at Trey Research elected to cache the product catalog by using a shared Windows Azure cache in each datacenter, while they decided that caching order and customer details would bring few benefits.

Defining and Configuring the Windows Azure Cache

The Windows Azure Caching service runs in the cloud, and an application should really connect only to an instance of the Windows Azure Caching service located in the same datacenter that hosts the application code. Therefore, Trey Research used the Windows Azure Caching service to create separate caches in the US North and US South datacenters, called **TreyResearchCacheUSN** (for the US North datacenter) and **TreyResearchCacheUSS** (for the US South datacenter). This ensures that each cache has a unique and easily recognizable name. The developers estimated that a 128MB cache (the minimum size available, with the cheapest cost) would be sufficient. However, the caches can easily be increased in size if necessary, without impacting the operation of the Orders application.

The web application, implemented in the **Orders.Website** project, defines the configuration parameters for accessing the cache in the service configuration file for the solution (ServiceConfiguration.csfg).

> *The Trey Research example application provided in the sample application is only deployed to a single datacenter, and the cache is named **TreyResearchCache**.*

```xml
<?xml version="1.0" encoding="utf-8"?>
<ServiceConfiguration serviceName="Orders.Azure" ...>
  ...
  <Role name="Orders.Website">
    ...
    <ConfigurationSettings>
      ...
      <Setting name="CacheHost"
        value="TreyResearchCache.cache.windows.net" />
      <Setting name="CachePort" value="22233" />
      <Setting name="CacheAcsKey" value="[data omitted]" />
      <Setting name="IsLocalCacheEnabled" value="false" />
```

```xml
      <Setting name="LocalCacheObjectCount" value="1000" />
      <Setting name="LocalCacheTtlValue" value="60" />
      <Setting name="LocalCacheSync"
        value="TimeoutBased" />
      ...
    </ConfigurationSettings>
    ...
  </Role>
</ServiceConfiguration>
```

Synchronizing the Caches and Databases in the Orders Application

The Orders application was modified to retrieve and update data from the local instance of the Windows Azure Cache, only fetching data from the SQL Azure database if the data is not currently available in the cache. Any changes made to cached data are copied back to SQL Azure. The following subsections describe how Trey Research implemented this approach.

Trey Research also had to consider the effects of caching on their data synchronization strategy. Each datacenter has a copy of the SQL Azure database holding the customers, orders, and products data. The Orders application can amend customers and orders information, and when it does so the cached copy of this information is copied back to the local SQL Azure database. This database is subsequently synchronized with the SQL Azure databases located in the other datacenters, as described in Chapter 2, "Deploying the Orders Application and Data in the Cloud."

However, suppose that the details of an order or customer have been cached by the Orders application running in the US North datacenter, and the same details are queried and cached by the Orders application running in the US South datacenter. At this point the two caches hold the same data. If the information in the US North datacenter is changed and written back to the SQL Azure database in the US North datacenter, and this database is subsequently synchronized with the US South datacenter, then the cached data in the US South datacenter is now out of date. However, when the cached data held in the US South datacenter expires or is evicted, the cache will be populated with the fresh data the next time it is queried.

So, although caching can improve the response time for many operations, it can also lead to issues of consistency if two instances of an item of data are not identical. Consequently, applications that use caching should be designed to cope with data that may be stale but that eventually becomes consistent.

This issue can become more acute if the same cached data is updated simultaneously in the US North and US South datacenters; SQL Azure Data Sync will ensure consistency between the different databases, but at least one of the caches will hold inconsistent data. For more advice and guidance on how to address these problems refer to the section "Guidelines for Using Windows Azure Caching" in "Appendix E - Maximizing Scalability, Availability, and Performance."

Retrieving and Managing Data in the Orders Application

The Orders application uses a set of classes for storing and retrieving each of the types of information it references. These classes are located in the DataStores folder of the **Orders.Website** project. For example, the **ProductStore** class in the ProductStore.cs file provides methods for querying products. These methods are defined by the **IProductsStore** interface:

C#
```
public interface IProductStore
{
  IEnumerable<Product> FindAll();
  Product FindOne(int productId);
}
```

The **FindAll** method returns a list of all available products from the SQL Azure database, and the **FindOne** method fetches the product with the specified product ID. In a similar vein, the **OrderStore** class implements the **IOrdersStore** interface which defines methods for retrieving and managing orders. None of these classes implements any form of caching.

Implementing Caching Functionality for the Products Catalog

The **Orders.Website** project contains a generic library of classes for caching data, located in the DataStores\Caching folder. This library is capable of caching any of the data items defined by the types in the DataStores folder, but for the reasons described earlier caching is only implemented for the **ProductStore** class.

The DataStores\Caching folder contains the **ICachingStrategy** interface, the **CachingStrategy** class, and the **ProductStoreWithCache** class. The following sections describe these classes.

The ICachingStrategy Interface

This is a simple interface that abstracts the caching functionality implemented by the library. It exposes a property named **DefaultTimeout** and a method called **Get**, as follows:

C#
```
public interface ICachingStrategy
{
  TimeSpan DefaultTimeout
  {
    get;
    set;
  }

  object Get<T>(string key, Func<T> fallbackAction,
           TimeSpan? timeout) where T : class;
}
```

The **key** parameter of the **Get** method specifies the unique identifier of the object to retrieve from the cache. If the object is not currently cached, the **fallbackAction** parameter specifies a delegate for a method to run to retrieve the corresponding data, and the **timeout** parameter specifies the lifetime of the object if it is added to the cache. If the **timeout** parameter is null, an implementation of this interface should set the lifetime of the object to the value specified by the **DefaultTimeout** property.

The CachingStrategy Class

This class implements the **ICachingStrategy** interface. The constructor for this class uses the Windows Azure caching APIs to authenticate and connect to the Windows Azure cache using the values provided as parameters (the web application retrieves these values from the service configuration file, and invokes the constructor by using the Unity framework as described later in this chapter, in the section "Instantiating and Using a ProductsStoreWithCache Object.")

The **Get** method of the **CachingStrategy** class queries the cache using the specified key, and if the object is found it is returned. If the object is not found, the method invokes the delegate to retrieve the missing data and adds it to the cache, specifying either the timeout value provided as the parameter to the **Get** method (if it is not null) or the default timeout value for the **CachingStrategy** object. The following code sample shows the important elements of this class:

C#
```
public class CachingStrategy :
  ICachingStrategy, IDisposable
{
  private readonly RetryPolicy cacheRetryPolicy;
  private DataCacheFactory cacheFactory;

  ...
  private TimeSpan defaultTimeout =
    TimeSpan.FromMinutes(10);

  public CachingStrategy(string host, int port,
    string key, bool isLocalCacheEnabled,
    long objectCount, int ttlValue, string sync)
  {
    // Declare array for cache host.
    var servers = new DataCacheServerEndpoint[1];

    servers[0] = new DataCacheServerEndpoint(
      host, port);

    // Setup DataCacheSecurity configuration.
    var secureAcsKey = new SecureString();
    foreach (char a in key)
    {
      secureAcsKey.AppendChar(a);
    }
    secureAcsKey.MakeReadOnly();
    var factorySecurity =
      new DataCacheSecurity(secureAcsKey);

    // Setup the DataCacheFactory configuration.
    var factoryConfig =
      new DataCacheFactoryConfiguration
```

```
    {
      Servers = servers,
      SecurityProperties = factorySecurity
    };
  ...
  this.cacheFactory =
    new DataCacheFactory(factoryConfig);

  this.cacheRetryPolicy = RetryPolicyFactory.
    GetDefaultAzureCachingRetryPolicy();
  ...
}

public TimeSpan DefaultTimeout
{
  get { return this.defaultTimeout; }
  set { this.defaultTimeout = value; }
}

public virtual object Get<T>(string key,
  Func<T> fallbackAction, TimeSpan? timeout)
  where T : class
  {
    ...
    try
    {
      var dataCache =
        this.cacheFactory.GetDefaultCache();

      var cachedObject =
        this.cacheRetryPolicy.ExecuteAction(
          () => dataCache.Get(key));

      if (cachedObject != null)
      {
        ...
        return cachedObject;
      }
      ...
      var objectToBeCached = fallbackAction();

      if (objectToBeCached != null)
      {
        try
        {
          this.cacheRetryPolicy.ExecuteAction(() =>
```

```
              dataCache.Put(key, objectToBeCached,
                timeout != null ?
                  timeout.Value : this.DefaultTimeout));
            ...
            return objectToBeCached;
          }
          ...
        }
      }
    }
  }
  ...
}
```

Notice that this class traps transient errors that may occur when fetching an item from the cache, by using the Transient Fault Handling Application Block. The static **GetDefaultAzureCachingRetry-Policy** method of the **RetryPolicyFactory** class referenced in the constructor returns the default policy for detecting a transient caching exception, and provides a construct for indicating how such an exception should be handled. The default policy implements the "Fixed Interval Retry Strategy" defined by the Transient Fault Handling Block, and the web.config file configures this strategy to retry to the failing operation up to six times with a five second delay between attempts.

The **Get** property of the **CachingStrategy** class invokes the **ExecuteAction** method of the retry policy object, passing it a delegate that attempts to read the requested data from the cache (this is the code that may exhibit a transient error, and if necessary will be retried based on the settings defined by the retry policy object). If a non-transient error occurs or an attempt to read the cache fails after six attempts, the exception handling strategy in the **Get** method (omitted from the code above) will return the value from the underlying store, retrieved by calling the **fallbackAction** delegate.

The ProductStoreWithCache Class

This class provides the caching version of the **ProductStore** class. It implements the **IProductsStore** interface, but internally employs an **ICachingStrategy** object to fetch data in the **FindAll** and **FindOne** methods, as shown by the following code sample:

```C#
public class ProductStoreWithCache : IProductStore
{
  private readonly IProductStore productStore;

  private readonly ICachingStrategy cachingStrategy;

  public ProductStoreWithCache(
    IProductStore productStore,
    ICachingStrategy cachingStrategy)
  {
    this.productStore = productStore;
    this.cachingStrategy = cachingStrategy;
  }
```

```
public IEnumerable<Product> FindAll()
{
  ...
  return (IEnumerable<Product>)
    this.cachingStrategy.Get(
    "ProductStore/FindAll",
    () => this.productStore.FindAll(),
    TimeSpan.FromMinutes(10));
}

public Product FindOne(int productId)
{
  ...
  return (Product)this.cachingStrategy.Get(
    string.Format(
      "ProductStore/Product/{0}", productId),
    () => this.productStore.FindOne(productId),
    TimeSpan.FromMinutes(10));
}
}
```

Instantiating and Using a ProductsStoreWithCache Object

The Orders application creates a **ProductsStoreWithCache** object by using the Unity Application Block. The static **ContainerBootstrapper** class contains the following code:

C#
```
public static class ContainerBootstrapper
{
  public static void RegisterTypes(
    IUnityContainer container)
  {
    ...
    container.RegisterType<IProductStore,
      ProductStoreWithCache>(
        new InjectionConstructor(
          new ResolvedParameter<ProductStore>(),
          new ResolvedParameter<ICachingStrategy>()));
    container.RegisterType<ProductStore>();

    // To change the caching strategy, replace the
    // CachingStrategy class with the strategy that
    // you want to use instead.
    var cacheAcsKey = CloudConfiguration.
      GetConfigurationSetting("CacheAcsKey", null);
```

```
    var port = Convert.ToInt32(CloudConfiguration.
      GetConfigurationSetting("CachePort", null));
    var host = CloudConfiguration.
      GetConfigurationSetting("CacheHost", null);

    var isLocalCacheEnabled = Convert.ToBoolean(
      CloudConfiguration.GetConfigurationSetting(
        "IsLocalCacheEnabled", null));
    var localCacheObjectCount = Convert.ToInt64(
      CloudConfiguration.GetConfigurationSetting(
        "LocalCacheObjectCount", null));
    var localCacheTtlValue = Convert.ToInt32(
      CloudConfiguration.GetConfigurationSetting(
        "LocalCacheTtlValue", null));
    var localCacheSync =
      CloudConfiguration.GetConfigurationSetting(
        "LocalCacheSync", null);

    container.RegisterType<ICachingStrategy,
                          CachingStrategy> (
      new ContainerControlledLifetimeManager(),
      new InjectionConstructor(host, port, cacheAcsKey,
        isLocalCacheEnabled, localCacheObjectCount,
        localCacheTtlValue, localCacheSync));
  }
}
```

These statements register the **ProductStore** and **CachingStrategy** objects, and the Unity Application Block uses them to create a **ProductStoreWithCache** object whenever the application instantiates an **IProductStore** object. Notice that the **CachingStrategy** class is configured to use the **ContainerControlledLifetimeManager** class of the Unity framework. This effectively ensures that the **CachingStrategy** object used by the application is created as a singleton that spans the life of the application. This is useful as the **DataCache-Factory** object that the **CachingStrategy** class encapsulates is very expensive and time consuming to create, so it is best to create a single instance of this class that is available throughout the duration of the application. Additionally, the parameters for the constructor for the **CachingStrategy** object are read from the configuration file and are passed to the **CachingStrategy** class by using a Unity **InjectionConstructor** object.

> The **RegisterTypes** method of the **ContainerBootstrapper** class is called from the **SetupDependencies** method in the Global.asax.cs file when the Orders application starts running. The **SetupDependencies** method also assigns the dependency resolver for the Orders application to the Unity container that registered these types. For more information about using the Unity Application Block see *"Unity Application Block"* on MSDN.

The **StoreController** class calls the **FindAll** method of the **ProductStoreWithCache** object when it needs to fetch and display the entire product catalog, and the **FindOne** method when it needs to retrieve the details for a single product:

```C#
public class StoreController : Controller
{
  private readonly IProductStore productStore;

  public StoreController(IProductStore productStore)
  {
    ...
    this.productStore = productStore;
  }

  public ActionResult Index()
  {
    var products = this.productStore.FindAll();
    return View(products);
  }

  public ActionResult Details(int id)
  {
    var p = this.productStore.FindOne(id);
    return View(p);
  }
}
```

This code transparently accesses the Windows Azure cache, populating it if the requested data is not currently available in the cache. You can change the caching configuration, and even elect not to cache data if caching is found to have no benefit, without modifying the business logic for the Orders application; all you need to do is switch the type for the **IProductsStore** interface in the **ContainerBootstrapper** class to **ProductStore**, as highlighted in bold in the following code example:

```C#
public static class ContainerBootstrapper
{
  public static void RegisterTypes(
    IUnityContainer container)
  {
    ...
    container.RegisterType<IProductStore, ProductStore>();
    ...
  }
}
```

Summary

This chapter has described the Windows Azure technologies that Trey Research used to improve the scalability, availability, and performance of the Orders application.

Windows Azure Traffic Manager can play an important role in reducing the network latency associated with sending requests to a web application by transparently routing these requests to the most appropriate deployment of the web application relative to the location of the client submitting these requests. Traffic Manager can also help to maximize availability by intelligently detecting whether the application is responsive, and if not, re-routing requests to a different deployment of the application.

Windows Azure provides a highly scalable environment for hosting web applications and services, and the Enterprise Library Autoscaling Application Block implements a mechanism that can take full advantage of this scalability by monitoring web applications and automatically starting and stopping instances as the demand from clients requires.

Finally, Windows Azure caching is an essential element in improving the responsiveness of web applications and services. It enables Trey Research to cache data locally to these applications, in the same datacenter. This technique removes much of the network latency associated with remote data access. However, as Trey Research discovered, you must be prepared to balance this improvement in performance against the possible complexity introduced by maintaining multiple copies of data.

More Information

All links in this book are accessible from the book's online bibliography available at: *http://msdn.microsoft.com/en-us/library/hh968447.aspx*.

- "Autoscaling Application Block Logging" at *http://msdn.microsoft.com/en-us/library/hh680883(v=pandp.50).aspx*
- Chapter 5, "Making Tailspin Surveys More Elastic," of the Developer's Guide to the Enterprise Library 5.0 Integration Pack for Windows Azure at *http://msdn.microsoft.com/en-us/library/hh680942(PandP.50).aspx*
- "Pricing Overview" at *http://www.windowsazure.com/en-us/pricing/details/*.
- "Caching, based on cache size per month" at *http://www.windowsazure.com/en-us/pricing/details/#caching*
- "Windows Azure CDN" at *http://msdn.microsoft.com/en-us/gg405416*.
- Chapter 3, "Accessing the Surveys Application" in the guide "Developing Applications for the Cloud, 2nd Edition" *http://msdn.microsoft.com/en-us/library/ff966499.aspx*.
- "Unity Application Block" at *http://msdn.microsoft.com/en-us/library/ff647202.aspx*.
- "Windows Azure Traffic Manager" at *http://msdn.microsoft.com/en-us/gg197529*.
- "Windows Azure Service Instances Auto Scaling" at *http://azureautoscaling.codeplex.com/releases/view/62421*.
- "Windows Azure Caching Service" at *http://msdn.microsoft.com/en-us/library/gg278356.aspx*.
- "Windows Azure CDN" at *http://msdn.microsoft.com/en-us/gg405416*.

7 Monitoring and Managing the Orders Application

When the design and implementation of the hybrid Orders application was completed, Trey Research considered how to monitor and manage the application as it runs on the Windows Azure™ technology platform.

The Orders application comprises a number of components, built using a variety of technologies, and distributed across a range of sites and connected by networks of varying bandwidth and reliability. With this complexity, it was very important for Trey Research to be able to monitor how well the system is functioning, and quickly take any necessary restorative action in the event of failure. However, monitoring a complex system is itself a complex task, requiring tools that can quickly gather performance data to help analyze throughput and pinpoint the causes of any errors, failures, or other shortcomings in the system. The range of problems can vary significantly, from simple failures caused by application errors in a service running in the cloud, through issues with the environment hosting individual elements, to complete systemic failure and loss of connectivity between components whether they are running on-premises or in the cloud.

This chapter focuses on the challenges associated with monitoring the Orders application, and the decisions Trey Research made when tackling these challenges.

Scenario and Context

The hybrid Orders application has components running remotely from the on-premises services; including a website, background order processing code, and databases. The application also communicates with transport partners as it processes orders, listens for status messages from these partners, and sends messages to the on-premises Audit Log service.

The designers at Trey Research had to decide how to monitor the application as it runs so that administrators can measure performance, ensure it meets Service Level Agreements, and verify that it provides acceptable response times to visitors. Administrators must also be able to retrieve data about errors or exceptions that occur at runtime, and be able to trace operations to assist in debugging the application. Developers had to add some code to the application before it was deployed in order to accomplish many of these tasks.

Trey Research also had to consider how to deploy the application to Windows Azure, and how to manage factors such as re-configuration and management of the individual Windows Azure services it uses while the application is deployed and running. Trey Research created a set of scripts and other executable programs that allow these kinds of tasks to be performed repeatedly, accurately, and securely.

Monitoring Services, Logging Activity, and Measuring Performance

Even though the Orders application runs remotely from Trey Research's head office, it is still possible for Trey Research administrators to obtain the same kinds of information about its operation and any exceptions or errors that occur as they would when administering an application deployed locally in their own data center. However, the way that this data is collected and accessed is very different in Windows Azure compared to a local server deployment.

You can configure Windows Azure Diagnostics to collect performance and diagnostics information. This data is stored in memory in the worker or web role being monitored, but it can be transferred to Windows Azure storage on a scheduled basis or on demand, so that it can be accessed from on-premises applications and monitoring solutions.

Figure 1 shows a high-level view of the monitoring mechanism in Windows Azure and some of the ways that Trey Research considered using it. The Windows Azure Diagnostics mechanism can be configured to collect data from a range of sources, such as Windows event logs and performance counters. It is also possible to use a third party logging mechanism, such as the Enterprise Library Logging Application Block, or custom code that writes events to the Window Azure Diagnostics mechanism.

> You can configure the Windows Azure Diagnostics mechanism to collect the data you need to monitor and debug applications, and to transfer this data to Windows Azure storage so that you can access it.

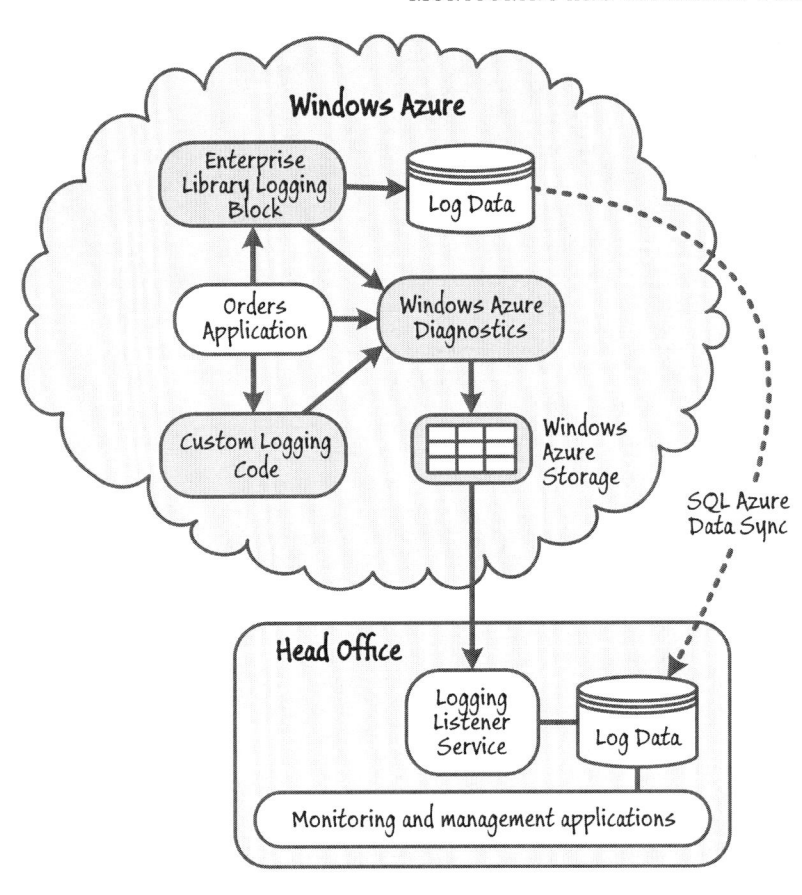

FIGURE 1
Monitoring approaches that Trey Research considered for the Orders application

Choosing a Monitoring and Logging Solution

Trey Research considered four ways of collecting information for monitoring services, logging activity, and measuring performance in the Orders application: Windows Azure Diagnostics, the Enterprise Library Logging Application Block, a third party monitoring solution, and using custom code in the application to generate logging messages. The following sections describe each of these options.

> A range of comprehensive ready-built monitoring solutions designed to work with applications deployed in Windows Azure is available. These products typically provide functions for collecting and analyzing monitoring information, displaying it in a dashboard, and notifying operators of significant events. Such solutions include Microsoft System Center Operations Manager and products from third parties.

Windows Azure Diagnostics

Windows Azure Diagnostics is the built-in mechanism for collecting all kinds of monitoring and diagnostic information in Windows Azure. It requires no additional code or assemblies. Windows Azure Diagnostics can collect data from Windows event logs and performance counters; the IIS log and failed request log; infrastructure logs; crash dump files; and custom error logs. Developers or administrators at Trey Research simply configure the diagnostics mechanism to collect the required data, and specify the intervals for this data to be transferred to Windows Azure storage. They can also use Windows Azure PowerShell cmdlets to reconfigure the diagnostics settings as the application runs, initiate a transfer of the data to Windows Azure storage on demand, and download the logged data to an on-premises store.

However, there is only a limited set of options for filtering and categorizing the logged information, and it can only be stored in Windows Azure storage. There are no opportunities to store the data in a database, or in a custom format or repository.

> *For more information about using Windows Azure Diagnostics, see "Appendix F - Monitoring and Managing Hybrid Applications" of this guide.*

Enterprise Library Logging Application Block

The Logging Application Block is a component of Enterprise Library, a framework of components for managing cross-cutting concerns in most types of applications. Trey Research could configure the Logging Application Block to send log entries to the Windows Azure Diagnostic trace listener, which is a component of the Windows Azure Diagnostics system that stores the log entries in memory so that they can be transferred to Windows Azure storage along with any other diagnostic data that the system collects.

Alternatively, Trey Research could configure the Logging Application Block to send log entries to other types of storage such as a database, text files in a range of formats, and XML files. One option that Trey Research considered was using the Logging Application Block to write log entries directly to a SQL Azure™ technology platform database located in the cloud, and then transfer this data back to an on-premises database for analysis. SQL Azure Data Sync could be used to simplify the task of synchronizing the data between the cloud and an on-premises database.

> You might also consider using the Enterprise Library Exception Handling Application Block to provide a structured policy-driven mechanism for collecting and managing exception information. The Exception Handling block can send its log entry messages to the Enterprise Library Logging Application Block for exposure through the Windows Azure Diagnostics mechanism.

The Logging Application Block is highly configurable and extensible, and includes a wide range of options for filtering and categorizing log messages. This would make it easy for developers at Trey Research to generate different types of log entries and provide useful additional support for administrators and operators.

The main limitation of the Enterprise Library Logging Application Block is that it cannot collect data from the host system; such as Windows event log entries, performance counters, or IIS log files. It is purely an activity logging mechanism where code generates the log entries in response to events occurring in the application. Using any of the Enterprise Library Application Blocks also means that external library assemblies must be uploaded and installed with the application code in Windows Azure.

For more information about the Enterprise Library Logging Application Block and Exception Handling Application Block, see "About This Release of Enterprise Library." There is also a whitepaper available that describes how you can use the Enterprise Library 5.0 application blocks with Windows Azure-hosted applications. You can download the whitepaper from the **Enterprise Library CodePlex site**.

Third Party Monitoring Solution

Trey Research could have adopted a third party monitoring solution. There are several solutions available that are aimed wholly or partly at monitoring Windows Azure applications and services. They include the following:

- *Windows Azure Management Pack for Microsoft System Center Operations Manager*
- *Azure Diagnostics Manager* from Cerebrata
- *AzureWatch* from Paraleap Technologies
- *ManageAxis* from Cumulux

These solutions can monitor role status, collect performance information, gather event data, and raise notifications to administration staff.

Custom Logging Solution

The developers at Trey Research considered building a custom logging and diagnostics solution for the Orders application. The Windows Azure Diagnostic trace listener exposes methods for creating and storing log entries, and so provides a way for Trey Research to monitor activity and expose these log entries through the standard Windows Azure Diagnostics mechanism. For example, the developers could add code to the Orders application that generates a message each time a visitor is initially authenticated and signs in. This code can call the methods of the diagnostic trace listener to store the message as a log entry. When the diagnostics data is later transferred to Windows Azure storage it will include entries created by the custom code.

> You can use the Windows Azure Diagnostic trace listener to generate log entries containing information you need for monitoring events and activity in your application. The data is exposed through the Windows Azure Diagnostics mechanism and can be transferred to Windows Azure storage for analysis as required.

It is also possible to create custom logging solutions that store data in other formats and locations. For example, like the Enterprise Library Logging Application Block, the code could store the log entries in a database, text file, or a repository in some other format. This approach will require a mechanism for accessing the data remotely from the on-premises applications and tools, or for transferring the data back to on-premises storage for future analysis.

The main limitation that Trey Research considered with using a custom logging solution is that, like the Logging Application Block, it cannot collect data from the host system; such as Windows event log entries, performance counters, or IIS log files. It is purely an activity logging mechanism where code generates the log entries in response to events occurring in the application.

How Trey Research Chose a Monitoring and Logging Solution

Trey Research wanted to be able to generate some monitoring and activity tracing information in the Orders application all of the time it is running. However, the administrators did not want to collect full tracing information or operating system diagnostics information all of the time. Instead, they want to be able to change the configuration so that additional information can be collected when required, such as when debugging a problem with the application.

After careful consideration, Trey Research decided to use a custom solution for activity tracing and recording specific errors by generating these log entries and then writing them to the Windows Azure Diagnostics mechanism. While the Enterprise Library Logging Application Block (and the Exception Handling Application Block) would have been suitable, the types of information that Trey Research collects are limited, and so the additional complexity of using these blocks was not felt to be an advantage in Trey Research's scenario.

How Trey Research Uses Windows Azure Diagnostics

Trey Research implements diagnostic logging, and downloads the information from the cloud to their on-premises servers. Trey Research traces the execution of each role instance, and also records the details of any exceptions raised by the role instances using a combination of a custom **TraceHelper** class and the standard Windows Azure Diagnostics mechanism. The data is initially stored in a table named WADLogsTable, in table storage at each datacenter. Trey Research considered the following two options for monitoring this data and using it to manage the system:

- Using System Center Operations Manager with the Windows Azure Management Pack, or another third party solution, to connect directly to each datacenter, examine the diagnostic data, generate the appropriate usage reports, and alert an operator if an instance failed or some other significant event occurred.
- Periodically transferring the diagnostic data to a secure on-premises location, and then reformatting this data for use by their own custom reporting and analytical tools.

Although System Center Operations Manager and other third party solutions provide many powerful features, the existing investment that Trey Research has already made in developing and procuring custom analytical tools led to the second option being more appealing. Additionally, it meant that Trey Research could more easily retain a complete audit log of all significant operations and events locally, which might be a requirement as ever-stricter compliance regulations become legally binding. However, this solution does not preclude Trey Research from deploying System Center Operations Manager or another third party solution in the future.

Selecting the Data and Events to Record

Trey Research decided to record different types of events using trace messages and Windows Azure Diagnostics. Under normal operation, Trey Research collects only trace messages that have a severity of **Warning** or higher. However, the mechanism Trey Research implemented allows administrators to change the behavior to provide more detail when debugging the application or monitoring specific activity.

The following table shows the logging configuration that Trey Research uses. Notice that Trey Research does not collect Windows event log events or Windows performance counter data. Instead, Trey Research captures information at all stages of the operation of the application and sends this information to the Windows Azure Diagnostics mechanism through a custom class named **TraceHelper**.

Event or tract type	Logging mechanism used	Type of event that initiates logging
Application-defined **Error**-level events	Collected by the custom **TraceHelper** class	Posting a message to a topic or queue fails after all retries. A job process task in a worker role fails. Failure to access Windows Azure Cache.
Application-defined **Warning**-level events	Collected by the custom **TraceHelper** class	Posting a message to a topic or queue fails but will be retried. Updating a database fails but will be retried.
Application-defined **Information**-level events	Collected by the custom **TraceHelper** class	Application startup. Starting a job process task in a worker role. Opening a view in the Orders website. Detailed information on a web role's interaction with Windows Azure Cache. Various events related to an order being placed, or customer data being added.
Application-defined **Verbose**-level events	Collected by the custom **TraceHelper** class	None defined; available for future extensions.
Windows event logs	Not collected by Trey Research.	Windows internal operating system or software events.
Windows performance counters	Not collected by Trey Research.	Performance counters implemented by Windows operating system and installed applications.

Configuring the Diagnostics Mechanism

The worker role in the **Orders.Workers** project and the web role in the **Orders.Website** project are both configured to use Windows Azure Diagnostics. The configuration file for both applications contains the following settings:

```xml
<?xml version="1.0" encoding="utf-8"?>
<configuration>
  ...
  <system.diagnostics>
    <sources>
      <source name="TraceSource">
        <listeners>
          <add type="Microsoft.WindowsAzure.Diagnostics
                    .DiagnosticMonitorTraceListener, ..."
              name="AzureDiagnostics">
            <filter type="" />
          </add>
        </listeners>
      </source>
    </sources>
  </system.diagnostics>
</configuration>
```

This configuration defines a diagnostic source listener named **TraceSource** that sends trace messages to the Windows Azure **DiagnosticsMonitorTraceListener** class. There is no filter level defined in the configuration because this will be set in the code that initializes the **TraceSource** listener.

To configure the diagnostics and schedule the transfer of diagnostic data to Windows Azure storage, Trey Research initially considered using an imperative approach by adding code such as that shown below to the **OnStart** methods of the classes implementing the web and worker roles.

```csharp
...
// Get default initial configuration.
var config =
    DiagnosticMonitor.GetDefaultInitialConfiguration();

// Update the initial configuration.
config.Logs.ScheduledTransferLogLevelFilter
        = LogLevel.Undefined;
config.Logs.ScheduledTransferPeriod
        = TimeSpan.FromSeconds(60);
```

> To collect Windows event logs and performance counter data you must configure Windows Azure Diagnostics to transfer the required data to Windows Azure Table storage. Windows event log entries are transferred to a table named **WADWindowsEventLogsTable**, and performance counter data is transferred to a table named **WADPerformanceCountersTable**. If Trey Research needs to capture this data in the on-premises management application, its developers must write additional code to download the data in these tables.

```
// Start the monitor with this configuration.
DiagnosticMonitor.Start("DiagnosticsConnectionString",
                        config);
...
```

However, although simple, Trey Research found that this approach could prove inflexible. If necessary, Trey Research needs to be able to modify the diagnostics configuration and transfer schedule remotely (by using Windows Azure Powershell cmdlets or one of the available third party monitoring solutions such as Cerebrata Diagnostics Manager listed earlier), but performing these tasks in code will cause any remote changes to the configuration to be lost if a role restarts.

Therefore Trey Research opted to configure the diagnostics by using the diagnostics configuration file, diagnostics.wadcfg. This file can be held in blob storage (and therefore survive role restarts) and read by the Windows Azure Diagnostics monitor when the role starts. For more information, see *"Using the Windows Azure Diagnostics Configuration File"* on MSDN.

Implementing Trace Message Logging and Specifying the Level of Detail

Trey Research collects trace messages generated by a custom class named **TraceHelper** (located in the Helpers folder of the **Orders.Shared** project). The **TraceHelper** class instantiates a **TraceSource** instance and exposes a set of static methods that make it easy to write trace messages with different severity levels.

C#
```
public class TraceHelper
{
  private static readonly TraceSource Trace;

  static TraceHelper()
  {
    Trace = new TraceSource("TraceSource",
                            SourceLevels.Information);
  }

  [EnvironmentPermissionAttribute(
    SecurityAction.LinkDemand, Unrestricted = true)]
  public static void Configure(SourceLevels sourceLevels)
  {
    Trace.Switch.Level = sourceLevels;
  }

  public static void TraceVerbose(string format,
                       params object[] args)
  {
    Trace.TraceEvent(TraceEventType.Verbose, 0,
                     format, args);
  }
```

```csharp
public static void TraceInformation(string format,
                        params object[] args)
{
  Trace.TraceEvent(TraceEventType.Information, 0,
            format, args);
}

public static void TraceWarning(string format,
                        params object[] args)
{
  Trace.TraceEvent(TraceEventType.Warning, 0,
            format, args);
}

public static void TraceError(string format,
                        params object[] args)
{
  Trace.TraceEvent(TraceEventType.Error, 0,
            format, args);
}
}
```

Data recorded in this way is directed to the Windows Azure Diagnostics monitor trace listener (due to the configuration shown in the section "Configuring the Diagnostics Mechanism" earlier in this chapter), and subsequently into the **WADLogsTable**. By default the **TraceHelper** class captures messages with the severity filter level of **Information**. However, this setting can be changed by calling the **Configure** method the **TraceHelper** class exposes, and supplying a value for the severity level of messages to trace. The worker roles and web roles both configure this setting in the **OnStart** method by reading it from the service configuration file.

C#
```csharp
public override bool OnStart()
{
  ...
  ConfigureTraceListener(
    RoleEnvironment.GetConfigurationSettingValue(
      "TraceEventTypeFilter"));
  ...
}

private static void ConfigureTraceListener(
                    string traceEventTypeFilter)
{
  SourceLevels sourceLevels;
  if (Enum.TryParse(traceEventTypeFilter, true,
              out sourceLevels))
  {
```

```
      TraceHelper.Configure(sourceLevels);
   }
}
```

The roles also set up handlers for the **RoleEnvironmentChanging** and **RoleEnvironmentChanged** events. These handlers reconfigure the **TraceHelper** class for the role when the configuration changes. This enables administrators to change the severity filter level to obtain additional information for debugging and monitoring while the application is running.

Writing Trace Messages

The web and worker roles use the **TraceHelper** class to record information about events, errors, and other significant occurrences. For example, exceptions are captured using code such as that shown in the following example taken from the **ReceiveNextMessage** method of the **ServiceBusReceiverHandler** class in the **Orders.Shared** project. Note that this code calls the **TraceError** method of the **Trace-Helper** class to write a trace message with severity "Error".

> The mechanism that Trey Research implements for specifying the trace level offers a high degree of control over the volume and nature of data that is captured. However, an alternative approach is to capture data for all events and then apply filters when transferring the trace data to Windows Azure storage by setting the **ScheduledTransferLog-LevelFilter** property of the diagnostic monitor configuration; this property can be specified as part of the Windows Azure Diagnostics monitor configuration stored in the diagnostics.wadcfg and can be updated remotely without requiring the roles to be restarted.

C#
```csharp
private void ReceiveNextMessage(
  CancellationToken cancellationToken)
{
  ...
  this.ReceiveNextMessage(cancellationToken);

  if (taskResult.Exception != null)
  {
    TraceHelper.TraceError(taskResult.Exception.Message);
    throw taskResult.Exception;
  }
  ...
}
```

The **TraceHelper** class is also used in the web role. Code in the Custom-Attributes folder of the **Orders.Website** project defines a custom attribute called **LogActionAttribute** that calls the **TraceInformation** method of the **TraceHelper** class to write a trace message with severity "Information".

C#
```csharp
public class LogActionAttribute : ActionFilterAttribute
{
  public override void OnActionExecuting(
    ActionExecutingContext filterContext)
  {
    ...
```

```csharp
    TraceHelper.TraceInformation(
      "Executing Action '{0}', from controller '{1}'",
      filterContext.ActionDescriptor.ActionName,
      filterContext.ActionDescriptor.
        ControllerDescriptor.ControllerName);
  }

  public override void OnActionExecuted(
    ActionExecutedContext filterContext)
  {
    ...
    TraceHelper.TraceInformation(
      "Action '{0}', from controller '{1}'
        has been executed",
      filterContext.ActionDescriptor.ActionName,
      filterContext.ActionDescriptor.
        ControllerDescriptor.ControllerName);
      }
   }
}
```

The controller classes in the **Orders.Website** project are tagged with this attribute. The following code shows the **StoreController** class, which retrieves products for display.

```csharp
[LogAction]
public class StoreController : Controller
{
  ...
  public ActionResult Index()
  {
    var products = this.productStore.FindAll();
    return View(products);
  }

  public ActionResult Details(int id)
  {
    var p = this.productStore.FindOne(id);
    return View(p);
  }
}
```

This feature enables the application to generate a complete record of all tagged actions performed on behalf of every user simply by changing the **TraceEventTypeFilter** setting in the **ServiceConfiguration.cscfg** file to Information.

Transferring Diagnostics Data from the Cloud

Trey Research uses a custom mechanism for collating and analyzing diagnostics information. It requires that all applications store event and trace messages in an on-premises database named DiagnosticsLog that the monitoring and analysis mechanism queries at preset intervals.

Trey Research could use a third-party tool to download the data from the WADLogsTable, or write scripts that use the Windows Azure PowerShell cmdlets (see *http://wappowershell.codeplex.com*). However, the Windows Azure SDK provides classes that make it easy to interact with Windows Azure storage through the management API using the .NET Framework. This is the approach that Trey Research chose.

The on-premises monitoring and management application (implemented in the **HeadOffice** project of the example) contains a page that administrators use to download and examine the diagnostics data collected in the **Orders** application.

The code that interacts with Windows Azure storage and updates the on-premises DiagnosticsLog database table is in the **DiagnosticsController** class, located in the Controllers folder of the **HeadOffice** project. The **DiagnosticsController** class uses the Enterprise Library Transient Fault Handling Block to retry any failed connection to Windows Azure storage and the on-premises database. The constructor of the **DiagnosticsController** class reads the retry policy from the application configuration file.

In a real application the diagnostics data would be downloaded automatically at preset intervals and stored in the on-premises database by a Windows service or other background application. For simplicity in the example application, the data is only downloaded when you open the Diagnostics page of the HeadOffice application.

```C#
this.storageRetryPolicy
  = RetryPolicyFactory.GetDefaultAzureStorageRetryPolicy();
```

When an administrator opens the Diagnostics page of the **HeadOffice** application, the **TransferLogs** action is executed. This action extracts a list of the datacenters from which it will download data from the application configuration, and then reads the corresponding account details (from the same configuration) for each datacenter. As the code iterates over the list of datacenters it creates a suitable **CloudStorageAccount** instance using the credentials collected earlier, and then calls a method named **TransferLogs** to download the data from this datacenter.

C#
```
[HttpPost]
public ActionResult TransferLogs(
                    FormCollection formCollection)
{
  var deleteEntries
    = formCollection.GetValue("deleteEntries") != null;
  var dataCenters
    = WebConfigurationManager.AppSettings["dataCenters"]
                        .Split(',');
  ...
  // Get account details for accessing each datacenter.
  var dataCenters2 = dataCenters.Select(
      dc => dc.Trim()).Where(dc =>
          !string.IsNullOrEmpty(dc.Trim()));
  var accountNames = dataCenters2.Select(
    dc => string.Format(CultureInfo.InvariantCulture,
        "diagnosticsStorageAccountName.{0}", dc));
  var accountKeys = dataCenters2.Select(
    dc => string.Format(CultureInfo.InvariantCulture,
        "diagnosticsStorageAccountKey.{0}", dc));

  for (var i = 0; i < dataCenters2.Count(); i++)
  {
    // Create credentials for this datacenter.
    var cred = new StorageCredentialsAccountAndKey(
        WebConfigurationManager.AppSettings[
                    accountNames.ElementAt(i)],
        WebConfigurationManager.AppSettings[
                    accountKeys.ElementAt(i)]);
    var storageAccount = new CloudStorageAccount(cred,
                                                true);

    // Download the data from this datacenter.
    this.TransferLogs(dataCenters2.ElementAt(i),
              storageAccount, deleteEntries);
  }
  ...
}
```

The **TransferLogs** method uses the **CreateCloudTableClient** class to access Windows Azure Table storage. The code accesses the table service context and generates a query over the **WADLogsTable** in Windows Azure storage. For each entry returned from the query (each row in the table) it creates a new **DiagnosticsLog** instance and saves this instance in the DiagnosticsLog database by using the **DiagnosticsLogStore** repository class. Notice how this method can also delete the entries in the **WADLogsTable** in Windows Azure storage at the same time to reduce storage requirements in the cloud.

```C#
private void TransferLogs(string dataCenter,
            CloudStorageAccount storageAccount,
            bool deleteWADLogsTableEntries)
{
  var tableStorage
     = storageAccount.CreateCloudTableClient();
  ...
  var context = tableStorage.GetDataServiceContext();

  if (!deleteWADLogsTableEntries)
  {
    context.MergeOption = MergeOption.NoTracking;
  }

  IQueryable<WadLog> query
    = this.storageRetryPolicy.ExecuteAction(() =>
          context.CreateQuery<WadLog>("WADLogsTable"));

  foreach (var logEntry in query)
  {
    var diagLog = new DiagnosticsLog
      {
        Id = Guid.NewGuid(),
        PartitionKey = logEntry.PartitionKey,
        RowKey = logEntry.RowKey,
        DeploymentId = logEntry.DeploymentId,
        DataCenter = dataCenter,
        Role = logEntry.Role,
        RoleInstance = logEntry.RoleInstance,
        Message = logEntry.Message,
        TimeStamp = logEntry.Timestamp
      };
    this.store.Save(diagLog);

    if (deleteWADLogsTableEntries)
    {
      context.DeleteObject(logEntry);
      this.storageRetryPolicy.ExecuteAction(() =>
                          context.SaveChanges());
    }
  }
}
```

> When accessing and performing operations on the Windows Azure Tables that store diagnostics information, consider the transaction charges that these operations will incur. It may be better to pay for storage than to pay for a large number of transactions that delete individual rows; then drop and recreate the table at appropriate intervals.

The strategy that Trey Research adopted for deleting diagnostic data from Windows Azure storage after it has been downloaded prevents this data from growing indefinitely, but it does come at a cost. Each record being deleted is counted as a single transaction against Windows Azure storage and Trey Research is billed accordingly. As the number of customers increase, the volume of diagnostics data that Trey Research captures will increase as well, and eventually the transaction charges associated with deleting each record individually as it is downloaded may become prohibitive.

To counter this overhead, Trey Research are currently evaluating an alternative approach; rather than deleting individual records from tables in Windows Azure storage, simply drop and recreate the tables themselves at an appropriate juncture, after downloading the data. This approach comprises far fewer transactions, but adds complexity to the code in the role that downloads the data; it may need to implement a locking mechanism to prevent a scheduled transfer of diagnostics data to a table that has just been dropped but has yet to be recreated. Additionally, dropping and creating tables may be more time consuming than removing individual records from an existing table, so this functionality may need to be implemented as a background task in a web or worker role.

Deployment and Management

Trey Research wanted to be able to configure and manage all of the services within its Windows Azure account that are used by the Orders application. Trey Research required that the configuration of features such as ACS and Service Bus be automated and reliably repeatable. This configuration is necessarily complex, and includes configuration of services in multiple datacenters where the application is deployed.

Trey Research also wanted to automate the deployment and redeployment of the application as it is updated and extended. Automating the deployment reduces the chances of errors, and helps to control the permissions for the employees that can perform these tasks.

Choosing Deployment and Management Solutions

Trey Research considered a range of solutions for deploying and managing the Orders application. These options included using the Windows Azure Management Portal, the Windows Azure Service Management REST API and Windows Azure SDK, and the Windows Azure PowerShell Cmdlets. The following sections describe each of these options.

Windows Azure Management Portal

The Management Portal is the primary location for creating service namespaces, and can also be used to configure all of the Windows Azure services for a subscription. It provides a graphical and intuitive interface that is easy to use, and provides feedback on the state of each service. However, all users of the Management Portal must access it by providing the administrative credentials for the Trey Research Windows Azure subscription, which means that they will have access to all features of the subscription.

The administrators at Trey Research are aware that using the Management Portal is the only way to create new namespaces for the services such as Service Bus, ACS, and Traffic Manager, although these namespaces can be configured afterwards using the portal, scripts, or code.

Windows Azure Service Management REST API and Windows Azure SDK
With the exception of creating namespaces for services, all of the features of the Windows Azure services can be accessed using the Windows Azure Service Management REST API and the Windows Azure SDK. The Windows Azure SDK contains assemblies used for performing service management tasks against the Service Management REST API. Alternatively, you can use third party tools or create your own code that accesses the REST interfaces of the Service Management API to automate management tasks. This approach is useful if you are building a solution based on a language that is not supported by the Windows Azure SDK; for example, you can install the Windows Azure SDK for Java and use the Java programming language.

Administrators and developers at Trey Research realized that they could use code inside their applications and management tools to perform complex tasks by using the Service Management API, including creation of setup programs and tools for managing most aspects of the application and the services it uses.

The major limitation with the Windows Azure Service Management REST API is that it cannot be used directly in scripts.

> *For more information about the Windows Azure Service Management REST API, see "About the Service Management API" on MSDN.*

Windows Azure PowerShell Cmdlets
The Windows Azure PowerShell cmdlets library that you can download from the Codeplex website contains almost one hundred PowerShell cmdlets that can accomplish most common Windows Azure management and configuration tasks.

These cmdlets are extremely useful for performing a wide range of management tasks, and they can be used within other scripts as well as being executed directly from the command line. The administrators at Trey Research realized that the cmdlets provide an ideal solution for accomplishing simple, everyday tasks.

However, while the Windows Azure PowerShell cmdlets can be used in scripts for complex management tasks, this can be more difficult than using the Windows Azure Service Management REST API.

> *For more information about the Windows Azure PowerShell cmdlets, see "Windows Azure PowerShell Cmdlets" at http://wappowershell.codeplex.com.*

How Trey Research Chose Deployment and Management Solutions
Managing Windows Azure applications and services typically involves two types of tasks:
- Occasional or infrequent tasks, such as configuring namespaces and features such as ACS, Windows Azure Service Bus, and Windows Azure Traffic Manager. These tasks are often complex.
- Frequent tasks, such as deploying and updating services, downloading logging data, adding and removing certificates, manipulating storage services, setting firewall rules, and interacting with SQL Azure. These tasks are typically more straight-forward.

For the occasional tasks, Trey Research decided to create applications and tools that use the Windows Azure Management REST API through objects exposed by the Windows Azure SDK and by a separate library. For example, Trey Research decided to create a setup program that can be executed to set many of the configuration options in Windows Azure instead of using the Management Portal.

For the more frequently performed tasks, administrators at Trey Research decided to use the Windows Azure PowerShell cmdlets within scripts to provide a repeatable, reliable, and automated process. For example, Trey Research uses a PowerShell script that is executed before deployment to set the appropriate values for namespaces, user names, passwords, and keys in the source files. These items are different for each datacenter in which the application is deployed.

Administrators at Trey Research also use PowerShell scripts to perform tasks such as changing the Windows Azure Diagnostics configuration for tracing and debugging, managing certificates, starting and stopping instances of the application roles, and managing SQL Azure databases.

How Trey Research Deploys and Manages the Orders Application

The following sections describe how Trey Research uses the Windows Azure Management REST API though a management wrapper library and directly to automate much of the configuration of Windows Azure services such as ACS and Service Bus.

Configuring Windows Azure by Using the Service Management Wrapper Library

Trey Research uses a library of functions that was originally developed by the Windows Azure team to help automate configuration of Windows Azure namespaces through the REST-based Management API. The library code is included in the **ACS.ServiceManagementWrapper** project of the sample code, and you can reuse this library in your own applications.

The setup program Trey Research created instantiates a **ServiceManagementWrapper** object and then calls several separate methods within the setup program to configure ACS and Service Bus for the Orders application. The Service Bus configuration Trey Research uses depends on ACS to authenticate the identities that post messages to, or subscribe to, queues and topics.

> The workings of the Service Management wrapper library are not described here, but you can examine the source code and modify it if you wish. It is a complex project, and exposes a great deal of functionality that makes many tasks for configuring Windows Azure much easier that writing your own custom code.

```C#
internal static void Main(string[] args)
{
  try
  {
    var acs = new ServiceManagementWrapper(
                  acsServiceNamespace,
                  acsUsername, acsPassword);

    Console.WriteLine("Setting up ACS namespace:"
                      + acsServiceNamespace);

    // ACS namespace setup for the Orders Website
    CleanupIdenityProviders(acs);
    CleanupRelyingParties(acs);
    CreateIdentityProviders(acs);
    CreateRelyingPartysWithRules(acs);

    // Create Service Bus topic, subscriptions and queue.
    SetupServiceBusTopicAndQueue();
  }
  catch (Exception ex)
  {
    ... display exception information ...
  }
    Console.ReadKey();
}
```

The values used by the setup program for namespace names, account IDs, passwords, and keys are stored in the App.config file of the setup program project named **TreyResearch.Setup**. Notice that the code first cleans up the current ACS namespace by removing any existing settings so that the new settings replace the old ones. If not, ACS may attempt to add duplicate settings or features such as identity providers or rule sets, which could cause an error.

To illustrate how easy the Service Management wrapper library is to use, the following code from **CleanupRelyingParties** method of the setup program removes all existing relying parties with the name "AccessControlManagement" from the current ACS namespace.

```C#
var rps = acsWrapper.RetrieveRelyingParties();
foreach (var rp in rps)
{
  if (rp.Name != "AccessControlManagement")
  {
    acsWrapper.RemoveRelyingParty(rp.Name);
  }
}
```

The setup program creates service identities by first removing any identity with the same name, and then adding a new one with the specified name and password. The values used in this code come from the App.config file.

C#
```
acswrapper.RemoveServiceIdentity(ContosoDisplayName);
acswrapper.AddServiceIdentity(ContosoDisplayName,
                              ContosoPassword);
```

The setup program also uses the Service Management wrapper library to create rules that map claims from identity providers to the claims required by the Orders application. For example, the following code creates a pass-through rule for the Windows Live ID® identity provider that maps the **Name-Identifier** claim provided by Windows Live ID to a new **Name** claim.

C#
```
var identityProviderName
    = SocialIdentityProviders.WindowsLiveId.DisplayName;

// pass nameidentifier as name
acsWrapper.AddPassThroughRuleToRuleGroup(
        defaultRuleGroup.RuleGroup.Name,
        identityProviderName,
        ClaimTypes.NameIdentifier, ClaimTypes.Name);
```

Configuring Windows Azure by Using the Built-in Management Objects

It is also possible to use the built-in objects that are part of the Windows Azure SDK to configure Windows Azure namespaces. For example, the setup program configures the access rules for the Service Bus endpoints in ACS. To do this it uses the classes in the **Microsoft.ServiceBus.AccessControlExtensions.AccessControlManagement** namespace.

The following code shows how the setup program creates a rule group for the Orders Statistics service. You can see that it sets ACS as the claim issuer, and adds two rules to the rule group. The first rule allows the authenticated identity **externaldataanalyzer** (a small and simple demonstration program that can display order statistics) to send requests. The second rule allows the authenticated identity **headoffice** (the on-premises management and monitoring application) to listen for requests. The code then adds the rule group to the **OrdersStatisticsService** relying party, and saves all the changes.

C#
```
var settings = new AccessControlSettings(
                ServiceBusNamespace, DefaultKey);
ManagementService serviceClient = ManagementServiceHelper
                .CreateManagementServiceClient(settings);

serviceClient.DeleteRuleGroupByNameIfExists(
            "Rule group for OrdersStatisticsService");
serviceClient.SaveChanges(SaveChangesOptions.Batch);

var ruleGroup = new RuleGroup {
    Name = "Rule group for OrdersStatisticsService" };
serviceClient.AddToRuleGroups(ruleGroup);

// Equivalent to selecting "Access Control Service" as
// the input claim issuer in the Management portal.
var issuer = serviceClient.GetIssuerByName(
                                "LOCAL AUTHORITY");

serviceClient.CreateRule(
    issuer,
    "http://schemas.xmlsoap.org/ws/2005/05/identity/
       claims/nameidentifier",
    "externaldataanalyzer",
    "net.windows.servicebus.action",
    "Send",
    ruleGroup,
    string.Empty);

serviceClient.CreateRule(
    issuer,
    "http://schemas.xmlsoap.org/ws/2005/05/identity/
       claims/nameidentifier",
    "headoffice",
    "net.windows.servicebus.action",
    "Listen",
    ruleGroup,
    string.Empty);

var relyingParty = serviceClient.GetRelyingPartyByName(
                        "OrdersStatisticsService", true);
var relyingPartyRuleGroup = new
  Microsoft.ServiceBus.AccessControlExtensions
   .AccessControlManagement.RelyingPartyRuleGroup();
relyingParty.RelyingPartyRuleGroups.Add(
                                relyingPartyRuleGroup);
```

> There is a great deal of code in the **TreyResearch. Setup** project. You may find it useful to examine this project when you create your own setup programs, and reuse some of the generic routines it contains.

```
serviceClient.AddToRelyingPartyRuleGroups(
                              relyingPartyRuleGroup);

serviceClient.AddLink(relyingParty,
    "RelyingPartyRuleGroups", relyingPartyRuleGroup);
serviceClient.AddLink(ruleGroup,
    "RelyingPartyRuleGroups", relyingPartyRuleGroup);

serviceClient.SaveChanges(SaveChangesOptions.Batch);
```

Summary

This chapter described how Trey Research tackled the issues surrounding deploying, configuring, monitoring and managing a hybrid application. The important point to realize is that the complexity of the environment and its distributed nature means it is inevitable that performance issues and failures will occur in such a system. The key to maintaining a good level of service is detecting these issues and failures, and responding quickly in a controlled, secure, and repeatable manner.

Windows Azure Diagnostics provides the basic tools to enable you to detect and determine the possible causes of errors and performance problems, but it is important that you understand how to relate the diagnostic information that is generated to the structure of your application. Analyzing this information is a task for a domain expert who not only understands the architecture and business operations of your application, but also has a thorough familiarity with the way in which this architecture maps to the services provided by Windows Azure.

The Windows Azure Service Management API provides controlled access to Windows Azure features, enabling you to build scripts and applications that an operator can use to deploy and manage the elements that comprise your application. This approach eliminates the need for an operator to have the same level of expertise with Windows Azure as the solution architect, and can also reduce the scope for errors by automating and sequencing many of the tasks involved in a complex deployment.

More Information

All links in this book are accessible from the book's online bibliography available at: *http://msdn.microsoft.com/en-us/library/hh968447.aspx* .

- "About This Release of Enterprise Library" at *http://msdn.microsoft.com/en-us/library/ff664636(v=PandP.50).aspx.*
- Whitepaper on the Enterprise Library Codeplex site that describes how you can use the Enterprise Library 5.0 application blocks with Windows Azure-hosted applications at *http://entlib.codeplex.com/releases/view/75025#DownloadId=336804.*
- Windows Azure Management Pack for Microsoft System Center Operations Manager at *http://pinpoint.microsoft.com/en-us/applications/system-center-monitoring-pack-for-windows-azure-applications-12884907699.*
- Azure Diagnostics Manager from Cerebrata at *http://www.cerebrata.com/Products/AzureDiagnosticsManager/Default.aspx.*
- AzureWatch from Paraleap Technologies at *http://www.paraleap.com/.*

- "Using the Windows Azure Diagnostics Configuration File" at *http://msdn.microsoft.com/en-us/library/gg604918.aspx*.
- "About the Service Management API" at *http://msdn.microsoft.com/en-us/library/windowsazure/ee460807.aspx*.
- "Collecting Logging Data by Using Windows Azure Diagnostics" at *http://msdn.microsoft.com/en-us/library/windowsazure/gg433048.aspx*.
- "Monitoring Windows Azure Applications" at *http://msdn.microsoft.com/en-us/library/windowsazure/gg676009.aspx*.
- "Windows Azure Service Management REST API Reference" at *http://msdn.microsoft.com/en-us/library/windowsazure/ee460799.aspx*.
- "Take Control of Logging and Tracing in Windows Azure" at *http://msdn.microsoft.com/en-us/magazine/ff714589.aspx*.
- "Windows Azure PowerShell Cmdlets" at *http://wappowershell.codeplex.com*.

APPENDIX A

Replicating, Distributing, and Synchronizing Data

All but the most trivial of applications have a requirement to store and retrieve data. For many systems, this aspect employs a database to act as a reliable repository. Modern database management systems, such as Microsoft SQL Server, provide multiuser access capable of handling many thousands of concurrent connections if the appropriate hardware and network bandwidth is available. However, to support highly-scalable data storage that reduces the need to install and maintain expensive hardware within an organization's data center, solutions such as the SQL Azure™ technology platform provide a cloud-based database management system that implements many of the same features.

Using SQL Azure, you can deploy a database to the same datacenter hosting the cloud-based applications and services that use it, which helps to minimize the network latency frequently associated with remote database access. However, in a hybrid system that spans applications running in a variety of distributed locations, using a single instance of SQL Azure in isolation may not be sufficient to ensure a good response time. Instead, an organization might decide to maintain a copy of the database at each location. In this scenario, it is necessary to ensure that all instances of the database contain the same data. This can be a non-trivial task, especially if the data is volatile.

Additionally, you may decide to store data in a different repository; for example, you may choose to use a different database management system or implement a different form of data source. In these cases, you may need to implement your own custom strategies to synchronize the data that they contain.

This appendix examines the issues concerned with distributing and replicating data the between services running in the cloud and across the cloud/on-premises divide by using the technologies available on the Windows Azure™ technology platform. It describes some possible solutions that an organization can implement to keep the data sources synchronized.

Use Cases and Challenges

A primary motivation for replicating data in a hybrid cloud-based solution is to reduce the network latency of data access by keeping data close to the applications and services that use it, thereby improving the response time of these applications and services. As described previously, if a service running in a datacenter uses data held in a local database stored in the same datacenter then it has eliminated the latency and reliability issues associated with sending and receiving messages across the Internet. However, these benefits are not necessarily cost-free as you must copy data held in a local database to other datacenters, and ensure that any changes to this data are correctly synchronized across all datacenters.

You must also consider that replicating data can introduce inconsistencies. When you modify data, the same modification must be made to all other copies of that data and this process may take some time. Fully transactional systems implement procedures that lock all copies of a data item before changing them, and only releasing this lock when the update has been successfully applied across all instances. However, in a globally distributed system such an approach is impractical due to the inherent latency of the Internet, so most systems that implement replication update each site individually. After an update, different sites may see different data but the system becomes "eventually consistent" as the synchronization process ripples the data updates out across all sites.

Consequently, replication is best suited to situations where data changes relatively infrequently or the application logic can cope with out-of-date information as long as it is eventually updated, possibly minutes or even hours later. For example, in an application that enables customers to place orders for products, the application may display the current stock level of each product. The number displayed is likely to be inaccurate if a product is popular; other concurrent users may be placing orders for the same product. In this case, when a user actually submits an order, the stock level should be checked again, and if necessary the customer can be alerted that there may be a delay in shipping if there are none left.

Depending on the strategy you choose to implement, incorporating replication and managing eventual consistency is likely to introduce complexity into the design, implementation, and management of your system. When you are considering replicating data, there are two main issues that you need to focus on:

- Which replication topology should you use?
- Which synchronization strategy should you implement?

The selection of the replication topology depends on how and where the data is accessed, while the synchronization strategy is governed by the requirements for keeping data up-to-date across replicas. The following sections describe some common use cases for replicating, distributing, and synchronizing data and summarize the key challenges that each use case presents.

Replicating Data across Data Sources in the Cloud and On-Premises

Description: Data must be positioned close to the application logic that uses it, whether this logic is running at a datacenter in the cloud or on-premises.

Replication is the process of copying data, and the problems associated with replication are those of managing and maintaining multiple copies of the same information. Choosing an appropriate replication topology can have a major impact on how you address these problems.

In its simplest form, the implementation of this use case copies all data in a data source to all other instances of the same data source, whether these data sources are located in the cloud or on-premises. In this scenario, applications running on-premises and services running in the cloud may be able to query and modify any data. They connect to the most local instance of the data source, perform queries, and make any necessary updates. At some point, these updates must be transmitted to all other instances of the data source, and these updates must be applied in a consistent manner. Figure 1 illustrates this topology, referred to as "Topology A" throughout this appendix.

In the diagrams in this section, the bold arrows indicate the synchronization paths between databases.

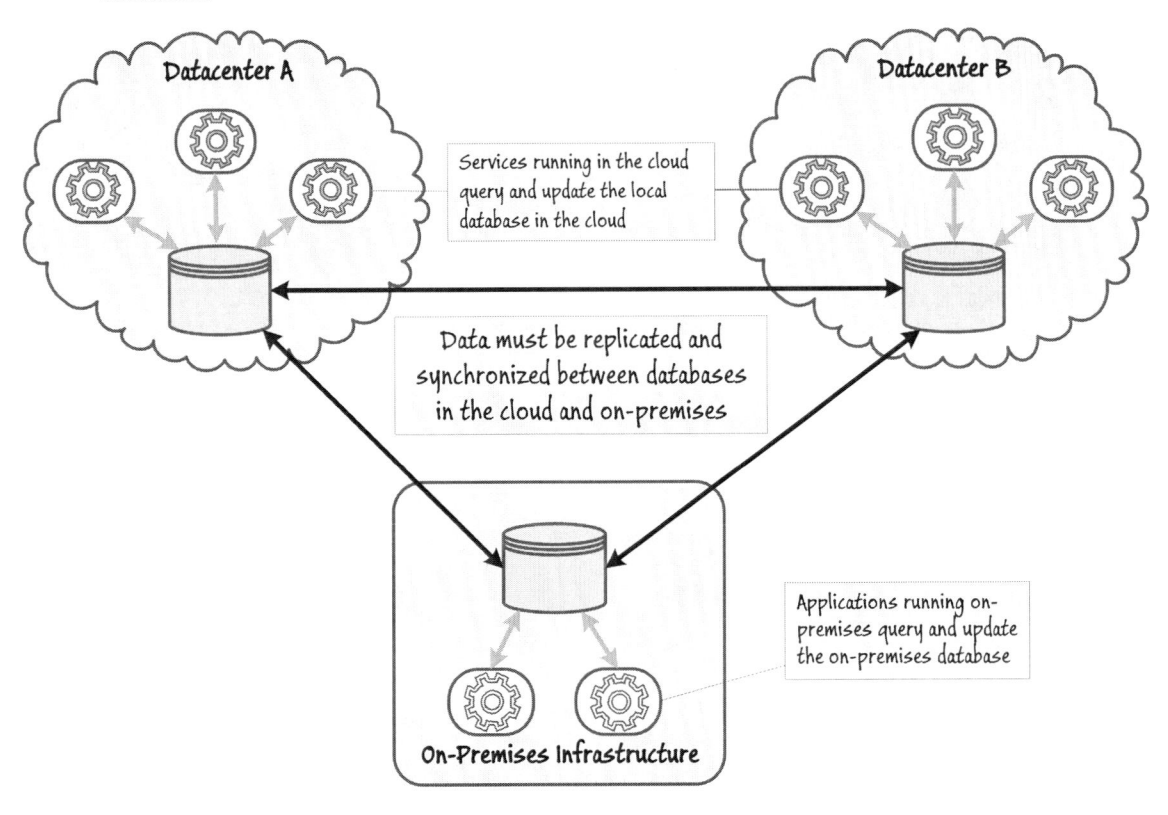

FIGURE 1
Topology A: Bidirectional synchronization across all databases on-premises and in the cloud

If your principal reason for moving data services to the cloud is purely for scalability and availability you might conclude that the data sources should just be removed from your on-premises servers, relocated to the cloud, and duplicated across all datacenters. Such a strategy might be useful if the bulk of the application logic that accesses the data has previously been migrated to the cloud. The same concerns surrounding updating data and propagating these changes consistently, as described in Topology A still apply, the only difference is that there is no data source located on-premises. Figure 2 shows this scenario.

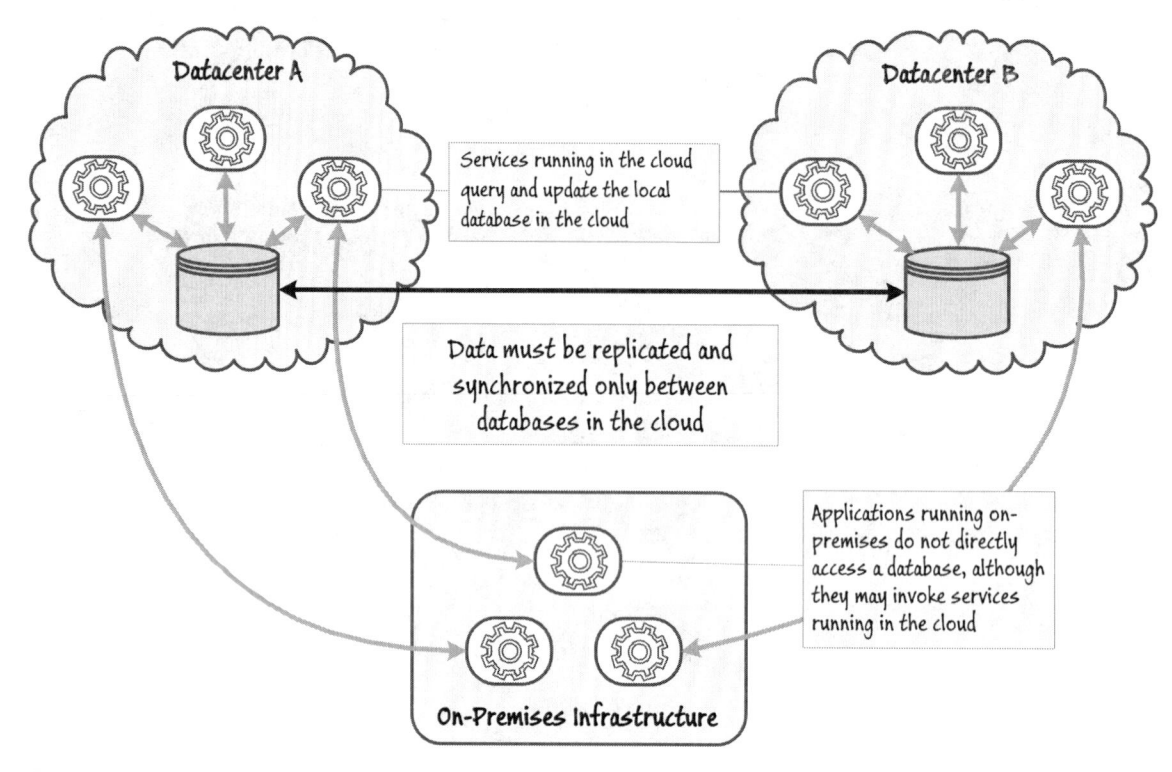

FIGURE 2
Topology B: Bidirectional synchronization only across databases in the cloud

Although Topology A and Topology B are simple to understand, such blanket strategies might not always be appropriate, especially if the data naturally partitions itself according to the location of the services that most use it. For example, consider a stock control system for an organization that maintains several warehouses at different geographical locations. A service running at a datacenter in the same geographical region as a warehouse might be responsible for managing only the data in that warehouse. In this case, it may be sensible to replicate just the data pertaining to that warehouse to the datacenter hosting the corresponding service instance, while retaining a copy of the entire database on-premises.

When a service modifies data in its local database, it can arrange to make the same change to the on-premises database. If a service should need to access data held elsewhere, it can query the on-premises database for this information. The on-premises database effectively acts as a master repository for the entire system, while the databases running at each datacenter act as a cache holding just the local data for that datacenter.

> *"Appendix E - Maximizing Scalability, Availability, and Performance"* describes additional ways to implement a cache by using the Windows Azure Caching.

This approach reduces the need to copy potentially large amounts of data that is rarely used by a datacenter at the cost of the developing the additional logic required in the code for the service to determine the location of the data. Additionally, if a service regularly requires query access to non-local data or makes a large number of updates then the advantages of adopting this strategy over simply accessing the on-premises data source for every request are reduced. This approach also assumes that each item in a data source is managed exclusively by one and only one datacenter, otherwise there is a risk of losing data updates (services running at two datacenters might attempt to update the same item).

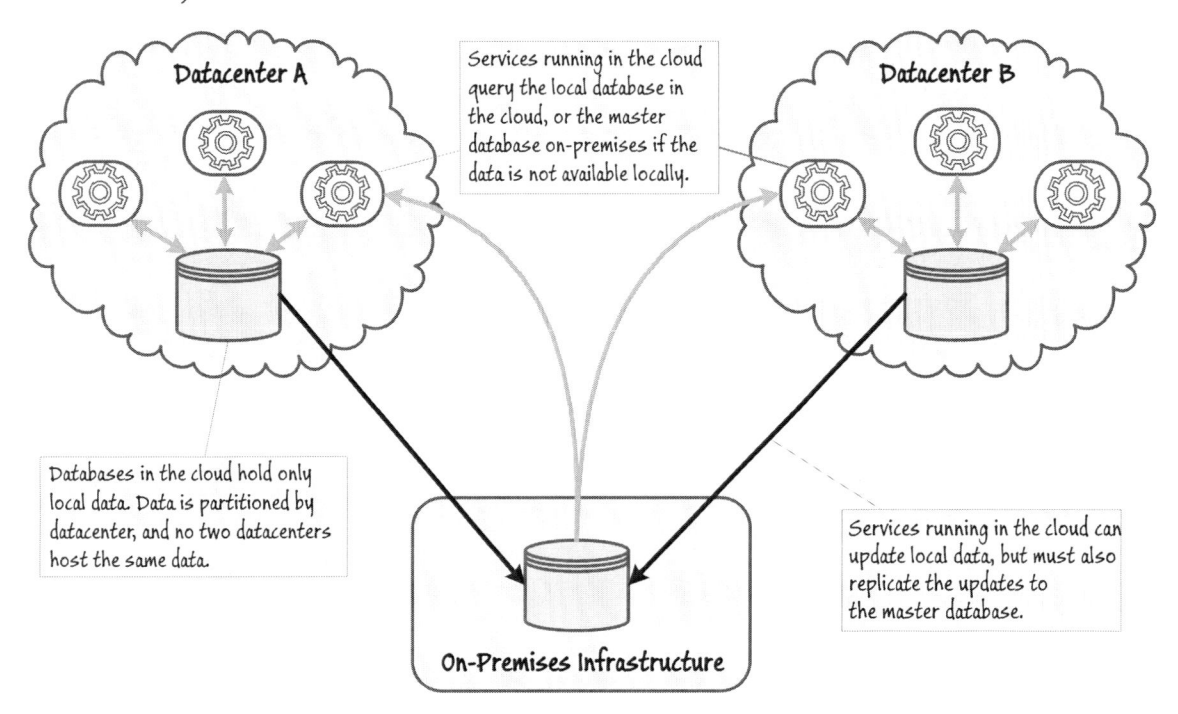

FIGURE 3
Topology C: On-premises master repository with one-way synchronization from the cloud

In a variation on this scenario, services running in the cloud primarily query the data, and send all updates to an application running on-premises; if a service running in the cloud needs to modify data, it sends a request to the on-premises application, which is designed to listen for and handle such requests. The on-premises application can modify the information in the master data source, also hosted on-premises, and then arrange for the corresponding changes to be copied out to the appropriate databases running in the cloud. This approach keeps the logic for the services in the cloud relatively straightforward (compared to Topology C) at the expense of providing an on-premises application to manage updates. This topology, shown in Figure 4, also ensures that all data sources in the cloud are eventually consistent by virtue of the replication process from the master data source.

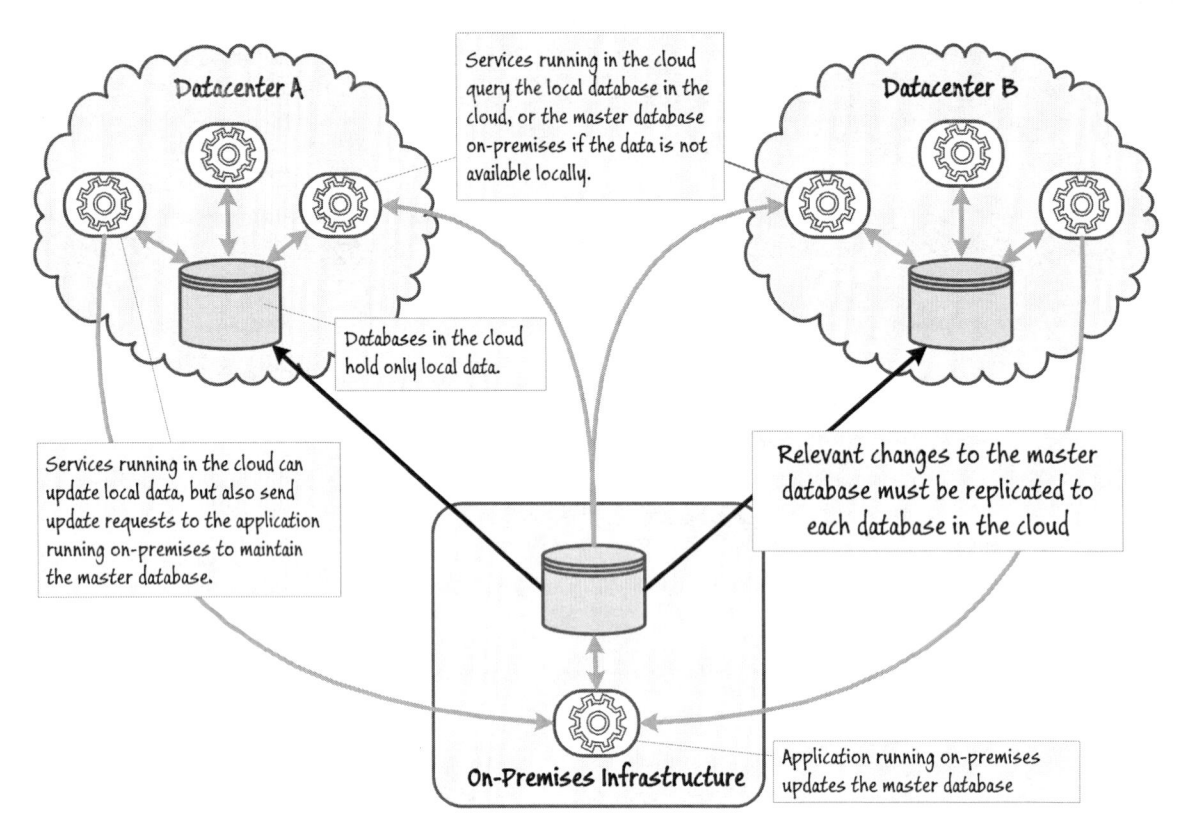

FIGURE 4
Topology D: On-premises master repository with one-way synchronization to the cloud

In a simple variant of Topology D, the application running on-premises updates the master database of its own volition rather than in response to requests from services in the cloud. The application may be performing some data maintenance tasks under the direct control of a user at the host organization. In this scenario, the updated data is simply replicated to each of the databases in the cloud.

A final use case concerns an organization spread across multiple sites not in the cloud. The organization retains the data and the applications that use it on-premises, but replicates this data between sites through the cloud. In this case, the cloud simply acts as a conduit for passing data between sites. You can apply this technique to situations such as remote branch offices which may require either a complete copy of the data, or just the subset that relates to the branch office. In either case, applications running at each branch office access the local data source hosted at the same site, and any updates are propagated through the cloud. A copy of the data source in the cloud through which all updates pass can act as the replication hub which gathers all updates and redistributes them, as well as performing the role of a master repository if a branch office requires access to non-local data.

REPLICATING, DISTRIBUTING, AND SYNCHRONIZING DATA 199

Datacenter

Master database in the cloud holds a copy of all data and acts as the replication hub

Changes to local databases must be replicated through the master database in the cloud

Databases at head office and each branch office may hold the entire data for the organization or just local data for the office

Local applications access and update data in local databases

Head Office

Branch Office A

Branch Office B

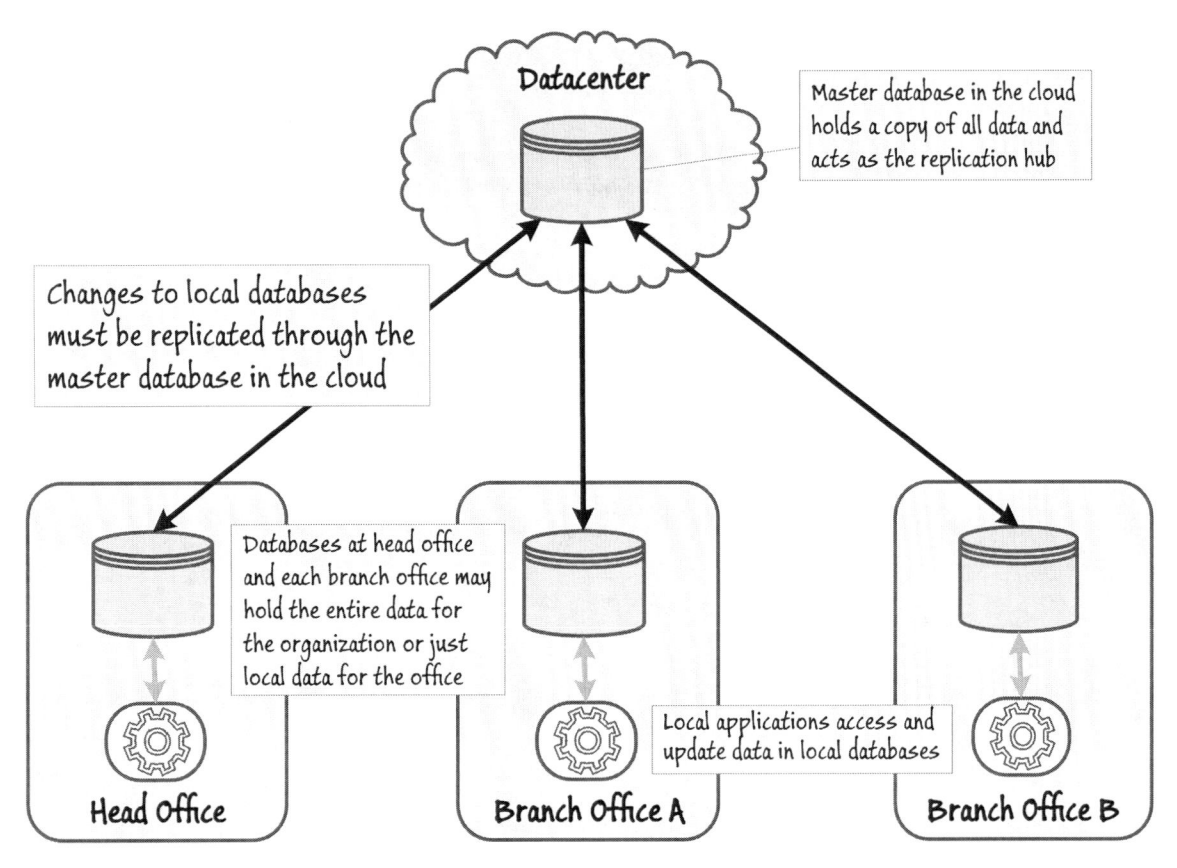

FIGURE 5
Topology E: On-premises databases synchronized through the cloud

Synchronizing Data across Data Sources
Description: Applications and services modify data, and these modifications must be propagated across all instances of the database.

Data changes, it is rarely entirely static. Applications inevitably insert, update, and delete records. In a replicated environment you must ensure that all such changes are propagated to all appropriate instances of a data source. Synchronizing data can be expensive in terms of network bandwidth requirements, and it may be necessary to implement the synchronization process as a periodic task that performs a batch of updates. Therefore you must be prepared to balance the requirements for data consistency against the costs of performing synchronization and ensure that your business logic is designed with eventual rather than absolute consistency in mind, as descried earlier in this appendix.

In determining your synchronization strategy you should consider the following questions:

- What data do you need to synchronize and how will you handle synchronization conflicts?

    The answer to this question depends largely on when and where the data is updated, as described by the topologies listed in the previous section. For example, in Topology D the master data source is only modified by applications running on-premises, so synchronization will be a matter of copying the changes made on-premises to each datacenter in the cloud. This is a one-way operation (on-premises out to the cloud) with little or no possibility of synchronization conflicts; the on-premises master database holds the definitive copy of the data, overwriting data held by datacenters in the cloud.

    Where data is modified by services in the cloud but not by applications running on-premises (Topology C), if the data is partitioned by datacenter again there is no possibility of conflicts (services at two different datacenters will not update the same data) and the synchronization process is effectively one-way, and instant. In this case, the datacenters in the cloud hold the definitive data and overwrite the data held on-premises.

    Where applications and services may modify data located anywhere in the cloud or on-premises (Topologies A, B, and E) the synchronization process is multi-way as each database must be synchronized with every other database. The data is not partitioned, so there is a possibility that conflicting changes can occur, and you must define a strategy for handling this situation.

- What are the expected synchronization data volumes? If there is a large amount of volatile data then replicating the effects of every insert, update, and delete operation may generate a lot of network traffic and consume considerable processing power, impacting the performance of each data source, and possibly nullifying the reason for replicating data in the first place. For example, in Topology C, if each service running in the cloud performs a large number of updates then maintaining a replica in the cloud becomes an overhead rather than a performance asset.

> Full multi-way synchronization between replicated relational databases can be a resource-intensive operation and the latency associated with transmitting and applying large numbers of updates across a number of sites may mean that some of the data held in one or more databases is inconsistent until all the sites are synchronized. To minimize the time taken and resources required to synchronize databases, you should carefully consider which data your applications and services need to replicate, whether your applications and services can live with potentially stale data, and whether any data should be read-only at each site.

- When do you need to synchronize data? Does every instance of the database have to be fully up-to-date all of the time; does your system depend on complete transactional integrity all of the time? If so, then replication might not be the most appropriate solution as synchronization will necessarily be a continuous process, circulating all changes immediately that they occur and locking resources in each data source while it does so to prevent inconsistency from occurring. In this case, using a single centralized data source is a better choice than implementing replication.

Cross-Cutting Concerns

Effective data replication has a high level of dependency on the network in terms of security, reliability, and performance. System requirements and application design can also have a significant bearing on how well your chosen replication approach functions. The following sections provide a summary of the possible issues.

Data Access Security

Each data source, whether it is a SQL Azure database or some other repository, must protect the data that it contains to prevent unauthorized access. This requirement applies during the synchronization process as well as during the regular data access cycle. The network packets containing the data being replicated must also be protected as a security breach at this point could easily propagate corrupted information to multiple instances of your precious data.

Data Consistency and Application Responsiveness

Data consistency and application responsiveness are conflicting requirements that you must balance to address the needs of your users.

If you require a high level of consistency across all replicated databases, then you must take steps to prevent competing applications from accessing data that may be in a state of flux somewhere in the system. This approach depends on application logic locking data items and their replicas before changing them and then releasing the locks. While the data is locked, no other applications can access it, adversely affecting the responsiveness of the system from the users' perspective. As mentioned elsewhere in this appendix, in many distributed systems immediate and absolute consistency may not be as important as maintaining application responsiveness; users want to be able to use the system quickly, and as long as information is not lost and eventually becomes consistent then they should be satisfied.

Integrity and Reliability

Even in a solution where immediate data consistency is not a critical requirement the system must still update data in a reliable manner to preserve the integrity of the information the application presents. For example, in the Orders application cited earlier, if the system accepts an order from a customer then that order should be fulfilled; the data comprising the order should not be lost and the order process must be completed. Therefore, any solution that replicates data between databases must implement a reliable mechanism for transporting and processing this data. If some aspect of the system handling this synchronization process fails, it should be possible to restart this process without losing or duplicating any data.

> Don't forget that the network is a critical component that impacts reliability. A reliable solution is one that is resilient in the case of network failure.

Windows Azure and Related Technologies

If you are implementing databases in the cloud using SQL Azure, you can configure replication and manage synchronization between these databases and SQL Server databases running on-premises by using SQL Azure Data Sync. This technology is a cloud-based synchronization service based on the Microsoft Sync Framework. Using the Windows Azure Management Portal you can quickly configure synchronization for the most common scenarios between your on-premises SQL Server databases and SQL Azure databases running in the cloud. Additionally, SQL Azure Data Sync is compatible with the Microsoft Sync Framework 2.1, so you can use the Sync Framework SDK to implement a custom synchronization strategy and incorporate additional validation logic if necessary.

SQL Azure Data Sync is compatible with SQL Server 2005 Service Pack 2 and later.

The Sync Framework SDK is also useful for scenarios where you need to implement a custom synchronization approach that you cannot configure easily by using the Management Portal. For example, you can build your own synchronization services if you need to synchronize data between databases located on-premises and mobile devices for roaming users.

Another approach is to implement a custom mechanism that passes updates between databases using messaging, with sender and listener applications applying the logic to publish updates and synchronize them with the exiting data. Service Bus topics and subscriptions provide an ideal infrastructure for implementing this scenario.

The following sections provide more information on how to use SQL Azure Data Sync, the Sync Framework SDK, and Service Bus topics and subscriptions for implementing replication in some common scenarios.

Replicating and Synchronizing Data Using SQL Azure Data Sync

Using SQL Azure Data Sync to implement SQL Server synchronization provides many benefits, including:

- **Elastic scalability**. The SQL Azure Data Sync service runs in the cloud and scales automatically as the amount of data and number of sites participating in the synchronization process increases.
- **Simple configuration**. You can use the Management Portal to define the synchronization topology. The portal provides wizards that step through the configuration process and enable you to specify the data to be synchronized. You can also indicate whether replication should be one-way or bidirectional. The portal provides a graphical view of your topology and its current health status through the Sync Group dashboard.
- **Scheduled synchronization**. You can specify how frequently the synchronization process occurs, and you can easily modify this frequency even after synchronization has been configured. Using the Management Portal you can also force an immediate synchronization.
- **Preconfigured conflict handling policies**. SQL Azure Data Sync enables you to select how to resolve any conflicts detected during synchronization by selecting from a set of built-in conflict resolution policies.
- **Comprehensive logging features**. SQL Azure Data Sync logs all events and operations. The Management Portal enables you to examine this information, and filter it in a variety of ways, enabling you to quickly determine the cause of any problems and take the necessary corrective action.

The following sections provide information about the way in which SQL Azure Data Sync operates, and include guidance on using SQL Azure Data Sync in a number of common scenarios.

Guidelines for Configuring SQL Azure Data Sync

When you configure SQL Azure Data Sync, you must make a number of decisions concerning the definition of the data that you want to replicate, and the location of the databases holding this data. This section provides guidance for defining the key elements of a synchronization architecture.

Defining a Sync Group and Sync Dataset

SQL Azure Data Sync organizes the synchronization process by defining a *sync group*. A sync group is a collection of member databases that need to be synchronized, together with a hub database that acts as a central synchronization point. All member databases participating in a topology synchronize through the hub; they send local updates to the hub and receive updates made by other databases from the hub.

When you define a sync group, you also define a *sync dataset* that specifies the tables, rows, and columns to synchronize. You do not have to select every table in a database, and you can define filters to restrict the rows that are synchronized. However, every table that participates in a sync dataset must have a primary key; otherwise synchronization will fail. Additionally, although you do not need to include every column in each participating table, you must include all columns that do not allow null values; again synchronization will fail otherwise.

> *SQL Azure Data Sync creates triggers on each table in a sync group. These triggers track the changes made to the data in each table in the sync group. For more information about the triggers that SQL Azure Data Sync generates, see "Considerations for Using Azure Data Sync" on MSDN.*

It is important to understand that SQL Azure Data Sync imposes some constraints on the column types in the tables that participate in the synchronization process. These constraints are due to the Sync Framework on which SQL Azure Data Sync is based; the Sync Framework is designed to operate with a variety of database management systems, not just SQL Server, and so the types it supports are limited to those common across the major database management systems. For example, you cannot synchronize columns based on user-defined data types, spatial data types, or CLR types. For a full list of supported and unsupported types see *"SQL Azure Data Sync – Supported SQL Azure Data Types"* on MSDN.

> If you attempt to create a sync dataset that includes columns with unsupported types, these columns will be ignored and the data that they contain will not be replicated.

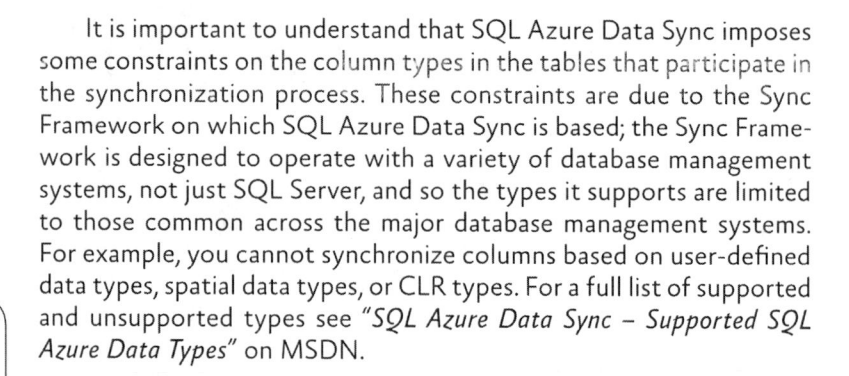

Implementing the Database Schema for Member Databases

In a typical scenario, the schema of the data that you want to replicate may already exist in an on-premises or SQL Azure database. When you deploy a sync group, if the necessary tables do not already exist in the other member databases or the hub, then SQL Azure Data Sync will automatically create them based on the definition of the sync dataset. In this case, the deployment process will only generate the columns specified by the sync dataset, and will add an index for the primary key of each table. While the deployment process does a reasonable job of replicating the schema for the sync dataset, it may not always be identical due to the differences between SQL Azure and SQL Server.

Additionally, any indexes other than that for the primary key will not be generated, and this may have an impact on the performance of queries performed against a replicated database. Therefore, to ensure complete accuracy and avoid any unexpected results, it is good practice to create a SQL script containing the commands necessary to create each table to be replicated, together with the appropriate indexes. You can also define any views and stored procedures that each member database may require as these cannot be replicated automatically. You can then run this script against each database in turn before provisioning replication.

> The same sync dataset applies globally across all member databases in the sync group. You define the sync dataset when you add the first member database to the sync group, and if necessary, the tables that underpin the sync dataset will be automatically added to subsequent member databases when they are enrolled in the sync group. However, once you have defined the sync dataset for a sync group you cannot modify the definition of this dataset; you must drop the sync group and build a new one with the new sync dataset.

Managing Synchronization Conflicts

During the synchronization process, SQL Azure Data Sync connects to each member database in turn to retrieve the updates performed in that database and applies them to the hub. Any updates previously applied to the hub from another member database are transmitted to the database and applied.

The hub is the focus for detecting and resolving conflicts. SQL Azure Data Sync enables you to select from two conflict resolution policies:

- **Hub Wins**. If the data at the hub has already been changed, then overwrite changes to this data made at the member database with the data at the hub. In effect, this means that the first member database to synchronize with the hub predominates.

- **Client Wins**. If the data has been changed at the member database, this change overwrites any previous changes to this data at the hub. In contrast to the Hub Wins policy, in this case the last member database to synchronize with the hub predominates.

During synchronization, each batch of updates is applied as a transaction; either all the updates in a batch are applied successfully or they are rolled back. However, these batch transactions do not necessarily reflect the business transactions performed by your system. For example, a business transaction that modifies data in two tables may have these updates propagated by different batches when these changes are synchronized.

Additionally, each synchronization applies only the changes in effect at that time to each database. If a row undergoes several updates between synchronizations, only the final update will be replicated; SQL Azure Data Sync does not keep a log of every change made between synchronizations.

> Use a tool such as Microsoft SQL Server Management Studio to generate and edit the SQL scripts that create the tables, views, and stored procedures for each member database. If you have the appropriate credentials, you can also connect to each member database using SQL Server Management Studio and run these scripts.

> The synchronization process visits each member database in turn in a serial manner and applies the necessary updates to synchronize that member database and the hub. Databases visited earlier will not incorporate the changes resulting from the synchronization with databases visited later. For member databases to be fully synchronized with each other, you need to perform two synchronizations across the sync group.

> You should give the conflict resolution policy careful thought as the same policy applies across all databases in a sync group. Additionally, you specify this policy when you first create the sync group and you cannot change it without dropping and recreating the sync group.

Note that, although you can select the conflict resolution policy, you cannot currently influence the order in which databases are synchronized with the hub. Ideally, you should design your solution to minimize the chances of conflicts occurring; in a typical distributed scenario, applications running at different sites tend to manage their own subset of an organization's data so the chances of conflict are reduced. Remember that the primary purpose of replication is to propagate updates made at one site to all other sites so that they all have the same view of the data.

If you need to guarantee the effects of the conflict resolution policy, you can divide your replication topology into a series of sync groups with each sync group containing the hub and a single member database. The synchronization schedule for each sync group determines the order in which each member database is synchronized with the hub. The sync group for a high priority member database with updates that must always take precedence can select the Client Wins conflict resolution policy so that these changes are always replicated.

The policy for other sync groups can be set to Hub Wins, and in this way the changes made at the high priority database will always be replicated out to the other member databases. You can implement many variations on this topology. For example you can place several member databases into the Hub Wins sync group if none of these databases are likely to contain changes that conflict with each other.

Conflict is typically a result of bidirectional synchronization. To reduce the chances of a conflict occurring you can configure one-way replication and specify the synchronization direction for each member database in a sync group relative to the hub. For more information, see the section "Selecting the Synchronization Direction for a Database" later in this appendix.

> To avoid issues with conflicting primary key values, do not use columns with automatically generated key values in replicated tables. Instead use a value that is guaranteed to be unique, such as a GUID.

*Pay careful attention to the definition of the columns in replicated tables as this can have a significant impact on the likelihood of conflict. For example, if you define the primary key column of a replicated table with the SQL Server **IDENTITY** attribute, then SQL Server will automatically generate values for this column in a monotonic increasing sequence, typically starting at 1 and incrementing by 1 for each newly inserted row. If rows are added at multiple member databases in a sync group, several of these rows might be given the same primary key value and will collide when the tables are synchronized. Only one of these rows will win and the rest will be removed. The results could be disastrous if, for example, this data represented orders for different customers; you will have lost the details of all the orders except for the winner selected by the conflict resolution policy!*

To avoid situations such as this, do not use columns with automatically generated key values in replicated tables, but instead use a value that is guaranteed to be unique, such as a GUID.

Locating and Sizing the Hub Database

The hub must be a SQL Azure database. After synchronizing with all the member databases, it holds the definitive and most up-to-date version of the data. The location of this database is key to maintaining the performance of the synchronization process; you should store it at a datacenter that is geographically closest to the most active member databases, whether these databases are located on-premises or in the cloud. This will help to reduce the network latency associated with transmitting potentially large amounts of data across the Internet. If your databases are distributed evenly around the world, and the volume of database updates and query traffic is roughly the same for each one, then you should position the hub at the datacenter closest to your highest priority sites.

SQL Azure Data Sync replicates and synchronizes data between your databases through the hub. You can provision a single instance of the SQL Azure Data Sync Server for each Windows Azure subscription that you own, and you can specify the region in which to run this server. Ideally, you should locate this server in the same region that you plan to use for hosting the hub database.

You create the hub database manually, and it should be at least as big as the largest of the member databases. SQL Azure does not currently support automatic growth for databases, so if you make the hub database too small synchronization could fail. You should also note that when you configure synchronization, SQL Azure Data Sync creates additional metadata tables in your databases to track the changes made, and you must take these tables into account when sizing the hub database.

Apart from acting as the focus around which the synchronization process revolves, the hub contains exactly the same data as any other SQL Azure member database in the sync group. You can insert, update, and delete data in this database and these changes will be replicated throughout the sync group. In some situations, you can elect to use one of the SQL Azure databases originally intended as a member of the sync group as the hub. For example, you may opt to designate the SQL Azure database for the most active site as the hub. This strategy can help to minimize the network latency and thereby improve the performance of the synchronization process.

However, every other member database in the sync group will periodically synchronize with this database, and the work involved in performing the synchronization operations may impact the performance of this database, especially if the tables in the sync dataset contain a complex collection of indexes. You must strike a balance between the overhead associated with a database being the hub of a sync group against the time required to synchronize this database with a hub located elsewhere.

Specifying the Synchronization Schedule for a Sync Group

Synchronization is a periodic process rather than a continuous operation; you can specify a simple synchronization schedule for a sync group, and you can also force synchronization to occur manually by using the Management Portal. If you set a synchronization schedule, you must select a synchronization frequency that is appropriate to your solution; if it is too long, then member databases may contain outdated information for an extended period, while if it is too short a previous synchronization might not have completed and the attempt will fail. As described previously, the time taken to complete the synchronization process depends on the location of the hub database.

It will also depend on the volume of data to be synchronized; the longer the interval between synchronizations the more data will need to be synchronized and transmitted to and from the hub. Additionally, as the synchronization period increases, it is more likely that conflicts will occur and the synchronization process will have to expend effort resolving these conflicts, which will increase the time taken still further. You may need to determine the optimal synchronization period based on observations, and tune it accordingly as you establish the data timeliness requirements of your applications and services.

Finally, you should also consider that SQL Azure charges are applied to data that is moved in and out of SQL Azure datacenters; the more data you synchronize between datacenters and the more frequently you perform this synchronization, the higher the cost.

> As with the conflict resolution policy, the synchronization schedule applies globally across all databases in the sync group. However, you can modify this schedule at any time, so you can observe the effects of synchronizing data at different intervals and then select the period most appropriate to your requirements.

Selecting the Synchronization Direction for a Database

When you add a member database to a sync group, you specify the synchronization direction. Synchronization can be:

- **Bidirectional**. The member database can make changes and upload them to the hub, and it can also receive updates from the hub. This is likely to be the most common form of synchronization implemented by many organizations.

- **To the hub**. The member database can make changes and upload them to the hub, but it will not receive changes from the hub. This form of synchronization is useful for situations such as Topology D (on-premises master repository with one-way synchronization to the cloud) described earlier in this appendix. A service running in the cloud updates the local member database and also copies changes to the database running on-premises as they occur. The on-premises database can be configured to synchronize to the hub. When synchronization occurs, the changes made by each service in the cloud can be propagated out to the member databases for the other services via the hub. The on-premises database does not need to be synchronized as it already contains all the updates.

- **From the hub**. The member database can receive changes from the hub, but will not upload any local changes to the hub. Again, this form of synchronization is useful for implementing scenarios similar to Topology D. In this case, the member databases can be configured to synchronize from the hub; any changes made locally will have already been made to the on-premises database by the services running in the cloud, so the only changes that need to be replicated are those originating from other services located elsewhere that have also updated the on-premises database.

> The synchronization direction is an attribute of each member database; each database in a sync group can specify a different synchronization direction.

Figure 6 depicts an updated version of Topology D with the hub database and Data Sync Service required by SQL Azure Data Sync.

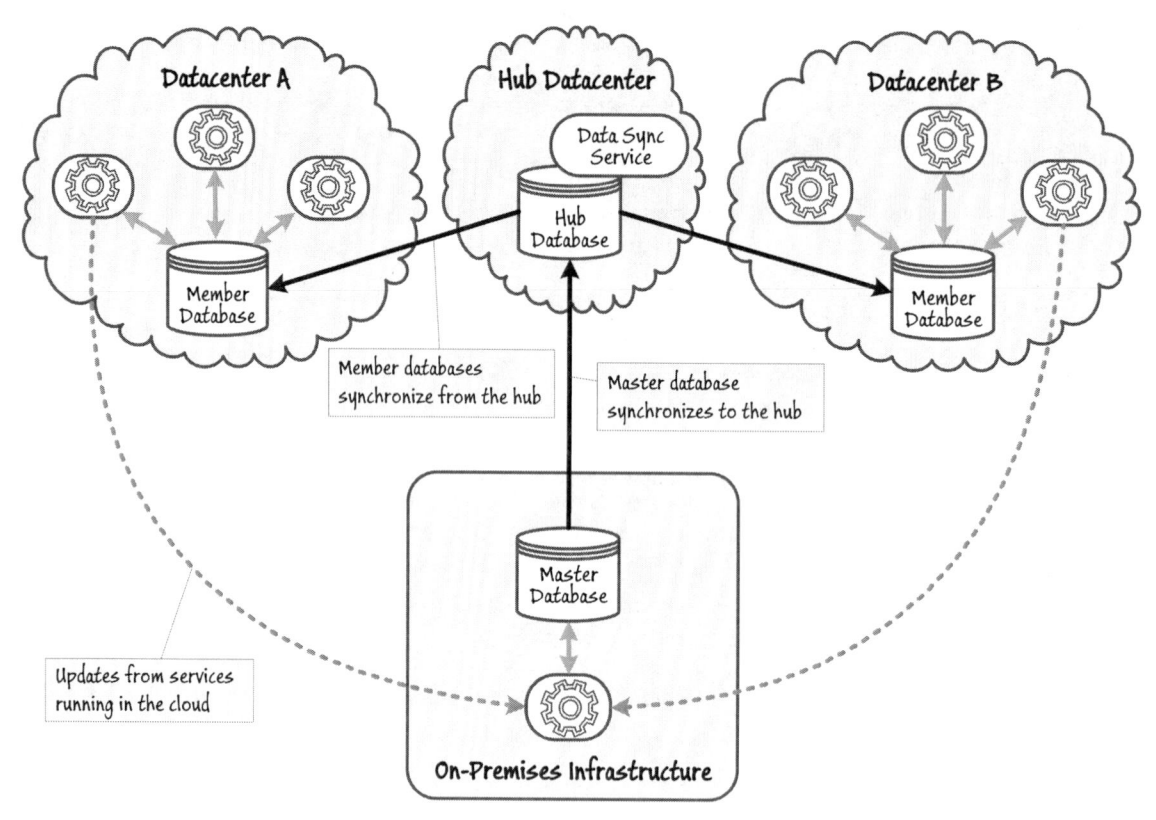

FIGURE 6
Specifying the synchronization direction for databases participating in Topology D

> *Although Figure 6 shows the hub as a separate database, except for the circumstances described at the end of the section "Locating and Sizing the Hub Database" it is likely that one of the member databases in the cloud would perform this role. The examples in the section "Guidelines for Using SQL Azure Data Sync" illustrate this approach.*

Avoiding Sync Loops

A member database can participate in more than one sync group. However, such a configuration can result in a sync loop. A sync loop occurs when the synchronization process in one sync group results in synchronization being required in another sync group, and when this second synchronization occurs the configuration of this sync group results in synchronization being required again in the first group, which again may render synchronization necessary in the second group, and so on. Sync loops are self-perpetuating and can result in large amounts of data being repeatedly written and rewritten, resulting in degraded performance and increased costs.

When you define a sync group, you must be careful to ensure that sync loops cannot exist by evaluating the role of any databases that participate in multiple sync groups, selecting the appropriate conflict resolution policy for each sync group, using row filtering to prevent the same rows in a table participating in different sync groups, and by carefully setting the synchronization direction for each database. For a more detailed description of sync loops and the circumstances under which they can occur, see *"Synchronization Loops"* on MSDN.

Guidelines for Using SQL Azure Data Sync

You can use SQL Azure Data Sync to implement the replication topologies described earlier in this appendix. You can apply these topologies with SQL Azure Data Sync to many common scenarios, as described in the following list.

- **Applications running on-premises access a SQL Server database also held on-premises. Services running in the cloud use a copy of the same data. Any changes made at any site must eventually be propagated to all other sites, although these updates do not have to occur immediately.**

    This is possibly the most common scenario for using SQL Azure Data Sync, and describes the situation covered by Topology A (bidirectional synchronization across all databases on-premises and in the cloud). As an example of this scenario, consider a database holding customer and order information. An application running on-premises maintains customer information, while customers use a web application running in the cloud that creates new orders. The web application requires access to the customer information managed by the on-premises application, and the code running on-premises frequently queries the order details to update the status of orders when they are delivered and paid for.

    In this example, response time is important, but neither the application running on-premises nor the web application running in the cloud requires access to completely up-to-date information. As long as the data is available at some near point in the future, that is good enough. Therefore, to minimize the effects of network latency and ensure that it remains responsive, the web application employs a SQL Azure database hosted in the same datacenter as the application and the on-premises application uses a SQL Server database also located on-premises.

If you are using Windows Azure Traffic Manager to route requests to a datacenter, be aware that services running at different datacenters may see different data if each datacenter has its own replica database. This is because the synchronization process may not have been completed at all sites, so updates visible in one datacenter might not have been propagated to other datacenters.

For further guidance about using Windows Azure Traffic Manager, see "Appendix E - Maximizing Scalability, Availability, and Performance." For an example describing how Trey Research used Windows Azure Traffic Manager, refer to Chapter 6, "Maximizing Scalability, Availability, and Performance in the Orders Application."

SQL Azure Data Sync enables you to replicate and share the customer and order information between the on-premises application and the cloud by using bidirectional synchronization, as shown in Figure 7. Note that, in this diagram, the SQL Azure database in Datacenter A also acts as the synchronization hub running the Data Sync service.

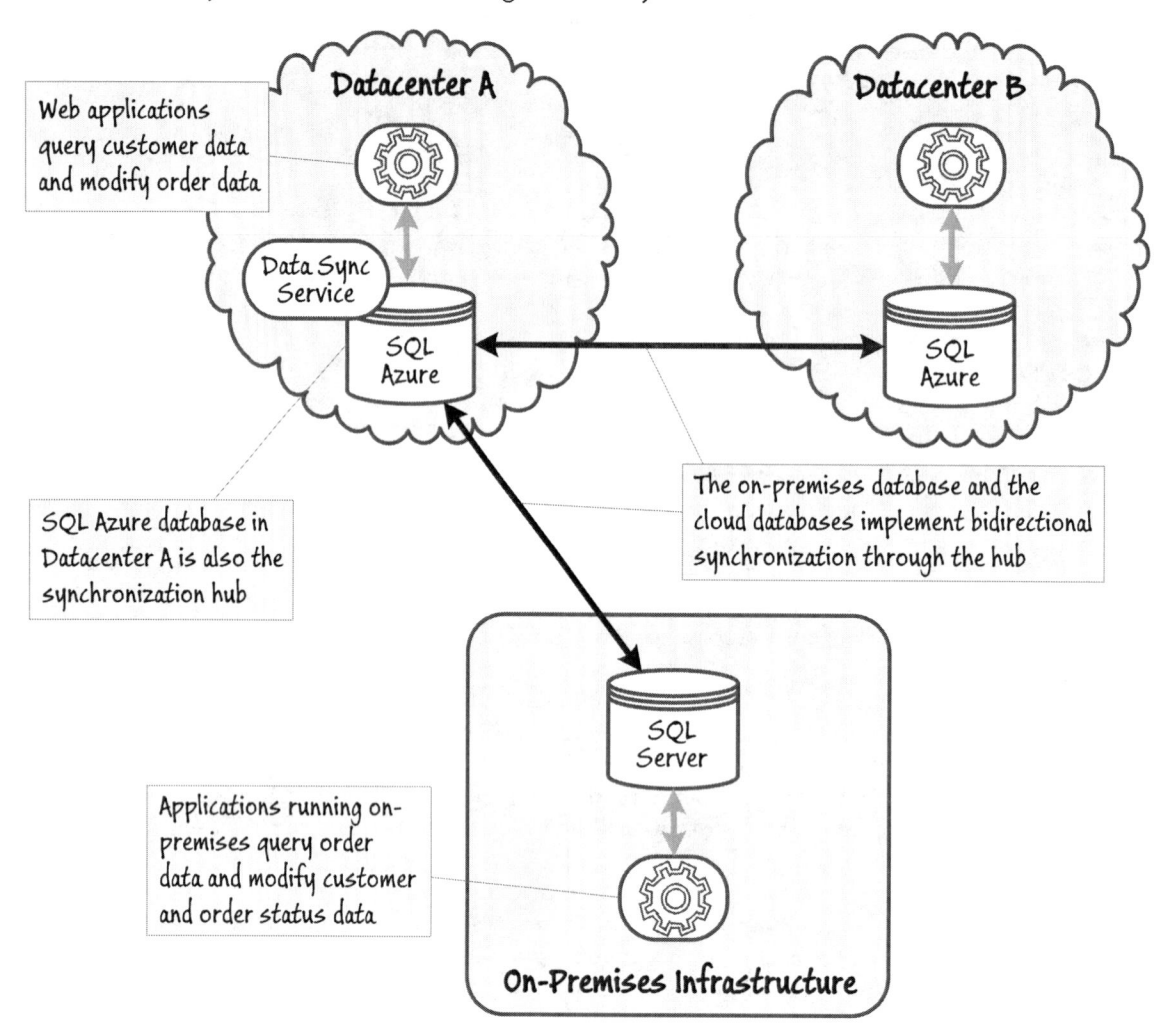

FIGURE 7
Sharing data between the applications running in the cloud and on-premises

- **You have relocated the logic for your business applications to services running in the cloud. The business applications previously used data held in a SQL Server database. The services have been distributed across different datacenters, and the SQL Server database has been migrated to SQL Azure. To minimize network latency each data center has a replica of the SQL Azure database.**

    This is the scenario that compares to Topology B (bidirectional synchronization only across databases in the cloud). In this example, the application logic that accesses the database has been completely relocated to the cloud, so the on-premises database has been eliminated. However, the cloud based applications all require access to the same data, and may modify this information, so each instance of the SQL Azure database must be periodically synchronized with the other instances. This replication will be bidirectional. Figure 8 shows the structure of a possible solution, with the SQL Azure database in Datacenter A also performing the role of the synchronization hub. In this example, any of the applications may query and modify any data. Consequently, the application logic might need to be amended to handle data that may be out of date until the next synchronization cycle.

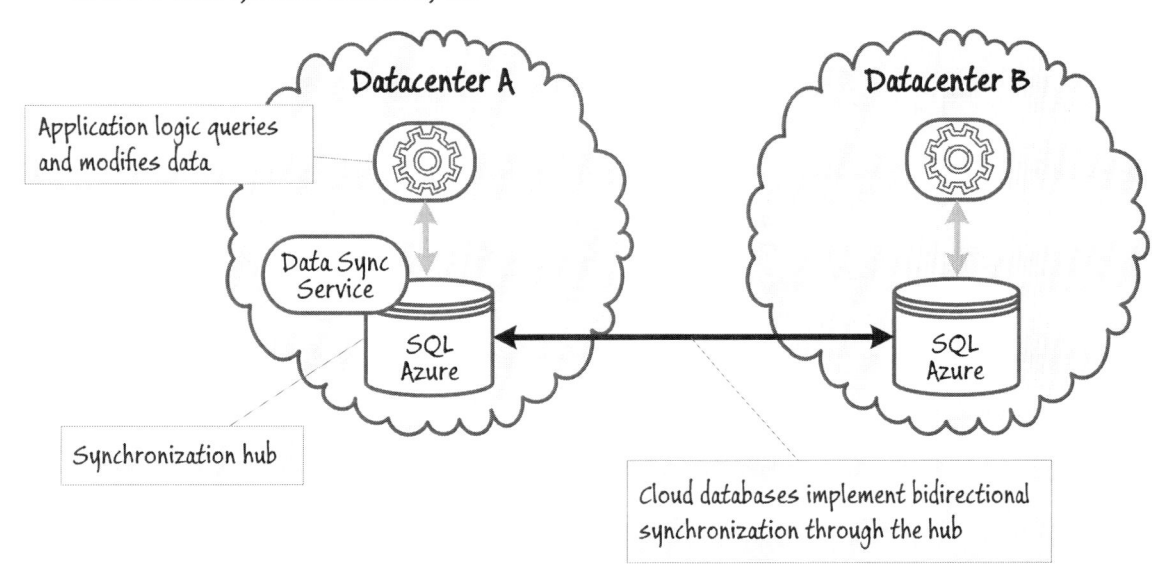

FIGURE 8
Replicating data between data centers in the cloud

- **You need to make your data held in an on-premises SQL Server database available to services running in the cloud. These services only query data and do not modify it; all modifications are performed by applications running on-premises.**

    This is the simple variant of Topology D (on-premises master repository with one-way synchronization to the cloud) described earlier. In this scenario, the services that query the data execute remotely from your on-premises database. To minimize response times, you can replicate the data to one or more SQL Azure databases hosted in the same datacenters as each of the services. Using SQL Azure Data Sync, you can publish the data held on premises and periodically synchronize any updates made by the on-premises applications with the databases in the cloud. Figure 9 shows an example. This configuration requires one-way replication, with the on-premises database synchronizing to a hub database in the cloud and each of the SQL Azure member databases synchronizing from the hub.

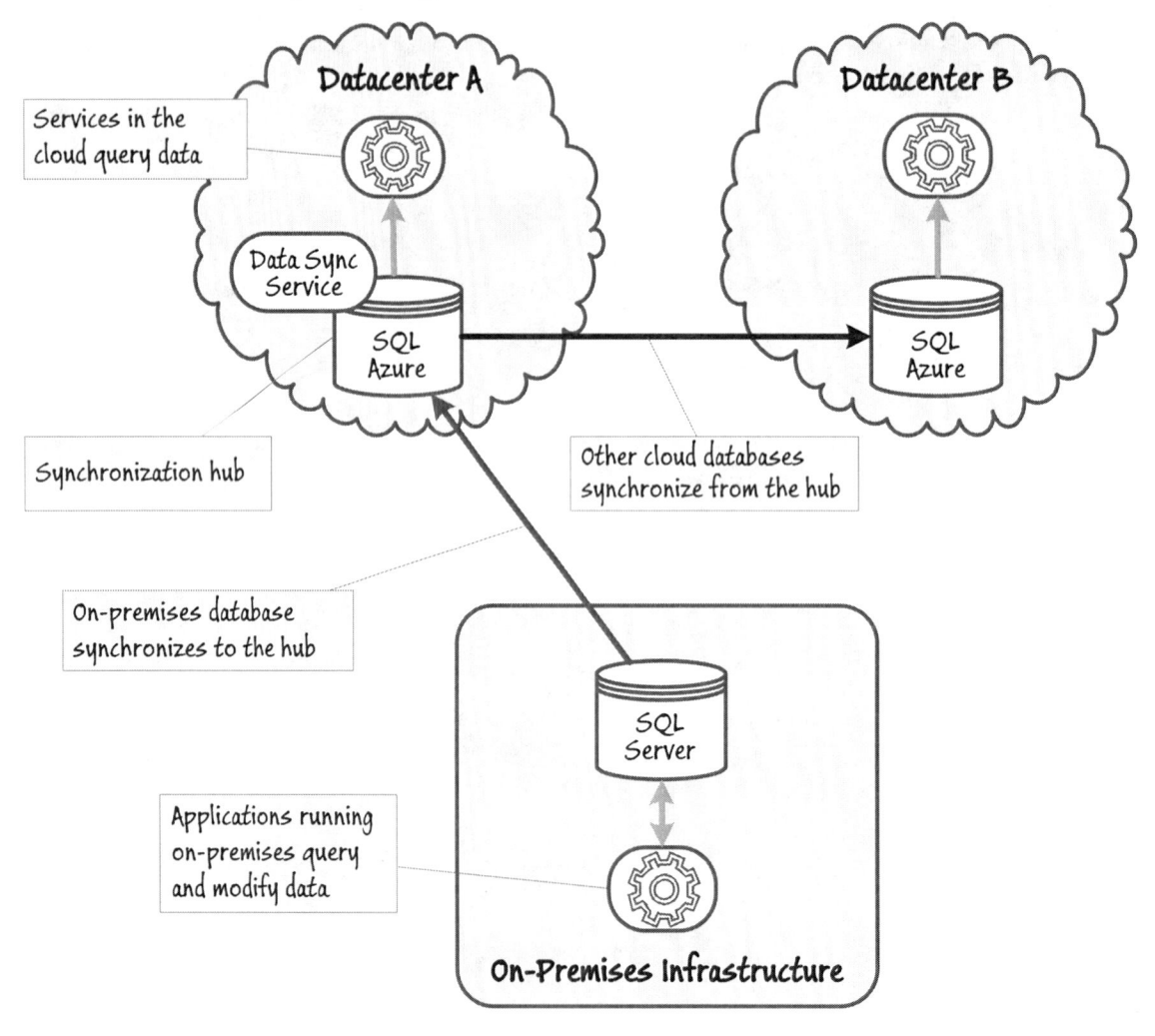

FIGURE 9
Publishing an on-premises database to the cloud

- **You have a number of applications and SQL Server databases running on-premises. However, you have migrated much of your business intelligence and reporting functionality to services running in the cloud. This functionality runs weekly, but to support your business operations it requires query access to your business data.**

    In this scenario, all the data modifications are performed against a number of SQL Server databases hosted within the organization by applications running on-premises. These applications may be independent from each other and operate by using completely different databases. However, assume that the business intelligence functionality performs operations that span all of these databases, querying data held across them all, and generating the appropriate reports to enable a business manager to make the appropriate business decisions for the organization. Some of these reports may involve performing intensive processing, which is why these features have been moved to the cloud.

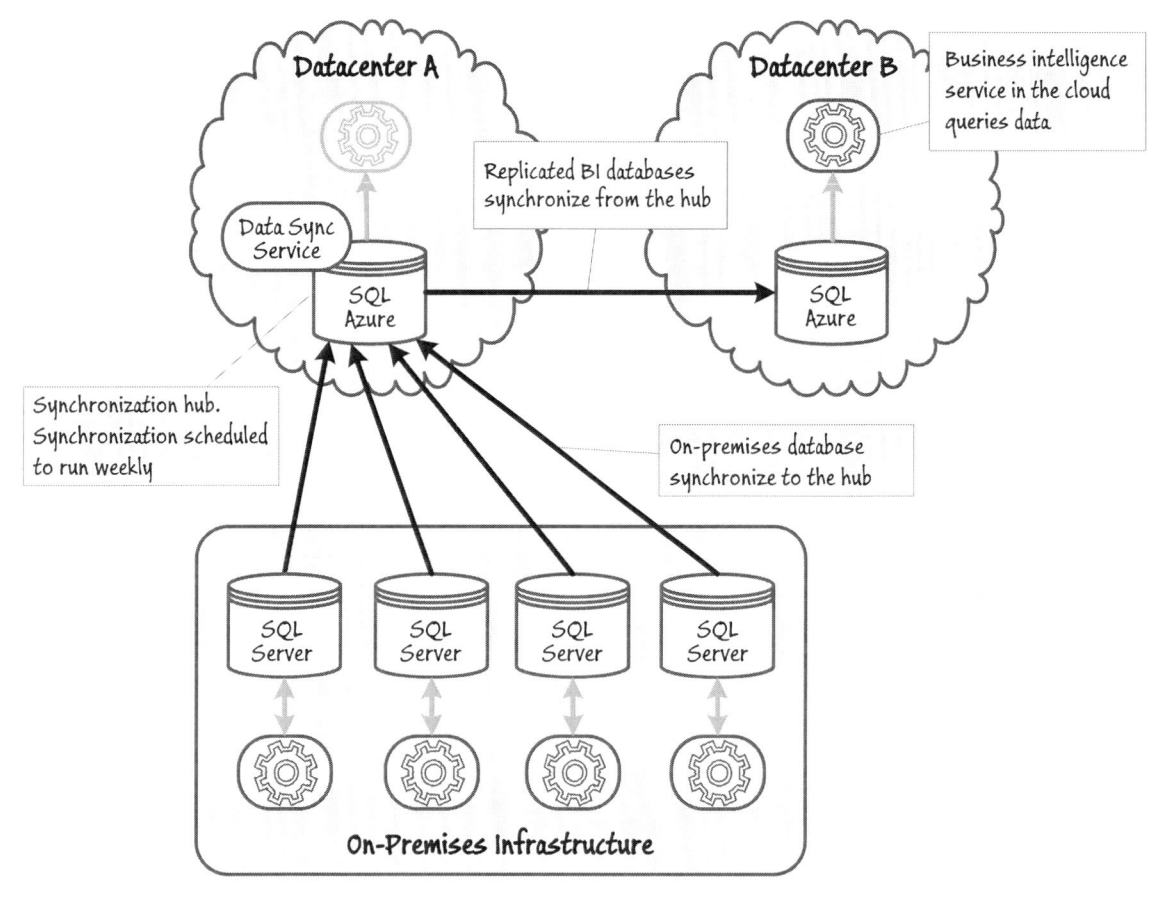

FIGURE 10
Aggregating and consolidating data in the cloud

You can use SQL Azure Data Sync to aggregate and consolidate data from the multiple on-premises databases into a single SQL Azure database in the cloud, possibly replicated to different datacenters as shown in Figure 10. The business intelligence service at each datacenter can then query this data locally. The synchronization process only needs to be one way, from the on-premises databases to the hub and then from the hub to each SQL Azure database; no data needs to be sent back to the on-premises database. Additionally, synchronization can be scheduled to occur weekly, starting a few hours before the business intelligence service needs to run (the exact schedule can be determined based on how much data is likely to be replicated and the time required for this replication to complete).

You can use the same approach to aggregate data from multiple offices to the cloud, the only difference being that the on-premises SQL Server databases are held at different locations.

- **You have a number of services running in the cloud at different datacenters. The services at each datacenter maintain a separate, distinct subset of your organization's data. However, each service may occasionally query any of the data, whether it is managed by services in that datacenter or any other datacenter. Additionally, applications running on-premises require access to all of your organization's data.**

    This situation occurs when the data is partitioned between different sites, as described in Topology C (on-premises master repository with one-way synchronization from the cloud). In this scenario, a SQL Server database running on-premises holds a copy of all the data for the organization, but each datacenter has a SQL Azure database holding just the subset of data required by the services running at that datacenter. This topology allows the services running at a datacenter to query and update just its subset of the data, and periodically synchronize this subset of the data with the on-premises database.

    If a service needs to query data that it does not hold locally, it can retrieve this information from the on-premises database. As described earlier, this mechanism necessitates implementing logic in each service to determine the location of the data, but if the bulk of the queries are performed against the local database in the same datacenter then the service should be responsive and maintain performance.

Implementing this system through SQL Azure Data Sync requires defining a separate sync group for each SQL Azure database. This is because the sync dataset for each SQL Azure database will be different; the data will be partitioned by datacenter. The on-premises database will be a member common to each sync group. To simplify the structure, you can specify that the SQL Azure database for each datacenter should act as its own synchronization hub. Figure 11 shows an implementation of this solution.

> Be careful to avoid introducing a sync loop if you follow this strategy; make sure that the different sync datasets do not overlap and include the same data.

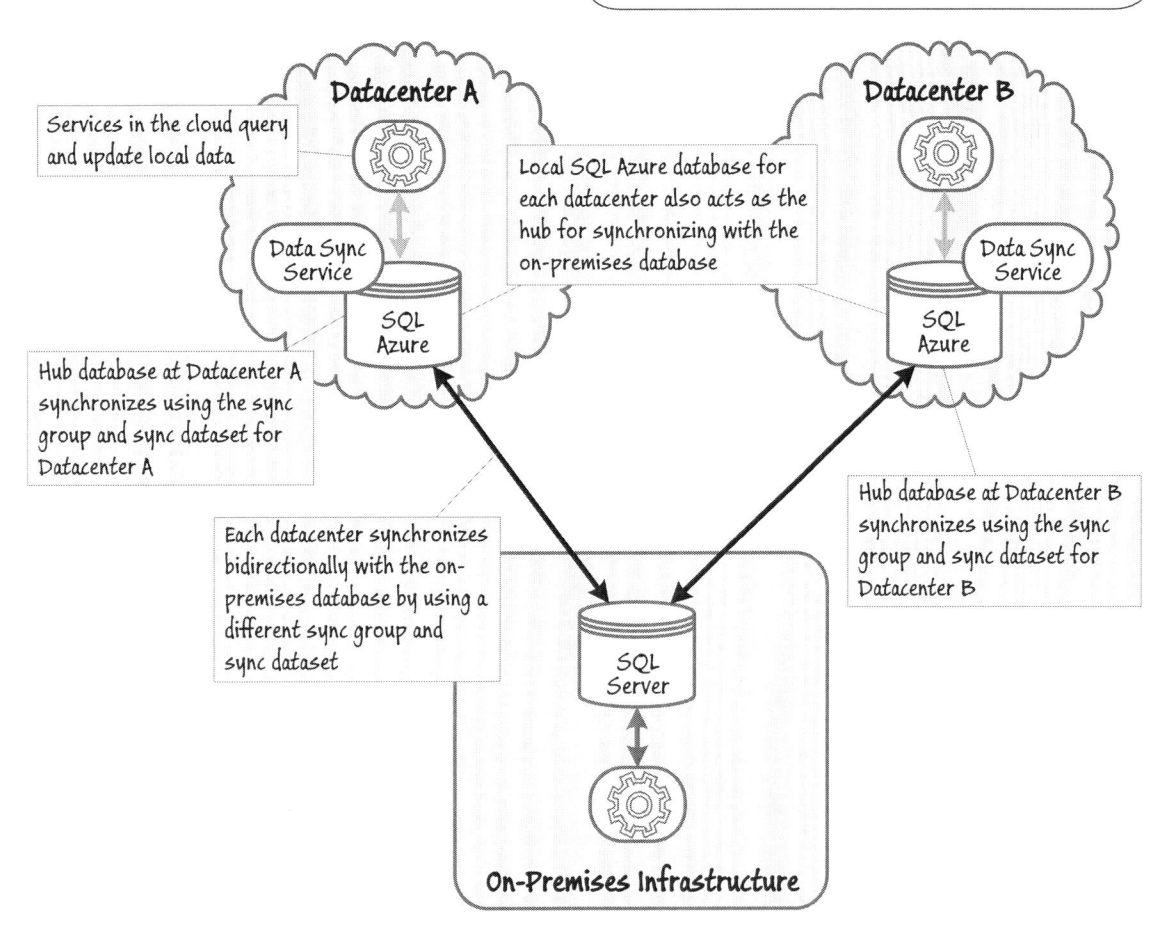

FIGURE 11
Using SQL Azure Data Sync to partition and replicate data in the cloud

- **Your organization comprises a head office and a number of remote branch offices. The applications running at head office require query access to all of the data managed by each of the branch offices, and may occasionally modify this data. Each branch office can query any data held in that branch office, any other branch office, or at head office, but can only modify data that relates to the branch office.**

    This is the scenario for Topology E (on-premises databases synchronized through the cloud). The data can be stored in a SQL Server database, and each branch office can retain a replica of this database. Other than the hub database, no data is stored in the cloud. The location of the hub should be close to the most active office (possibly the head office). Synchronization should be bidirectional, and can be scheduled to occur with a suitable frequency depending on the requirement for other branch offices to see the most recent data for any other branch. If each branch only stores its own local subset of the data, you will need to create a separate sync group for each branch database with the appropriate sync dataset, as described in the previous scenario. If the sync datasets for each branch do not overlap, it is safe to use the same SQL Azure database as the synchronization hub for each sync group. Figure 12 shows a possible structure for this solution.

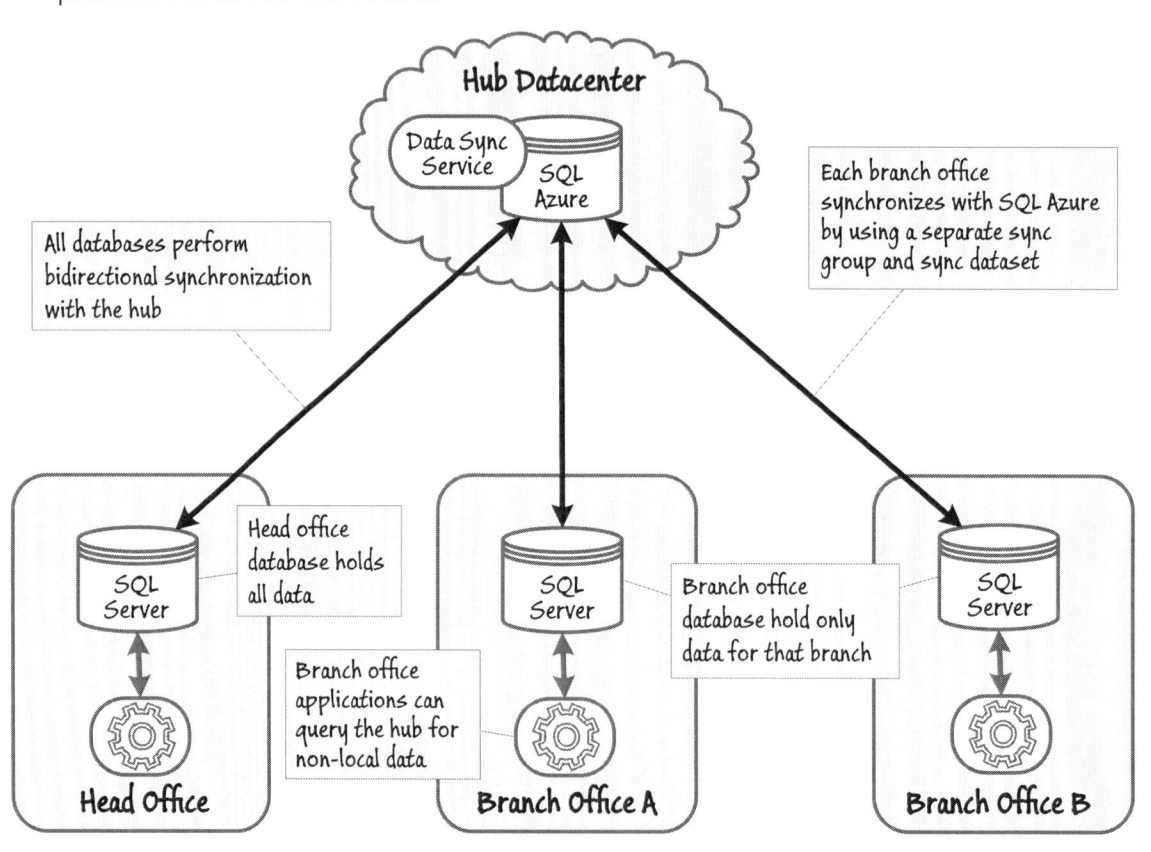

Figure 12
Using SQL Azure Data Sync to partition and replicate data across branch offices

- **Many of your services running in the cloud perform a large number of on-line transaction processing (OLTP) operations, and the performance of these operations is crucial to the success of your business. To maintain throughput and minimize network latency you store the information used by these services in a SQL Azure database at the same datacenter hosting these services. Other services at the same site generate reports and analyze the information in these databases. Some of this reporting functionality involves performing very complex queries. However, you have found that performing these queries causes conflict in the database that can severely impact the performance of the OLTP services.**

In this scenario, the solution is to replicate the database supporting the OLTP operations to another database intended for use by the reporting and analytical services, implementing a read scale-out strategy. The synchronization only needs to be performed one-way, from the OLTP database to the reporting database, and the schedule can be set to synchronize data during off-peak hours. The OLTP database can perform the role of the hub. Additionally, the reporting database can be optimized for query purposes; the tables can be configured with indexes to speed the various data retrieval operations required by the analytical services, and the data can be denormalized to reduce the processing requirements of complex queries. In contrast, the number of indexes in the OLTP database should be minimized to avoid the overhead associated with maintaining them during update-intensive operations. Figure 13 shows this solution.

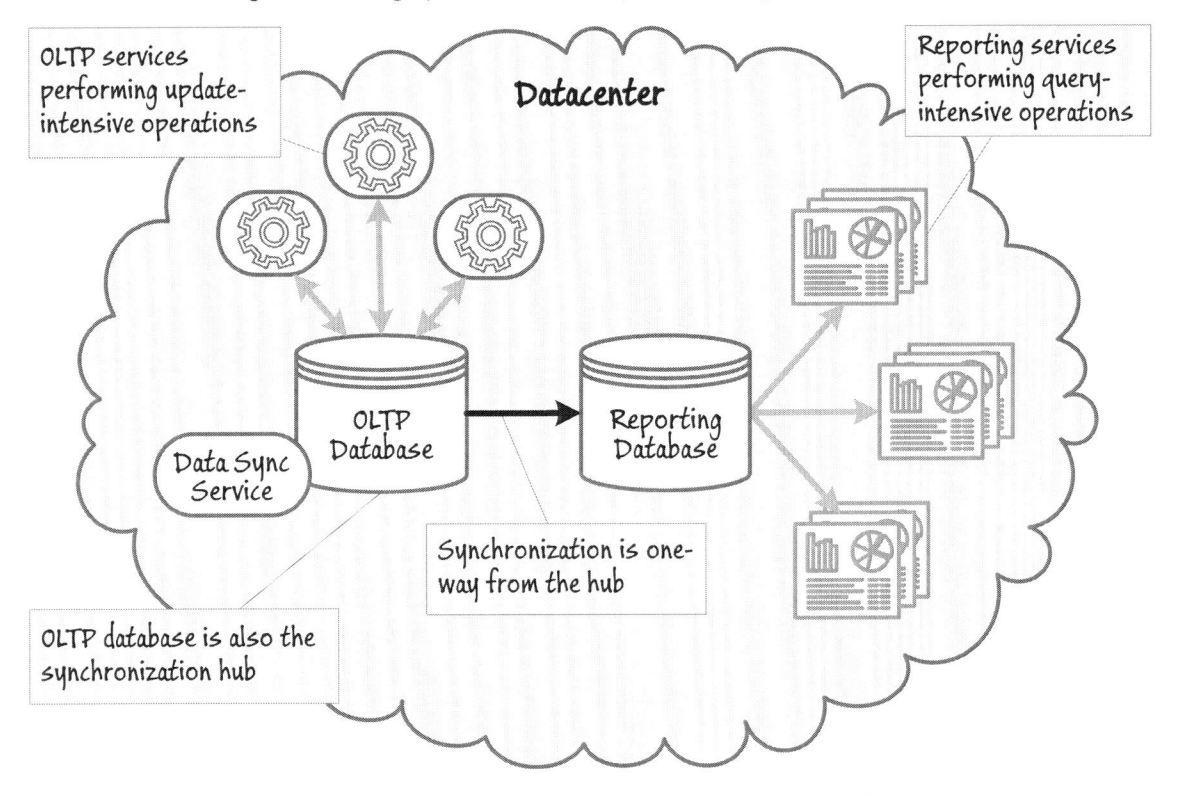

Figure 13
Replicating a database to implement read scale-out

SQL Azure Data Sync Security Model

SQL Azure Data Sync uses a piece of software called the Data Sync client agent to communicate between your on-premises instances of SQL Server and the SQL Azure Data Sync Server in the cloud; you must download and install the Data Sync client agent on one of your on-premises servers. The communications between the SQL Azure Data Sync Server and the Data Sync client agent are encrypted. The Data Sync client agent uses an outbound HTTPS connection to communicate with the SQL Azure Data Sync Server. Additionally, all sensitive configuration information used by the Data Sync client agent and SQL Azure Data Sync Server are encrypted, including the credentials used to connect to each on-premises database and SQL Azure database. The agent key defined in the Management Portal is used by the Data Sync Service for authentication.

The Data Sync client agent software consists of two elements:

- A Windows service that connects to the on-premises databases, and
- A graphical utility for configuring the agent key and registering on-premises databases with this service.

The client agent service must be configured to run using a Windows account that has the appropriate rights to connect to each server hosting on-premises databases to be synchronized; these databases do not have to be located on the same server as the client agent service. When you register an on-premises database with the client agent you must provide the login details for accessing the SQL Server hosting the database, and this information is stored (encrypted) in the client agent configuration file. When the SQL Azure Data Sync Server synchronizes with an on-premises database, the client agent uses these details to connect to the database.

For more information about the SQL Azure Data Sync security model, see *"Data Security"* on MSDN.

> For additional security, if the client agent is running on a different server from your databases, you can configure the client agent service to encrypt the connection with each on-premises database by using SSL. This requires that you have installed the appropriate SSL certificates installed in each instance of SQL Server. For more information, see *"Encrypting Connections to SQL Server"* on MSDN.

Implementing Custom Replication and Synchronization Using the Sync Framework SDK

The Management Portal enables you to replicate and synchronize SQL Server and SQL Azure databases without writing any code, and it is suitable for configuring many common replication scenarios. However, there may be occasions when you require more control over the synchronization process, for example a service may need to force a synchronization to occur at a specific time. For example, if you are caching data by using the Windows Azure Caching service, using SQL Azure Data Sync may render any cached data invalid after synchronization occurs. You may need to more closely coordinate the lifetime of cached data with the synchronization frequency, perhaps arranging to flush data from the cache when synchronization occurs, to reduce the likelihood of this possibility. You might also need to implement a different conflict resolution mechanism from the policies provided by SQL Azure Data Sync, or replicate data from a source other than SQL Server. You can implement a just such customized approach to synchronization in your applications by using the Sync Framework SDK.

Chapter 6, "Maximizing Scalability, Availability, and Performance in the Orders Application" describes how Trey Research implemented caching. "Appendix E - Maximizing Scalability, Availability, and Performance" provides further details and guidance on using Windows Azure Caching.

The Sync Framework 2.1 SDK includes support for building applications that can synchronize with SQL Azure. Using this version of the Sync Framework SDK, you can write your own custom synchronization code and control the replication process directly.

Using this approach, you can address a variety of scenarios that are not easy to implement by using SQL Azure Data Sync, such as building offline-capable/roaming applications. As an example, consider an application running on a mobile device such as a notebook computer used by a plumber or a building maintenance engineer. At the start of each day, he or she uses the application to connect to the local branch office and receive a work schedule with a list of customers' addresses and job details. As each job is completed, an application running on the mobile device is used to input the details, which are stored in a database on the mobile device. Between jobs, he or she can connect to the branch office again and upload the details of the work completed so far, and the application also downloads any amendments to the work schedule for that day. For example, he or she can be directed to attend an urgent job prior to moving on to the previously scheduled engagement. Every Friday afternoon, an administrator in the branch office generates a report detailing the jobs carried out by all workers reporting to that branch.

If the branch office database is implemented by using SQL Azure, the mobile application running on the mobile device can use the Sync Framework SDK to connect to the datacenter hosting this database and synchronize with the local database on the device. Figure 14 shows a simplified view of this architecture.

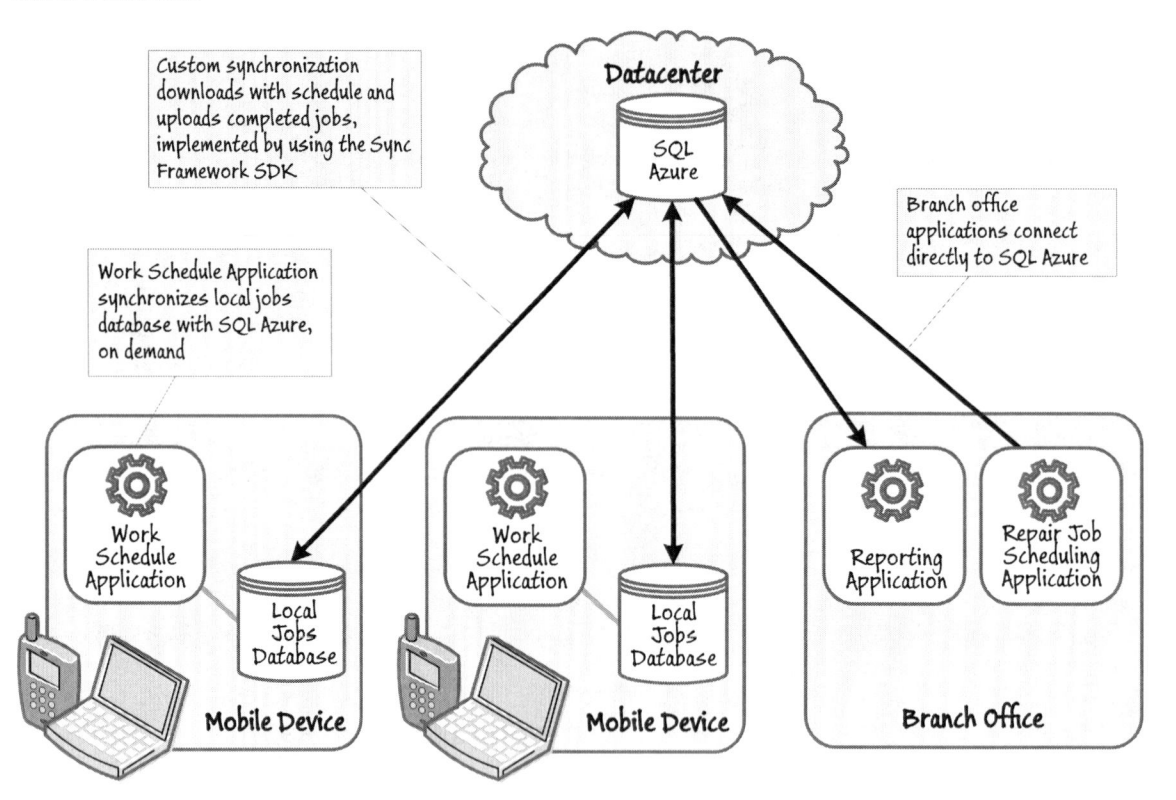

Figure 14
Using the Data Sync SDK to implement custom synchronization

For more information about using the Sync Framework SDK with SQL Azure, see *"SQL Server to SQL Azure Synchronization using Sync Framework 2.1"* on MSDN.

Replicating and Synchronizing Data Using Service Bus Topics and Subscriptions

SQL Azure Data Sync provides an optimized mechanism for synchronizing SQL Server and SQL Azure databases, and is suitable for an environment where changes can be batched together, propagated as a group, and any conflicts resolved quickly. In a dynamic environment such batch processing this may be inappropriate; it may be necessary to replicate changes as they are made rather than batching them up and performing them at some regular interval. In these situations, you may need to implement a custom strategy. Fortunately, this is a well-researched area, and several common patterns are available. This section describes two generic scenarios that cover most situations, and summarizes how you might implement solutions for these scenarios by using Service Bus topics and subscriptions.

Guidelines for Using Service Bus Topics and Subscriptions

You can use Service Bus topics and subscriptions to implement a reliable infrastructure for routing messages between sender and receiver applications. You can exploit this infrastructure to provide a basis for constructing a highly customizable mechanism for synchronizing data updates made across a distributed collection of data sources, as described by the following scenarios:

- **Concurrent instances of applications or services running as part of your system require read and write access to a set of distributed resources. To maintain responsiveness, read operations should be performed quickly, and so the resources are replicated to reduce network latency. Write operations can occur at any time, so you must implement controlled, coordinated write access to each replica to reduce the possibility of any updates being lost.**

    As an example of this scenario, consider a distributed system that comprises multiple instances of an application accessing a database. The instances run in various datacenters in the cloud, and to minimize network latency a copy of the database is maintained at each datacenter. If an application instance at datacenter A needs to modify an item in the database at datacenter A, the same change must be propagated to all copies of the database residing at other datacenters. If this does not happen in a controlled manner, application instances running in different datacenters might update the local copy of the same data to different values, resulting in a conflict, as shown in Figure 15.

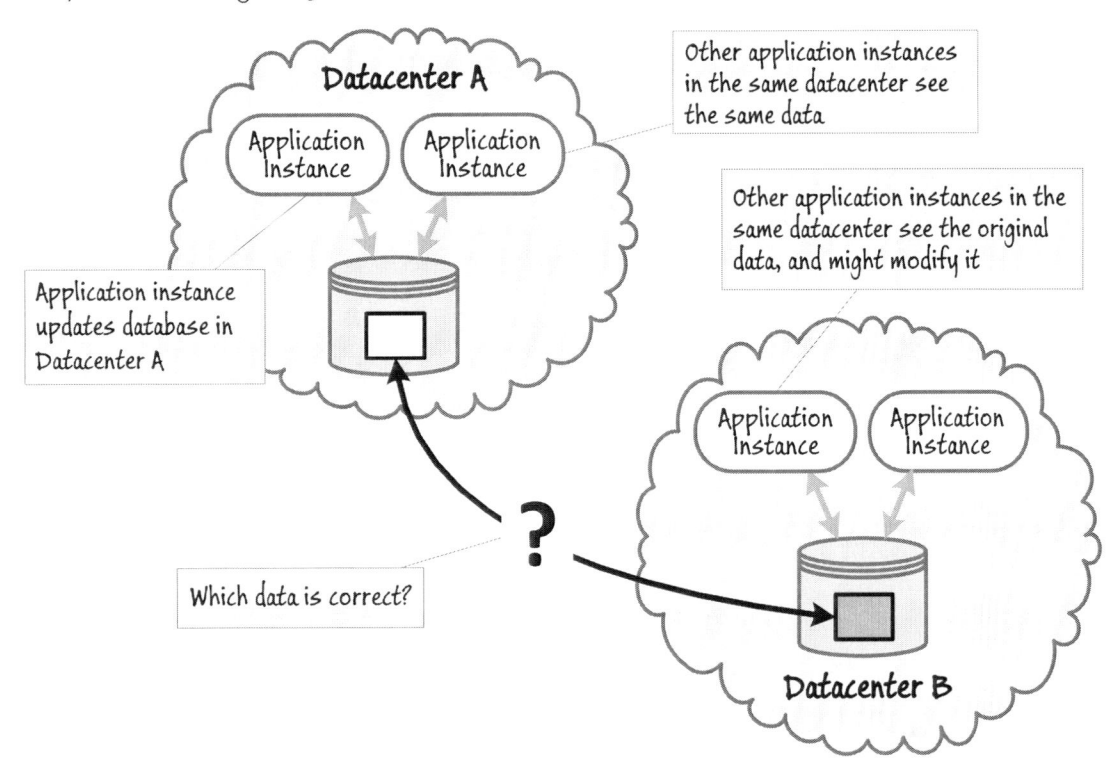

FIGURE 15
Conflict caused by uncontrolled updates to replicas of a database

In the classic distributed transaction model, you can address this problem by implementing transaction managers coordinating with each other by using the Two-Phase Commit protocol (2PC). However, although 2PC guarantees consistency, it does not scale well. In a global environment based on networks that are not always fully reliable this could lead to data being locked for excessively long periods, reducing the responsiveness of applications that depend on this data. Therefore you must be prepared to make some compromises between consistency and availability.

One way to approach this problem is to implement update operations as BASE transactions. BASE is an acronym for Basic Availability, Soft-state, and Eventual consistency, and is an alternative viewpoint to traditional ACID (Atomic, Consistent, Isolated, and Durable) transactions. With BASE transactions, rather than requiring complete and total consistency all of the time, it is considered sufficient for the database to be consistent eventually, as long as no changes are lost in the meantime. What this means in practice, in the example shown in Figure 15, is that an application instance running at Datacenter A can update the database in the same datacenter, and this update must be performed in such a way that it is replicated in the database a Datacenter B. If an application instance running at Datacenter B updates the same data, it must likewise be propagated to Datacenter A. The key point is that after both of the updates are complete, the result should be consistent and both databases should reflect the most recent update.

There may be a period during which the modification made at Datacenter A has yet to be copied to Datacenter B, and during this time an application instance running at Datacenter B may see old, stale data, so the application has to be designed with this possibility in mind; consider the orders processing system displaying the stock levels of products to customers cited earlier in this appendix as an example.

Service Bus topics and subscriptions provide one solution to implementing controlled updates in this scenario. Application instances can query data in the local database directly, but all updates should be formatted as messages and posted to a Service Bus topic. A receiving application located in each datacenter has its own subscription for this topic with a filter that simply passes all messages through to this subscription. Each receiver therefore receives a copy of every message and it uses these messages to update the local database. Figure 16 shows the basic structure of this solution.

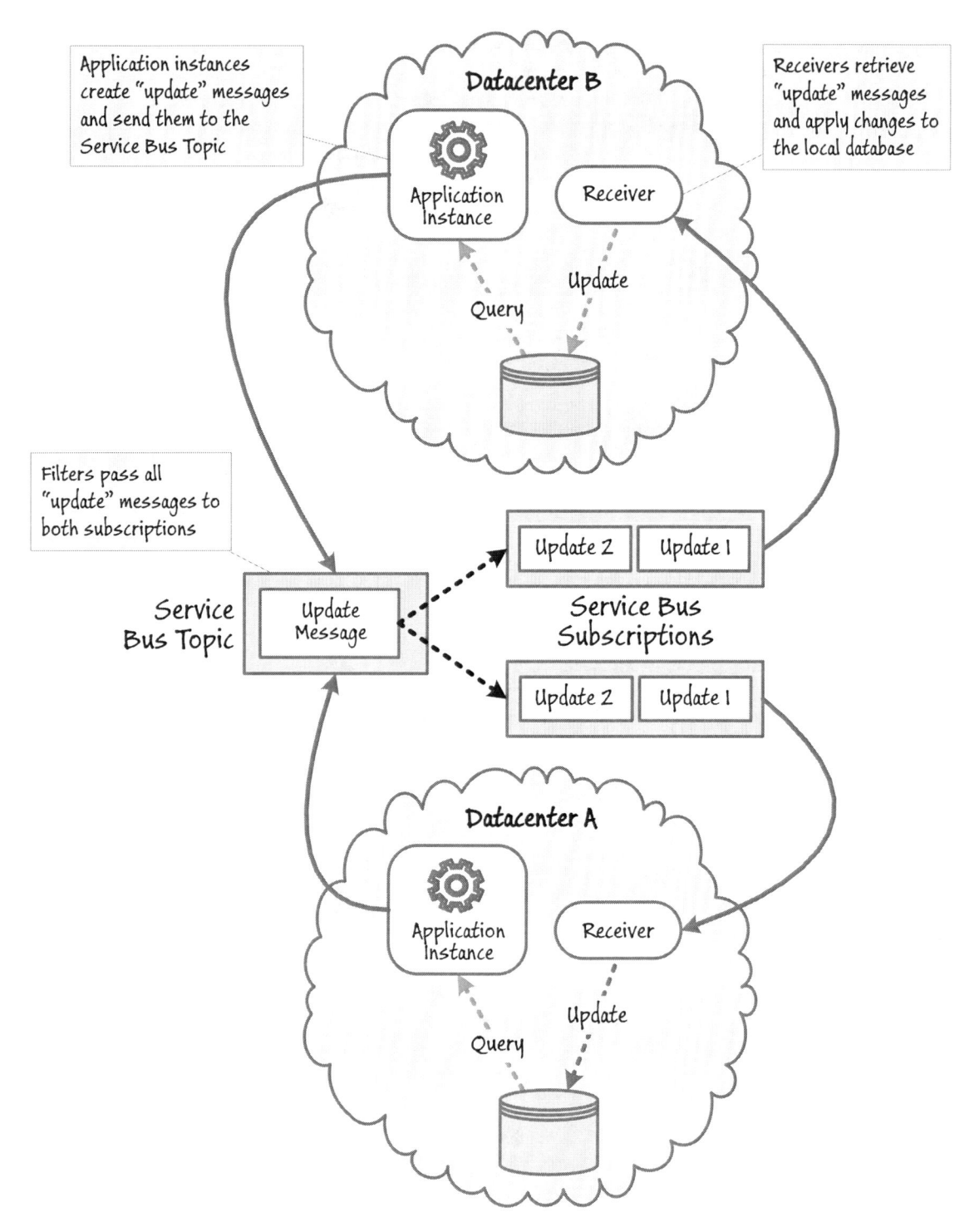

Figure 16
Routing update messages through a Service Bus topic and subscriptions

You can use the Enterprise Library Transient Fault Handling Block to provide a structure for posting messages reliably to a topic and handling any transient errors that may occur. As an additional safeguard, you should configure the topic with duplicate detection enabled, so if the same update message does get posted twice any duplicates will be discarded.

The receiver should retrieve messages by using the **PeekLock** receive mode. If the update operation succeeds, the receive process can be completed, otherwise it will be abandoned and the update message will reappear on the subscription. The receiver can retrieve the message again and retry the operation.

- **Instances of a long-running service executing in the cloud access a single remote data source, but to maintain response times each instance has a cache of the data that it uses. The volume of data is reasonably small and is held in-memory. The data acting as the source of the cache may be updated, and when this happens each service instance should be notified so that it can maintain its cached copy.**

    An example of this scenario is a job processor service. Client applications store information about business tasks that need to be performed in a database. The job processor service periodically polls the database looking for new tasks to perform, and then executes the appropriate actions when a task appears.

    In this scenario, each instance of the job processor service is seeded with data from the database when it starts up. When a new job is added by a client application it is stored in the database. However, the client application can also post a message to a Service Bus topic with the details of the new job. Each instance of the job processor service has a subscription to this topic, and uses the messages posted to this topic to update its local copy of the cached data and perform the appropriate processing. The job processor service no longer needs to poll the database which now acts purely as an audit log of jobs.

    This architecture optimizes the use of the database and also reduces the possibility of conflict occurring; if multiple instances of the job processor service have to poll the database, there is the possibility that they might all pick up the details of the same new job unless the database query logic includes a mechanism for exclusively locking data as it is retrieved, whereas a Service Bus subscription automatically prevents multiple instances from retrieving the same message. Additionally, this architecture can more easily spread the load evenly and is naturally scalable; you can partition the job messages to direct them to different subscriptions by defining an appropriate set of filters, and you can start and stop instances of the job processor service listening on each subscription as the queue length grows and shrinks. Figure 17 shows the structure of this solution.

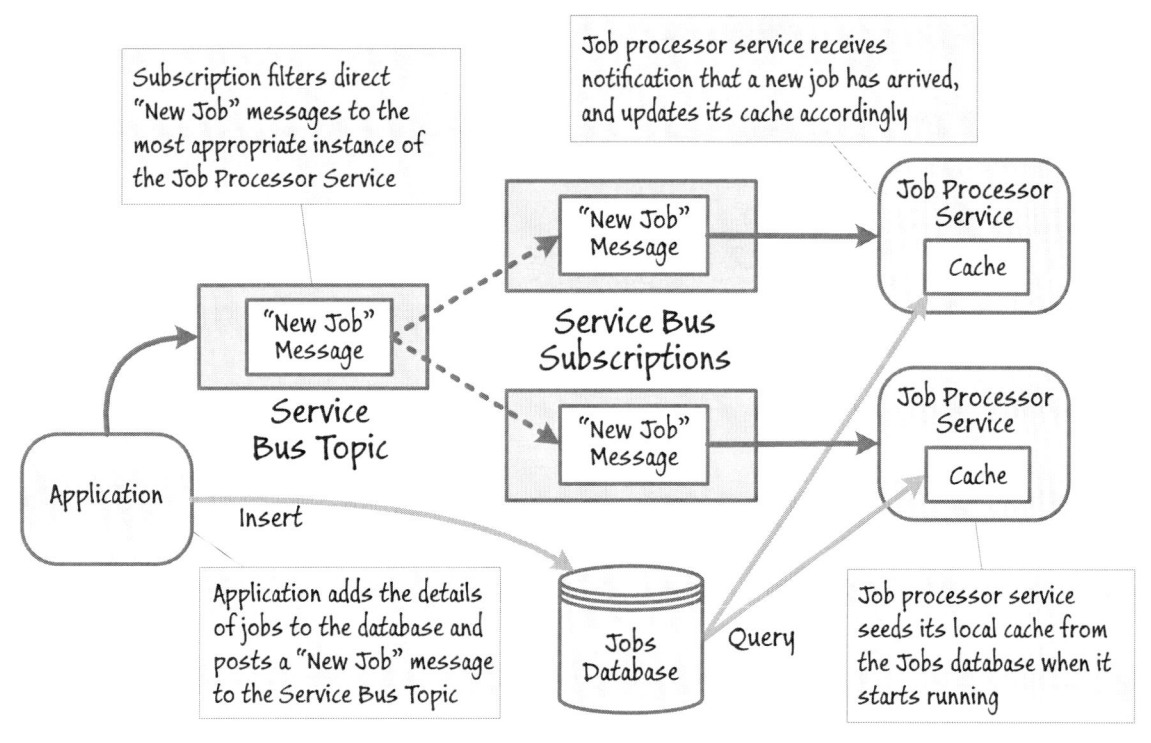

FIGURE 17
Implementing update notifications by using a Service Bus topic and subscriptions

The logic of the client application and job processor service can easily be extended if, for example, the client application wishes to cancel a job, or change some of the details for a job. In these cases, the client application can remove or update the job information in the database and post a job cancellation or job update message to the Service Bus topic.

More Information

All links in this book are accessible from the book's online bibliography available at: *http://msdn.microsoft.com/en-us/library/hh968447.aspx*.
- "Considerations for Using Azure Data Sync" at *http://sqlcat.com/sqlcat/b/technicalnotes/archive/2011/12/21/considerations-when-using-data-sync.aspx*.
- "SQL Azure Data Sync – Supported SQL Azure Data Types" at *http://msdn.microsoft.com/en-us/library/hh667319.aspx*.
- "Synchronization Loops" at *http://msdn.microsoft.com/en-us/library/hh667312.aspx*.
- "Encrypting Connections to SQL Server" at *http://msdn.microsoft.com/en-us/library/ms189067.aspx*.
- "Data Security" at *http://msdn.microsoft.com/en-us/library/hh667329.aspx*.
- "SQL Server to SQL Azure Synchronization using Sync Framework 2.1" at *http://blogs.msdn.com/b/sync/archive/2010/08/31/sql-server-to-sql-azure-synchronization-using-sync-framework-2-1.aspx*.

APPENDIX B Authenticating Users and Authoring Requests

Most applications will need to authenticate and authorize visitors or partners at some stage of the process. Traditionally, authentication was carried out against a local application-specific store of user details, but increasingly users expect applications to allow them to use more universal credentials; for example, existing accounts with social network identity providers such as Windows Live, Google, Facebook, and Open ID.

This process, called *federated authentication*, also offers the opportunity for applications to support single sign-on (SSO). With SSO, users that sign in to one application by means of, for example, their Windows Live® ID credentials are able to visit other sites that use Windows Live ID without being prompted to reenter their credentials.

Alternatively, applications may need to authenticate users with accounts defined within the corporate domain; or support a combination of federated and corporate credentials, and allow users to choose which they specify to sign in. In addition, when using Service Bus in the Windows Azure™ technology platform, access to Service Bus Relay, queue, and topic endpoints must be secured by requiring a token containing the appropriate claims to be presented.

All of these scenarios can be implemented using claims-based authentication, where a Security Token Service (STS) generates tokens that are stored in cookies in the user's web browser or presented by services when they make a request to a server. This appendix describes the Windows technologies that are available to help you implement claims-based authentication, federated authentication, single sign-on, and security for Service Bus queues.

> Federated claims-based authentication and single sign-on are powerful techniques that can simplify development as well as providing benefits to users by making it easier to access different applications without requiring them to reenter their credentials every time.

> *This appendix does not provide a full reference to claims-based authentication technologies and techniques. A detailed exploration of claims-based authentication, authorization, and Windows Azure Access Control Service can be found at "Claims-Based Identity and Access Control Guide" available at http://claimsid.codeplex.com/.*

Uses Cases and Challenges

Most business applications require that users are authenticated and authorized to perform the operations provided by the application. The following sections describe the most common generic use cases for authentication and authorization. The solutions you implement must be reliable, responsive, available, and secure. They must be able to uniquely and accurately identify users and provide information that the application can use to decide what actions that user can take. The solutions should also be as unobtrusive as possible so that the task of signing in is simple and painless.

> Poorly designed or badly implemented authentication can be a performance bottleneck in applications. Users expect authentication to be simple and quick; it should not get in the way of using the application.

Authenticating Public Users

Publicly available applications, such as online shopping sites or forums, typically need to authenticate a large number of users, of whom the application has no prior knowledge. Users register by providing information that the site requires, such as a name and address, but the key factor is to be able to uniquely identify each user so that the correct information can be mapped to them as they sign on next time.

Applications may store the users' credentials and prompt for users to enter these when signing on. However, for public applications, it is useful to allow users to sign in with credentials that they use for other websites and applications so that they do not need to remember another user name and password. By using federated authentication, an application can delegate the responsibility for this task to an external identity provider that handles storing the credentials and checking them when the user signs in. This approach also removes the responsibility for storing sensitive credentials from your application, as this is now the responsibility of the identity provider.

> *The section "Federated Authentication" later in this appendix provides more details of how this works.*

Authenticating Corporate Users and Users from Partner Organizations

Applications used by only a limited and known set of users generally have different requirements than public applications. Users are not usually expected to register before using their account; they expect the organization to already have an account configured for them. In addition, the account details will often be more comprehensive than those held by a public application or social identity provider, and not editable by the user. For example, the account will typically define the membership of security groups and the users' corporate email addresses.

This approach is generally used for "internal" or corporate accounts in most organizations, and is exposed to applications though the built-in operating system features and a directory service. However, where known users from partner organizations must be able to authenticate, an intermediary service that is available outside of the organization is required to generate the claims-based identity tokens. This can be achieved using the same federated authentication techniques as described later in this appendix.

Authorizing User Actions

Identifying individual users is only part of the process. After authentication, the application must be able to control the actions users can take. With claims-based authentication, the token the user obtained from their chosen identity provider may contain claims that specify the role or permissions for the user.

However, this is generally only the case when the user was authenticated by a trusted identity provider that is an enterprise directory service, and consequently contains role information for each user (such as "Manager" or "Accounting"). Social identity providers such as Windows Live ID and Google typically include only a claim that is the unique user identifier, and perhaps the name, so in this case you must implement a mechanism that matches the user identifier with data held by your application or located in a central user repository.

This repository must contain the roles, permissions, or rights information for each user that the application can use to authorize the users' actions. In the case of a public application such as a shopping website the repository will typically hold the information provided by users when they register, such as address and payment details, allowing you to match users with their stored account details when they sign in.

Authorizing Service Access for Non-Browser Clients

Claims-based identity techniques work well in web browsers because the automatic redirection capabilities of the browser make the process seamless and automatic from the user's point of view. For non-browser clients, such as other applications that make service requests to your application's web services, you must publish information that allows the client to discover how to obtain the required token, the accepted formats, and the claims that the token must contain.

The client is responsible for obtaining the required token from a suitable identity provider and STS, and presenting this token with the request. After the application receives the token, it can use the claims it contains to authorize access. If the token contains only a unique identifier, the application may need to map this to an existing account to discover the roles applicable to that client.

> Claims-based and federated identity solutions allow you to clearly separate the tasks of authentication and authorization, so that changes to one of these do not impact the other. This decoupling makes it easier to maintain and update applications to meet new requirements.

Authorizing Access to Service Bus Queues

Hybrid applications that use Windows Azure Service Bus queues must be able to authenticate users, client applications, and partner applications that access the queues. Clients accessing a Service Bus queue can have one of three permission levels: Send, Listen, or Manage. It is vital for security reasons that clients connecting to a queue have the appropriate permissions to prevent reading messages that should not be available to them, or sending messages that may not be valid.

Authorizing Access to Service Bus Relay Endpoints

Hybrid applications that use Windows Azure Service Bus Relay must be able to authenticate users, client applications, and partner applications that access Service Bus Relay endpoints. Clients accessing a Service Bus Relay endpoint must present a suitable token unless the endpoint is configured to allow anonymous (unauthenticated) access. The service is likely to be inside the corporate network or a protected boundary, so is vital for security reasons that clients accessing it through Service Bus Relay have the appropriate permissions in order to prevent invalid access.

> Service Bus topics and subscriptions behave in the same way as queues from a security perspective, and you authenticate users and authorize access to operations in the same way.

Service Bus integrates with Windows Azure Access Control Service, which acts as the default identity provider for Service Bus queues and Service Bus Relay endpoints. However, you can configure Service Bus to use other identity providers if you wish.

Cross-Cutting Concerns

Authentication and authorization mechanisms must, by definition, be secure and robust to protect applications from invalid access and illegal operations. However, other requirements are also important when considering how to implement authentication and authorization in Windows Azure hybrid applications.

Security

The identity of a user or client must be established in a way that uniquely identifies the user with a sufficiently high level of confidence, and is not open to spoofing or other types of attack that may compromise the accuracy of the result. When using federated authentication and delegating responsibility for validating user identity, you must be able to trust the identity provider and any intermediate services with the appropriate level of confidence.

Access to STSs and identity providers should take place over a secure connection protected by Secure Sockets Layer (SSL) or Transport Layer Security (TLS) to counter man-in-the-middle attacks and prevent access to the credentials or authentication tokens while passing over the network.

Many STSs can encrypt the authentication tokens they return, and this should be considered as an additional layer of protection, even when using secure connections.

If an in-house repository of user information is maintained, this must be protected from external penetration through unauthorized access over the network, and from internal attack achieved by means of physical access to the servers, by setting appropriate permissions on the repository tables and content. All sensitive information should be encrypted so that it is of no use if security is compromised.

You must be aware of legal and contractual obligations with regard to personally identifiable information (PII) about users that is held in the repository.

Responsiveness
Authentication mechanisms can be a bottleneck in applications that have many users signing in at the same time, such as the start of a working day. The implementation you choose must be able to satisfy requests quickly during high demand periods. Keep in mind that the process can involve several network transitions between identity providers and STSs, and any one of these may be a weak point in the chain of events.

Reliability
Authentication mechanisms are often a single point of failure. While it is easy to add more servers to an application to handle additional capacity, this approach is often more difficult with authentication mechanisms that must all access a single repository, or use an external identity provider over which you have no control. If the authentication mechanism fails, users will not be able to access the application.

Interoperability
The protocols and mechanisms for claims-based authentication are interoperable by default. They use standards-based protocols and formats for communication and security tokens. However, some STSs and identity providers may only provide one type of token, such as a Simple Web Token (SWT), whereas others might accept or return other types such as Security Assertions Markup Language (SAML) tokens. You must ensure that the external providers and services you choose are compatible with requirements and with each other.

Claims-Based Authentication and Authorization Technologies

This section provides a brief overview of the technologies for authentication and authorization that are typically used in Windows applications and applications hosted in Windows Azure. It focuses on claims-based authentication, which is likely to be the most appropriate approach for hybrid applications. It covers:
- Federated authentication, security token services, and identity providers
- Windows Identity Foundation (WIF)
- Windows Azure Access Control Service (ACS)

Federated Authentication

In traditional applications, authentication and authorization are typically operating system features that make use of enterprise directory services such as Microsoft Active Directory® and resource permissions settings such as Access Control Lists (ACLs). More recently, frameworks and technologies such as Microsoft ASP.NET have provided built in mechanisms that implement a similar approach for web-based applications.

However, all of these approaches require the maintenance of a user directory that defines which users can access the application or service, and what these users can do within the application or service. Maintaining such lists is cumbersome and can even be prone to errors or lack of security. For example, if you store the list of users in a partner organization that can access your application, you depend on that partner to tell you when a user leaves the company or when the permissions required by specific users change.

Instead, you can choose to trust the partner and allow that partner to maintain the list of users and their roles. You still retain control of authorization so you can manage what users can do within the application, but you are freed from the need to authenticate every user and allocate them to the appropriate authorization roles or groups.

This approach to delegated authentication requires a standard way of querying user repositories and distributing tokens that contain claims. Security Token Services (STSs) such as Microsoft Active Directory Federation Service (ADFS) and Windows Azure Access Control Service provide this capability, and compatible STSs are available for other platforms and operating systems.

Each STS uses Identity Providers (IdPs) that are responsible for authenticating users. After the identity provider authenticates the user, the STS will create a token containing the claims for this user. When using a web browser, this token is delivered in a cookie; for smart clients and service applications the token is delivered within the service response (depending on the protocol in use). An STS may also deliver a cookie to the web browser that indicates the user has been authenticated, allowing this user to access other applications without needing to reenter credentials. This provides a single sign-in (SSO) experience.

> ADFS is an IdP as well as an STS because it can use Active Directory to validate a user based on credentials the user provides. ACS uses external IdPs such as Windows Live ID, Google, Facebook, and OpenID, but you can also define service identities in ACS that allow clients to authenticate without using an external IdP.

In practice, you decide which STS your application will trust. This may be an STS that you own, such as ADFS connected to your corporate Active Directory. Alternatively, it may be an external STS such as ACS. You can also configure an STS to trust another STS, so that users can be authenticated by any of the STSs in the trust chain, as shown in Figure 1.

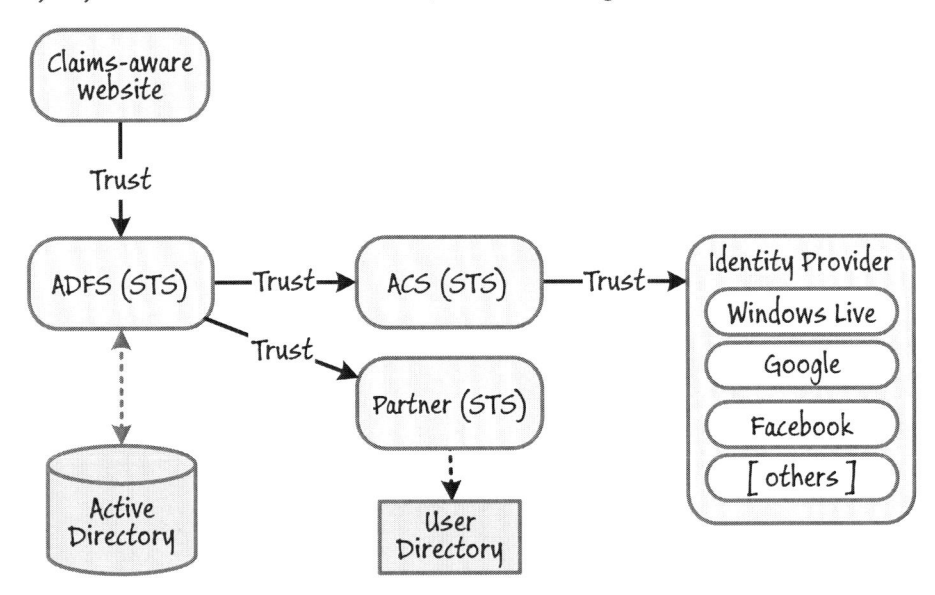

FIGURE 1
An authentication trust chain that can support federated identity and single sign-on

An Overview of the Claims-Based Authentication Process

An STS chooses an IdP that will authenticate the user based on the user's home realm or domain. For example, a user has a home realm of Google if this user has an identity that is managed and authenticated by Google. If a user's account is in a company's Active Directory, their home realm will be that company's corporate domain. For users authenticating through a web browser, the STS directs the user to the appropriate IdP. If the user can be successfully authenticated, the IdP returns a token containing claims (information) about that user to the STS. The STS then generates an application-specific token, which can contain these claims or some augmented version of the claims, and redirects the user to the application with this token. The application can use the claims in this token to authorize user actions. Figure 2 shows a simplified view of the process when using ACS as the token issuer.

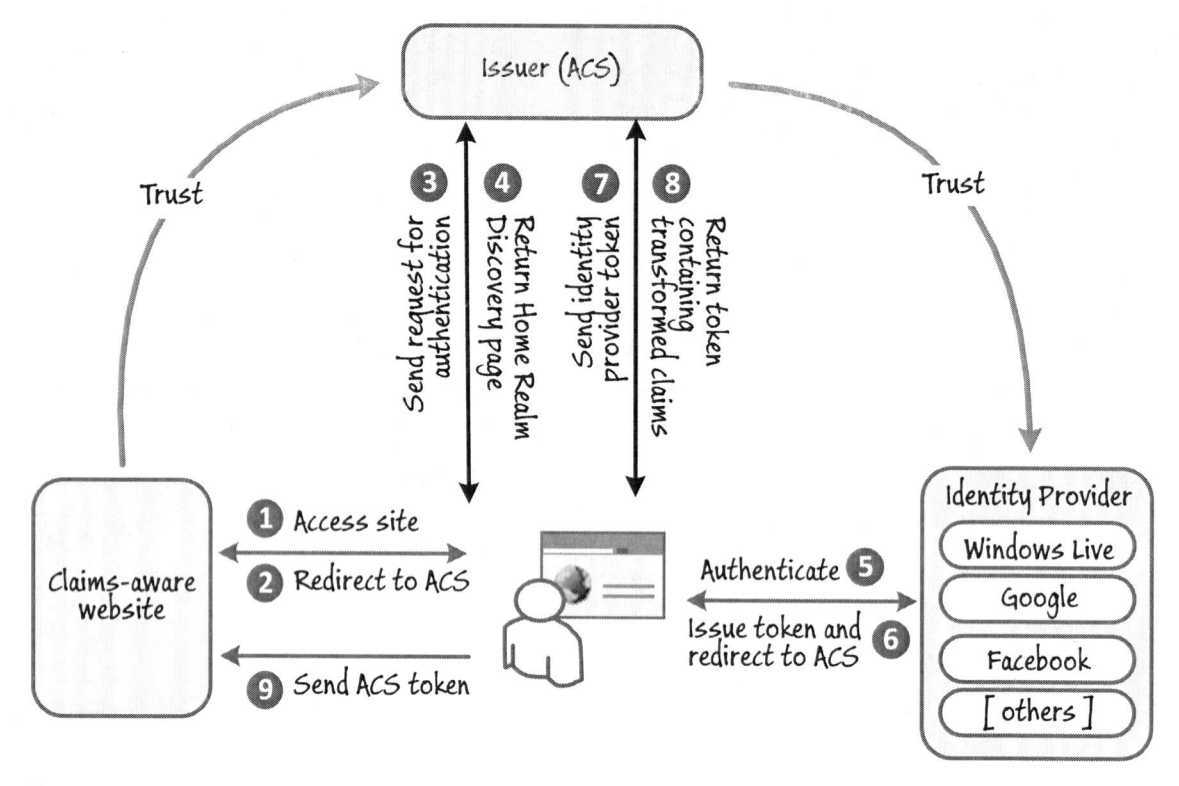

FIGURE 2
A simplified view of the browser authentication process using ACS and social identity providers

To support single sign-on, the STS can store an additional cookie containing an STS-specific token in the browser to indicate that the user was successfully authenticated. When the user accesses another application that trusts the same STS, this STS can generate a suitable token without requiring the user to authenticate again with the original IdP.

Authorizing Web Service Requests

The automatic redirection process described above is only applicable when authenticating requests that come from a web browser. Requests to a web service may be generated by code running in another application, such as a smart client application or service application. In an environment such as this, outside of a web browser, request redirection and cookies cannot be used to direct the request along the chain of STSs and IdPs.

When using claims authentication with a smart client application or a web service, the client must actively request the tokens it requires at each stage, and send these tokens along with the request to the next party in the authentication chain. Figure 3 shows a simplified view of the process for a smart client or service application when using ACS as the token issuer.

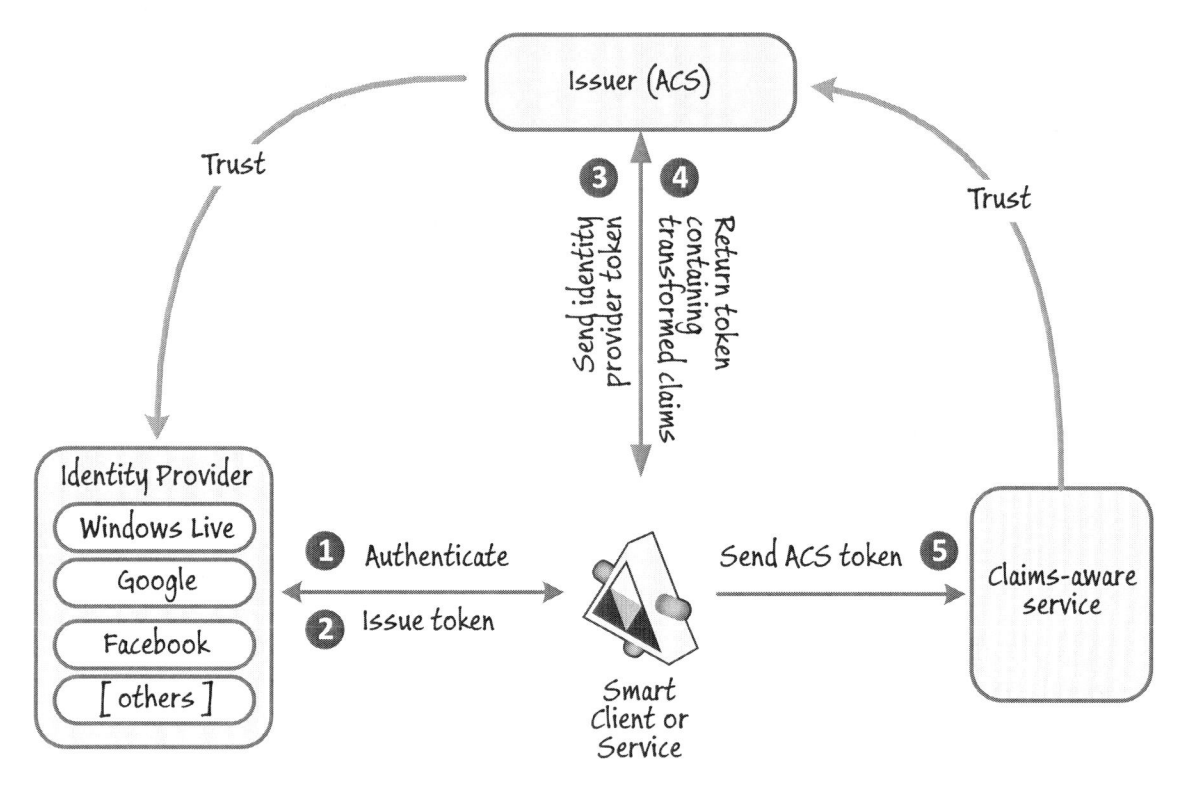

FIGURE 3
A simplified view of the smart client or service authentication process using ACS and social identity providers

WINDOWS IDENTITY FOUNDATION

Microsoft provides a framework that makes it easy to implement claims-based authentication and authorization in web applications and service applications. Windows Identity Foundation (WIF) is a core part of the Microsoft identity and access management framework based on Active Directory, Active Directory Federation Services, Windows Azure Access Control Services, and federated claims-based authentication.

WIF automatically checks requests for the presence of the required claims tokens, and performs redirection for web browsers to the specified STS where these can be obtained. It also exposes the claims in valid tokens to application code, so that the code can make authorization decisions based on these claims. WIF includes a wizard that developers can use in the Visual Studio® development system, or from the command line, to configure applications for claims-based identity.

Claims-based authentication for web browsers is known as passive authentication because the browser automatically handles the redirection and presentation of tokens at the appropriate points in the authentication chain of events. Authentication for service calls, where this process must be specifically managed by the application code, is referred to as active authentication. For more information about active authentication, see "Claims Enabling Web Services."

The WIF components expose events as they handle requests, allowing developers to control the process as required. WIF also includes controls that can be embedded in the application UI to initiate sign-in and sign-out functions.

> *For more information about Windows Identity Foundation, see the Identity Management home page and the patterns & practices "Claims Based Identity & Access Control Guide" at http://claimsid.codeplex.com/.*

Windows Azure Access Control Service

Windows Azure Access Control Service (ACS) is a cloud-based service that makes it easy to authenticate and authorize website, application, and service users and is compatible with popular programming and runtime environments. ACS integrates with the WIF tools and environments and Microsoft Active Directory Federation Services (ADFS), and supports a range of protocols that includes OAuth, OpenID, WS-Federation, and WS-Trust. It allows authentication to take place against many popular web and enterprise identity providers.

ACS accepts SAML 1.1, SAML 2.0, and SWT formatted tokens, and can issue a SAML 1.1, SAML 2.0, or SWT token.

When a user requests authentication from a web browser, ACS receives a request for authentication from the web application and presents a home realm discovery page that lists the identity providers the web application trusts. The user selects an identity provider, and ACS redirects the user to that identity provider's login page. The user logs in and is returned to ACS with a token containing the claims this user has agreed to share in that particular identity provider. ACS then applies the appropriate rules to transform the claims, and creates a new token containing the transformed claims. The rules configured within ACS can perform protocol transition and claims transformation as required by the web application. It then redirects the user back to the web application with the ACS token. The web application can use the claims in this token to apply authorization rules appropriate for this user.

The process for authentication of requests from smart clients and other service applications is different because there is no user interaction. Instead, the service must first obtain a suitable token from an identity provider, present this token to ACS for transformation, and then present the token that ACS issues to the relying party.

ACS also acts as a token issuer for Windows Azure Service Bus. You can configure roles and permissions for Service Bus queues and Service Bus Relay endpoints within ACS. Service Bus automatically integrates with ACS and uses it to validate access and operations on queues and endpoints.

ACS is configured through the service interface using an OData-based management API, or through the web portal that provides a graphical and interactive administration experience. You can also use Windows Azure PowerShell® command-line interface cmdlets to configure ACS. These cmdlets provide wrappers around the OData API and enable you to write scripts that you can use to help automate many ACS configuration tasks.

> *For more information about the Windows Azure PowerShell cmdlets visit the "Windows Azure PowerShell Cmdlets" page on CodePlex, at* http://wappowershell.codeplex.com/.

ACS and Unique User IDs

One point to be aware of if you decide to use ACS is that the user ID it returns after authenticating a user is unique not only to the user, but also to the combination of ACS instance and user. Each configured instance of an ACS namespace generates a different user ID for the same user. If a user is authenticated through Windows Live ID in one ACS instance, the ID it returns will be different from the ID returned when the same user was authenticated by Windows Live ID through a different ACS instance. This is done to protect user privacy; it prevents different applications from identifying users through their ID when they are authenticated by different ACS instances.

If you deploy multiple instances of ACS in different datacenters or change the namespace of your ACS instance you must implement a mechanism that matches multiple ACS-delivered unique IDs to each user in your user data store.

> *For more information about the Windows Azure PowerShell cmdlets visit the Windows Azure Download page at* http://www.windowsazure.com/en-us/manage/downloads/.

Windows Azure Service Bus Authentication and Authorization

Windows Azure Service Bus is a technology that allows you to expose internal services through a corporate firewall or router without having to open inbound ports or perform address mapping. It creates a virtual endpoint in the cloud that users connect to, and the requests and messages are passed to your service or message queue. "Appendix C - Implementing Cross-Boundary Communication" covers Service Bus in more detail; in this section you will see how the integration with ACS is used to authenticate Service Bus requests.

> To view and configure the settings and rules for an ACS namespace that authenticates Service Bus requests, navigate to the Service Bus namespace in the Azure portal and click the "Access Control" icon at the top of the window. You can also configure the ACS namespace programmatically using the Management API.

Service Bus integrates with ACS for authenticating requests. For every Service Bus namespace created in the Windows Azure Management Portal there is a matching ACS namespace, created automatically, which is used to authenticate Service Bus requests. This ACS namespace is the Service Bus namespace with the suffix "**-sb**". It has the Service Bus namespace as its realm, and generates SWTs with an expiration time of 1200 seconds. You cannot change these settings for the default internal identity provider, but you can add additional identity providers that have different behavior. You can also configure access rules and rule groups for your services within this ACS namespace.

CLIENT AUTHENTICATION

Figure 4 shows the overall flow of requests for authenticating a client and accessing an on-premises service through Service Bus. Clients accessing Service Bus Relay endpoints or Service Bus queues must obtain a token from ACS that contains claims defining the client's roles or permissions (**1** and **2**), and present this token to Service Bus (**3**) along with the request for the service or access to a message queue. Service Bus validates the token and allows the request to pass to the required service (**4**). The only exception is if you specifically configure a Service Bus Relay endpoint to all allow unauthenticated access. You cannot configure unauthenticated access for Service Bus queues.

> *To allow unauthenticated access to a Service Bus Relay endpoint you set the **Relay-ClientAuthenticationType** property of the WCF relay binding security element to **None**. For more information, see "Securing Services" in the MSDN WCF documentation and "Securing and Authenticating a Service Bus Connection."*

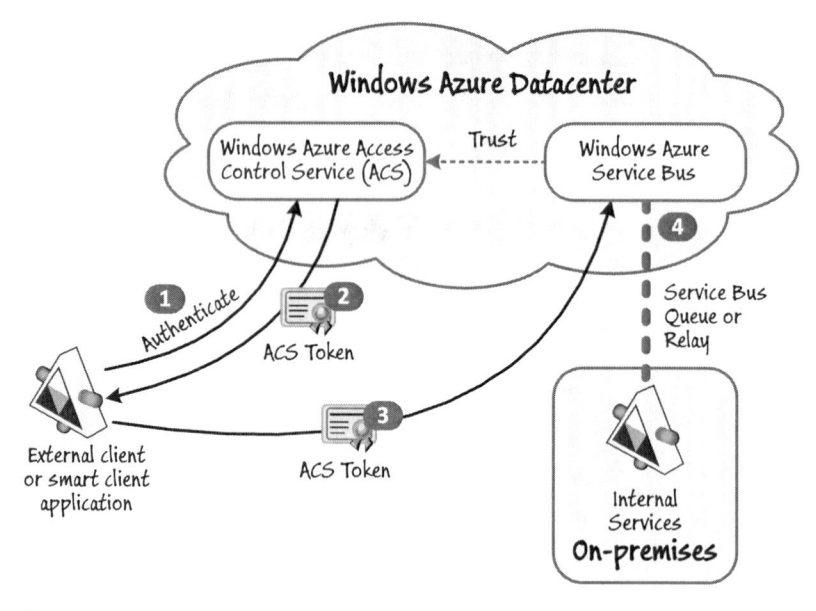

FIGURE 4
Authenticating a Service Bus request with ACS

The token issued by ACS can contain multiple claims. For example, when using Service Bus queues the token may contain claims that allow the client to send messages to a queue, listen on a queue, and manage a queue. When using Service Bus Relay, the claims can allow a client to register an endpoint to listen for requests, send requests to a service, and manage the endpoint. However, it is possible that the service itself (in Figure 4, the internal on-premises service) will require additional credentials not related to Service Bus. The ACS token simply allows access to the service through Service Bus; other authentication and message security issues are the same as when calling a service directly. For example, you may need to provide separate credentials within the service call headers that allow the client to access specific resources within the server that is hosting the service.

ACS is an STS that can be configured to trust other identity providers. This is useful where you want to be able to authenticate clients in a specific domain, such as your own corporate domain, to manage access to internal services exposed through Service Bus. For example, you may expose a service that external employees and partners use to submit expenses. Each valid user will have an existing account in your corporate directory.

> The authentication process for accessing a Service Bus Relay endpoint is entirely separate from any other authentication and security concern related to calling the service itself. Service Bus Relay simply provides a routing mechanism that can expose internal services in a secure way; the techniques you use to protect and secure the service itself and the resources it uses are no different from when you expose it directly over a network.

Figure 5 shows this approach. A client first obtains a token from the corporate ADFS and Active Directory service (**1** and **2**), then presents this token to ACS (**3**). ACS is configured to trust the ADFS instance as an identity provider. ACS validates the token presented by the user with ADFS, and exchanges it for an ACS token containing the appropriate claims based on transformations performed by ACS rules (**4**). The client can then present the ACS token with the service request to Service Bus (**5**). Service Bus validates the token and allows the request to pass to the required service (**6**).

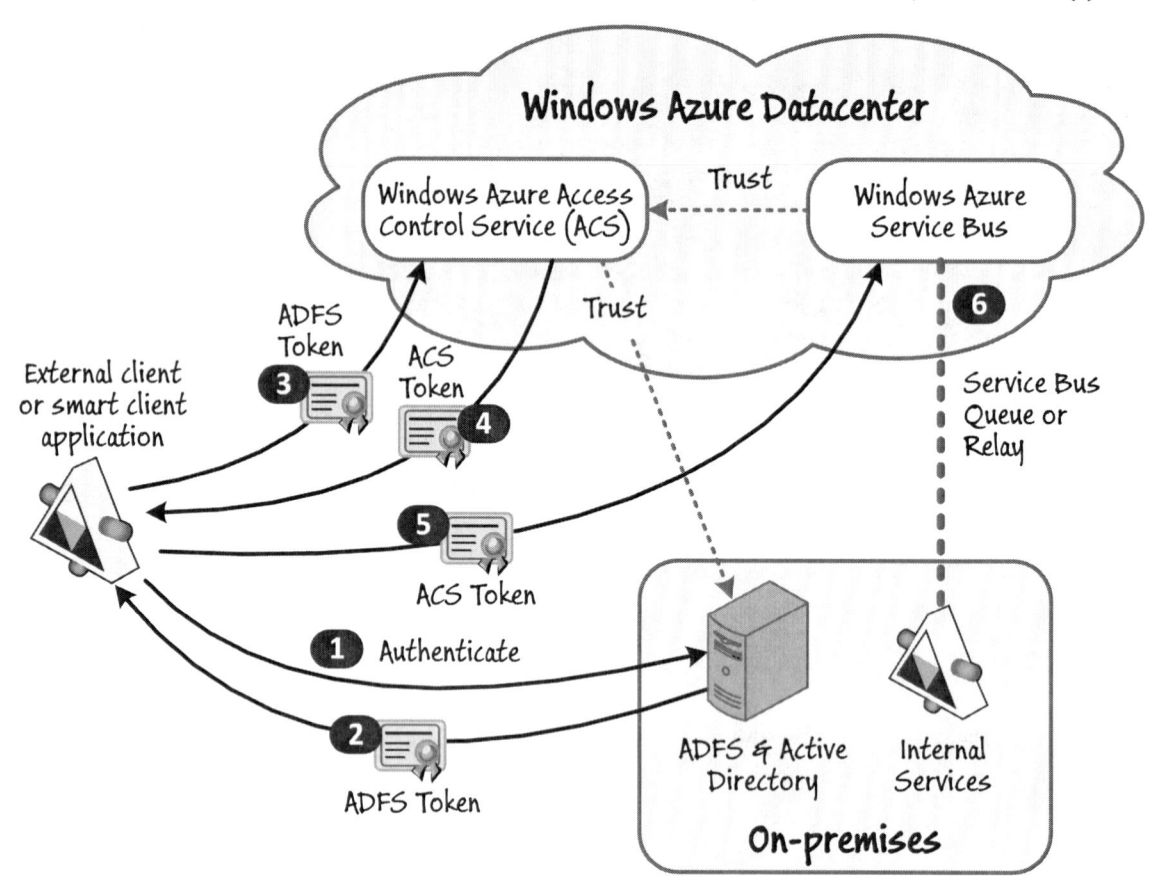

FIGURE 5
Authenticating a Service Bus request with ADFS and ACS

A similar approach can be taken with other identity providers that can generate SAML tokens. The ACS namespace used by Service Bus can also be configured with several identity providers, allowing clients to choose which one to use.

Service Bus Tokens and Token Providers

Unless the endpoint is configured for unauthenticated access, clients must provide a token to ACS when accessing Service Bus endpoints and queues. This token can be one of the following types:

- **Shared secret**. In this approach, the client presents a token containing the service identity and the associated key to ACS to obtain an ACS token to send with the request to Service Bus. The service identity is configured in the ACS Service Bus namespace, which can generate a suitable key that the client will use. This is the approach shown in Figure 4 earlier in this appendix.
- **Simple Web Token (SWT)**. In this approach, the client obtains an SWT from an identity provider and presents this to ACS to obtain an ACS token to send with the request to Service Bus. Figure 5 earlier in this appendix shows the overall process for this approach.
- **SAML Token**. In this approach, the client obtains a SAML token from an identity provider and presents this to ACS to obtain an ACS token to send with the request to Service Bus. Figure 5 earlier in this appendix also shows the overall process for this approach.

To include the required token with a request, the client uses the Service Bus **MessagingFactory**, **NamespaceManager**, or **TransportClientEndpointBehavior** class. All of these types contain methods that accept an instance of a concrete class that inherits the abstract **TokenProvider** base class, such as **SamlTokenProvider**, **SharedSecretTokenProvider**, and **SimpleWebTokenProvider**.

The concrete **TokenProvider** implementations contain methods that create the corresponding type of token from string values and byte arrays, or from existing tokens. You can create custom implementations of **TokenProvider** to perform other approaches to token creation; perhaps to implement a federated authentication mechanism if this is required.

For more information, see "TokenProvider Class" on MSDN.

Service Bus Endpoints and Relying Parties

ACS automatically generates a default relying party in the corresponding ACS namespace when you configure a namespace in Service Bus. This relying party has a realm value that encompasses all endpoint addresses you define within the namespace. It is effectively the root of your Service Bus namespace.

However, you can define additional relying parties within the ACS namespace that correspond to endpoint addresses you add to the Service Bus namespace. For example, if your root namespace in Service Bus is treyresearch, you might define additional endpoints such as the following:

- http://treyresearch.servicebus.windows.net/orders/
- http://treyresearch.servicebus.windows.net/orders/usnorth/partners
- http://treyresearch.servicebus.windows.net/orders/usnorth/partners/partner1
- http://treyresearch.servicebus.windows.net/orders/usnorth/partners/partner2

You can create multiple relying party definitions in ACS that correspond to any subsection of the endpoints defined in Service Bus. ACS matches a request with the longest matching definition, and applies the permissions specified for that definition. This allows you to set up granular control of permissions to each endpoint, or to groups of endpoints. For example, from the list of shown above, the definition for **http://treyresearch.servicebus.windows.net/orders/usnorth/partners** would be used for requests to **http://treyresearch.servicebus.windows.net/orders/usnorth/partners/partner3** because there is no more specific match available.

Authorization Rules and Rule Groups

The ACS Service Bus namespace contains rules and rule groups that specify the roles and permissions for clients based on their authenticated identity. ACS automatically creates a default rule that provides Send, Listen, and Manage permissions to the service owner and the root relying party definition. The default owner identity should be used only for administrative tasks. You should create additional service identities within ACS for clients, and assign the appropriate permissions (Send, Listen, and Manage) to each one to restrict access to the minimum required.

For each relying party definition you specify rules that transform the claims in the token received by ACS, and/or rules that add claims. When using the shared secret approach for authentication, there is no existing token and so there are no existing input claims. The output claims are of the type net.windows.servicebus.action, and have the values Send, Listen, and Manage. You can add more than one claim to any rule. To make configuration easier, you can create rule groups and apply these to multiple relying parties.

> *For a comprehensive list of tutorials on using ACS, including configuring identities and identity providers, see "ACS How Tos" on MSDN. For a step-by-step guide to configuring rules and rule groups, see "How To Implement Token Transformation Logic Using Rules" on MSDN.*

More Information

All links in this book are accessible from the book's online bibliography available at: *http://msdn.microsoft.com/en-us/library/hh968447.aspx*.

- " Claims-Based Identity and Access Control Guide" at *http://claimsid.codeplex.com/* and on MSDN at *http://msdn.microsoft.com/en-us/library/ff423674.aspx*
- "Claims Enabling Web Services" at *http://msdn.microsoft.com/en-us/library/hh446528.aspx*.
- Identity Management home page at *http://msdn.microsoft.com/en-us/security/aa570351.aspx*
- Windows Azure PowerShell Cmdlets are available from the Windows Azure Download page at *http://www.windowsazure.com/en-us/manage/downloads/*.
- "Access Control Service 2.0" at *http://msdn.microsoft.com/en-us/library/windowsazure/gg429786.aspx*
- "Securing Services" in the MSDN WCF documentation at *http://msdn.microsoft.com/en-us/library/ms734769.aspx*
- "Securing and Authenticating a Service Bus Connection" at *http://msdn.microsoft.com/en-us/library/dd582773.aspx*.
- "TokenProvider Class" at *http://msdn.microsoft.com/en-us/library/microsoft.servicebus.tokenprovider.aspx*.
- "ACS How Tos" at *http://msdn.microsoft.com/en-us/library/gg185939.aspx*.
- "How To Implement Token Transformation Logic Using Rules" at *http://msdn.microsoft.com/en-us/library/gg185955.aspx*.

APPENDIX C

Implementing Cross-Boundary Communication

A key aspect of any solution that spans the on-premises infrastructure of an organization and the cloud concerns the way in which the elements that comprise the solution connect and communicate. A typical distributed application contains many parts running in a variety of locations, which must be able to interact in a safe and reliable manner. Although the individual components of a distributed solution typically run in a controlled environment, carefully managed and protected by the organizations responsible for hosting them, the network that joins these elements together commonly utilizes infrastructure, such as the Internet, that is outside of these organizations' realms of responsibility.

Consequently the network is the weak link in many hybrid systems; performance is variable, connectivity between components is not guaranteed, and all communications must be carefully protected. Any distributed solution must be able to handle intermittent and unreliable communications while ensuring that all transmissions are subject to an appropriate level of security.

The Windows Azure™ technology platform provides technologies that address these concerns and help you to build reliable and safe solutions. This appendix describes these technologies.

Uses Cases and Challenges

In a hybrid cloud-based solution, the various applications and services will be running on-premises or in the cloud and interacting across a network. Communicating across the on-premises/cloud divide typically involves implementing one or more of the following generic use cases. Each of these use cases has its own series of challenges that you need to consider.

> Making the most appropriate choice for selecting the way in which components communicate with each other is crucial, and can have a significant bearing on the entire systems design.

Accessing On-Premises Resources From Outside the Organization

Description: Resources located on-premises are required by components running elsewhere, either in the cloud or at partner organizations.

The primary challenge associated with this use case concerns finding and connecting to the resources that the applications and services running outside of the organization utilize. When running on-premises, the code for these items frequently has direct and controlled access to these resources by virtue of running in the same network segment. However, when this same code runs in the cloud it is operating in a different network space, and must be able to connect back to the on-premises servers in a safe and secure manner to read or modify the on-premises resources.

Accessing On-Premises Services From Outside the Organization

Description: Services running on-premises are accessed by applications running elsewhere, either in the cloud or at partner organizations.

In a typical service-based architecture running over the Internet, applications running on-premises within an organization access services through a public-facing network. The environment hosting the service makes access available through one or more well-defined ports and by using common protocols; in the case of most web-based services this will be port 80 over HTTP, or port 443 over HTTPS. If the service is hosted behind a firewall, you must open the appropriate port(s) to allow inbound requests. When your application running on-premises connects to the service it makes an outbound call through your organization's firewall. The local port selected for the outbound call from your on-premises application depends on the negotiation performed by the HTTP protocol (it will probably be some high-numbered port not currently in use), and any responses from the service return on the same channel through the same port. The important point is that to send requests from your application to the service, you do not have to open any additional inbound ports in your organization's firewall.

> Opening ports in your corporate firewall without due consideration of the implications can render your systems liable to attack. Many hackers run automated port-scanning software to search for opportunities such as this. They then probe any services listening on open ports to determine whether they exhibit any common vulnerabilities that can be exploited to break into your corporate systems.

When you run a service on-premises, you are effectively reversing the communication requirements; applications running in the cloud and partner organizations need to make an inbound call through your organization's firewall and, possibly, one or more Network Address Translation (NAT) routers to connect to your services. Remember that the purpose of this firewall is to help guard against unrestrained and potentially damaging access to the assets stored on-premises from an attacker located in the outside world. Therefore, for security reasons, most organizations implement a policy that restricts inbound traffic to their on-premises business servers, blocking access to your services. Even if you are allowed to open up various ports, you are then faced with the task of filtering the traffic to detect and deny access to malicious requests.

The vital question concerned with this use case therefore, is how do you enable access to services running on-premises without compromising the security of your organization?

Implementing a Reliable Communications Channel across Boundaries

Description: Distributed components require a reliable communications mechanism that is resilient to network failure and enables the components to be responsive even if the network is slow.

When you depend on a public network such as the Internet for your communications, you are completely dependent on the various network technologies managed by third party operators to transmit your data. Utilizing reliable messaging to connect the elements of your system in this environment requires that you understand not only the logical messaging semantics of your application, but also how you can meet the physical networking and security challenges that these semantics imply.

A reliable communications channel does not lose messages, although it may choose to discard information in a controlled manner in well-defined circumstances. Addressing this challenge requires you to consider the following issues:

- How is the communications channel established? Which component opens the channel; the sender, the receiver, or some third-party?
- How are messages protected? Is any additional security infrastructure required to encrypt messages and secure the communications channel?
- Do the sender and receiver have a functional dependency on each other and the messages that they send?
- Do the sender and receiver need to operate synchronously? If not, then how does the sender know that a message has been received?
- Is the communications channel duplex or simplex? If it is simplex, how can the receiver transmit a reply to the sender?
- Do messages have a lifetime? If a message has not been received within a specific period should it be discarded? In this situation, should the sender be informed that the message has been discarded?
- Does a message have a specific single recipient, or can the same message be broadcast to multiple receivers?
- Is the order of messages important? Should they be received in exactly the same order in which they were sent? Is it possible to prioritize urgent messages within the communications channel?
- Is there a dependency between related messages? If one message is received but a dependent message is not, what happens?

Cross-Cutting Concerns

In conjunction with the functional aspects of connecting components to services and data, you also need to consider the common non-functional challenges that any communications mechanism must address.

SECURITY

The first and most important of these challenges is security. You should treat the network as a hostile environment and be suspicious of all incoming traffic. Specifically, you must also ensure that the communications channel used for connecting to a service is well protected. Requests may arrive from services and organizations running in a different security domain from your organization. You should be prepared to authenticate all incoming requests, and authorize them according to your organization's data access policy to guard your organization's resources from unauthorized access.

You must also take steps to protect all outgoing traffic, as the data that you are transmitting will be vulnerable as soon as it leaves the environs of your organization.

The questions that you must consider when implementing a safe communications channel include:

- How do you establish a communications channel that traverses the corporate firewall securely?
- How do you authenticate and authorize a sender to enable it to transmit messages over a communications channel? How do you authenticate and authorize a receiver?
- How do you prevent an unauthorized receiver from intercepting a message intended for another recipient?
- How do you protect the contents of a message to prevent an unauthorized receiver from examining or manipulating sensitive data?
- How do you protect the sender or receiver from attack across the network?

> Robust security is a vital element of any application that is accessible across a network. If security is compromised, the results can be very costly and users will lose faith in your system.

RESPONSIVENESS

A well designed solution ensures that the system remains responsive, even while messages are flowing across a slow, error prone network between distant components. Senders and receivers will probably be running on different computers, hosted in different datacenters (either in the cloud, on-premises, or within a third-party partner organization), and located in different parts of the world. You must answer the following questions:

- How do you ensure that a sender and receiver can communicate reliably without blocking each other?

- How does the communications channel manage a sudden influx of messages?
- Is a sender dependent on a specific receiver, and vice versa?

Interoperability

Hybrid applications combine components built using different technologies. Ideally, the communications channel that you implement should be independent of these technologies. Following this strategy not only reduces dependencies on the way in which existing elements of your solution are implemented, but also helps to ensure that your system is more easily extensible in the future.

Maintaining messaging interoperability inevitably involves adopting a standards-based approach, utilizing commonly accepted networking protocols such as TCP and HTTP, and message formats such as XML and SOAP. A common strategy to address this issue is to select a communications mechanism that layers neatly on top of a standard protocol, and then implement the appropriate libraries to format messages in a manner that components built using different technologies can easily parse and process.

Windows Azure Technologies for Implementing Cross-Boundary Communication

If you are building solutions based on direct access to resources located on-premises, you can use Windows Azure Connect to establish a safe, virtual network connection to your on-premises servers. Your code can utilize this connection to read and write the resources to which it has been granted access.

If you are following a service-oriented architecture (SOA) approach, you can build services to implement more functionally focused access to resources; you send messages to these services that access the resources in a controlled manner on your behalf. Communication with services in this architecture frequently falls into one of two distinct styles:

- **Remote procedure call (RPC) style communications.**

    In this style, the message receiver is typically a service that exposes a set of operations a sender can invoke. These operations can take parameters and may return results, and in some simple cases can be thought of as an extension to a local method call except that the method is now running remotely. The underlying communications mechanism is typically hidden by a proxy object in the sender application; this proxy takes care of connecting to the receiver, transmitting the message, waiting for the response, and then returning this response to the sender. Web services typically follow this pattern.

You can also deploy data to the cloud and store it in Windows Azure blob and table storage. The Windows Azure SDK provides APIs that enable you to access this data from applications running on-premises as well as other services running in the cloud or at partner organizations. These scenarios are described in detail in the patterns & practices guide "Developing Applications for the Cloud on the Microsoft Windows Azure Platform" (available at http://wag.codeplex.com) but are not covered in this guide.

> Older applications and frameworks also support the notion of *remote objects*. In this style of distributed communications, a service host application exposes collections of objects rather than operations. A client application can create and use remote objects using the same semantics as local objects. Although this mechanism is very appealing from an object-oriented standpoint, it has the drawback of being potentially very inefficient in terms of network use (client applications send lots of small, chatty network requests), and performance can suffer as a result. This style of communications is not considered any further in this guide.

This style of messaging lends itself most naturally to synchronous communications, which may impact responsiveness on the part of the sender; it has to wait while the message is received, processed, and a response sent back. Variations on this style support asynchronous messaging where the sender provides a callback that handles the response from the receiver, and one-way messaging where no response is expected.

You can build components that provide RPC-style messaging by implementing them as Windows Communication Foundation services. If the services must run on-premises, you can provide safe access to them using the Windows Azure Service Bus Relay mechanism described later in this appendix.

If these services must run in the cloud you can host them as Windows Azure roles. This scenario is described in detail in the patterns & practices guide *"Developing Applications for the Cloud (2nd Edition)"* on MSDN.

- **Message-oriented communications.**

    In this style, the message receiver simply expects to receive a packaged message rather than a request to perform an operation. The message provides the context and the data, and the receiver parses the message to determine how to process it and what tasks to perform.

    This style of messaging is typically based on queues, where the sender creates a message and posts it to the queue, and the receiver retrieves messages from the queue. It's also a naturally asynchronous method of communications because the queue acts as a repository for the messages, which the receiver can select and process in its own time. If the sender expects a response, it can provide some form of correlation identifier for the original message and then wait for a message with this same identifier to appear on the same or another queue, as agreed with the receiver.

    Windows Azure provides an implementation of reliable message queuing through Service Bus queues. These are covered later in this appendix.

The following sections provide more details on Windows Azure Connect, Windows Azure Service Bus Relay, and Service Bus queues; and describe when you should consider using each of them.

Accessing On-Premises Resources from Outside the Organization Using Windows Azure Connect

Windows Azure Connect enables you integrate your Windows Azure roles with your on-premises servers by establishing a virtual network connection between the two environments. It implements a network level connection based on standard IP protocols between your applications and services running in the cloud and your resources located on-premises, and vice versa.

Guidelines for Using Windows Azure Connect

Using Windows Azure Connect provides many benefits over common alternative approaches:

- Setup is straightforward and does not require any changes to your on-premises network. For example, your IT staff do not have to configure VPN devices or perform any complex network or DNS configuration, nor are they required to modify the firewall configuration of any of your on-premises servers or change any router settings.
- You can selectively relocate applications and services to the cloud while protecting your existing investment in on-premises infrastructure.
- You can retain your key data services and resources on-premises if you do not wish to migrate them to the cloud, or if you are required to retain these items on-premises for legal or compliance reasons.

Windows Azure Connect is suitable for the following scenarios:

- **An application or service running in the cloud requires access to network resources located on-premises servers.**

    Using Windows Azure Connect, code hosted in the cloud can access on-premises network resources using the same procedures and methods as code executing on-premises. For example, a cloud-based service can use familiar syntax to connect to a file share or a device such as a printer located on-premises. The application can implement impersonation using the credentials of an account with permission to access the on-premises resource. With this technique, the same security semantics available between on-premises applications and resources apply, enabling you to protect resources using access control lists (ACLs).

> If you are exposing resources from a VM role, you may need to configure the firewall of the virtual machine hosted in the VM role. For example, file sharing may be blocked by Windows Firewall.

- **An application running in the cloud requires access to data sources managed by on-premises servers.**

  Windows Azure Connect enables a cloud-based service to establish a network connection to data sources running on-premises that support IP connectivity. Examples include SQL Server, a DCOM component, or a Microsoft Message queue. Services in the cloud specify the connection details in exactly the same way as if they were running on-premises. For instance, you can connect to SQL Server by using SQL Server authentication, specifying the name and password of an appropriate SQL Server account (SQL Server must be configured to enable SQL Server or mixed authentication for this technique to work.)

  This approach is especially suitable for third-party packaged applications that you wish to move to the cloud. The vendors of these applications do not typically supply the source code, so you cannot modify the way in which they work.

- **An application running in the cloud needs to execute in the same Windows security domain as code running on-premises.**

  Windows Azure Connect enables you to join the roles running in the cloud to your organization's Windows Active Directory® domain. In this way you can configure corporate authentication and resource access to and from cloud-based virtual machines. This feature enables you take advantage of integrated security when connecting to data sources hosted by your on-premises servers. You can either configure the role to run using a domain account, or it can run using a local account but configure the IIS application pool used to run the web or worker role with the credentials of a domain account.

- **An application requires access to resources located on-premises, but the code for the application is running at a different site or within a partner organization.**

  You can use Windows Azure Connect to implement a simplified VPN between remote sites and partner organizations. Windows Azure Connect is also a compelling technology for solutions that incorporate roaming users who need to connect to on-premises resources such as a remote desktop or file share from their laptop computers.

> Using Windows Azure Connect to bring cloud services into your corporate Windows domain solves many problems that would otherwise require complex configuration, and that may inadvertently render your system open to attack if you get this configuration wrong.

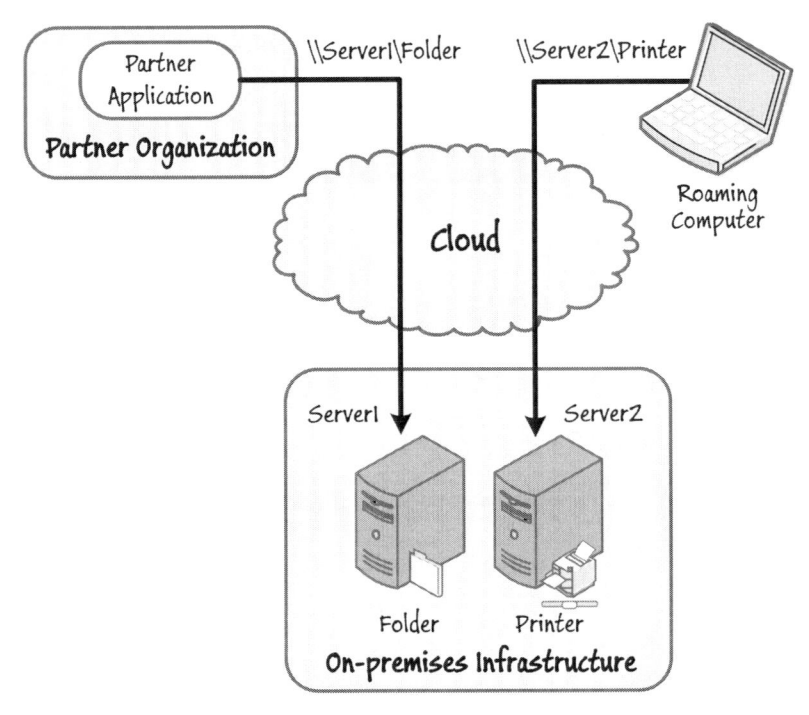

FIGURE 1
Connecting to on-premises resources from a partner organization and roaming computers

For up-to-date information about best practices for implementing Windows Azure Connect, visit the Windows Azure Connect Team Blog.

Windows Azure Connect Architecture and Security Model

Windows Azure Connect is implemented as an IPv6 virtual network by Windows Azure Connect endpoint software running on each server and role that participates in the virtual network. The endpoint software transparently handles DNS resolution and manages the IP connections between your servers and roles. It is installed automatically on roles running in the cloud that are configured as connect-enabled. For servers running on-premises, you download and install the Windows Azure Connect endpoint software manually. This software executes in the background as a Windows service. Similarly, if you are using Windows Azure Connect to connect from a VM role, you must install the Windows Azure Connect endpoint software in this role before you deploy it to the cloud.

You use the Windows Azure Management Portal to generate an activation token that you include as part of the configuration for each role and each instance of the Windows Azure Connect endpoint software running on-premises. Windows Azure Connect uses this token to link the connection endpoint to the Windows Azure subscription and ensures that the virtual network is only accessible to authenticated servers and roles. Network traffic traversing the virtual network is protected end-to-end by using certificate-based IPsec over Secure Socket Tunneling Protocol (SSTP). Windows Azure Connect provisions and configures the appropriate certificates automatically, and does not require any manual intervention on the part of the operator.

The Windows Azure Connect endpoint software establishes communications with each node by using Connect Relay service, hosted and managed by Microsoft in their datacenters. The endpoint software uses outbound HTTPS connections only to communicate with the Windows Azure Connect Relay service. However, the Windows Azure Connect endpoint software creates a firewall rule for Internet Control Message Protocol for IPv6 (ICMPv6) communications which allows Router Solicitation (Type 133) and Router Advertisement (Type 134) messages when it is installed. These messages are required to establish and maintain an IPv6 link. Do not disable this rule.

> Microsoft implements separate instances of the Windows Azure Connect Relay service in each region. For best performance, choose the relay region closest to your organization when you configure Windows Azure Connect.

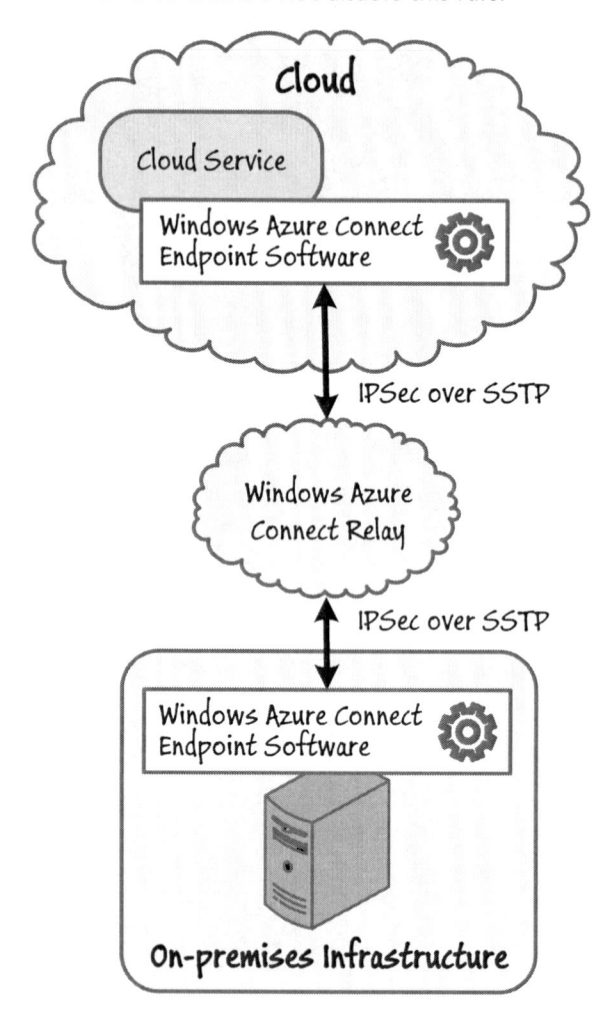

FIGURE 2
The security architecture of Windows Azure Connect

You manage the connection security policy that governs which servers and roles can communicate with each other using the Management Portal; you create one or more endpoint groups containing the host servers that comprise your solution (and that have the Windows Azure Connect endpoint software installed), and then specify the Windows Azure roles that they can connect to. This collection of host servers and roles constitutes a single virtual network.

For more information about configuring Windows Azure Connect and creating endpoint groups, see "Windows Azure Connect" on MSDN.

Limitations of Windows Azure Connect

Windows Azure Connect is intended for providing direct access to corporate resources, either located on-premises or in the cloud. It provides a general purpose solution, but is subject to some constraints as summarized by the following list:

- Windows Azure Connect is an end-to-end solution; participating on-premises computers and VM roles in the cloud must be defined as part of the same virtual network. Resources (on-premises and in a VM role) can be protected by using ACLs, although this protection relies on the users accessing those resources being defined in the Windows domain spanning the virtual network or in another domain trusted by this domain. Consequently, Windows Azure Connect is not suitable for sharing resources with public clients, such as customers communicating with your services across the Internet. In this case, Windows Azure Service Bus Relay is a more appropriate technology.

- Windows Azure Connect implements an IPv6 network, although it does not require any IPv6 infrastructure on the underlying platform. However, any applications using Windows Azure Connect to access resources must be IPv6 aware. Older legacy applications may have been constructed based on IPv4, and these applications will need to be updated or replaced to function correctly with Windows Azure Connect.

- If you are using Windows Azure Connect to join roles to your Windows domain, the current version of Windows Azure Connect requires that you install the Windows Azure Connect endpoint software on the domain controller. This same machine must also be a Windows Domain Name System (DNS) server. These requirements may be subject to approval from your IT department, and you organization may implement policies that constrain the configuration of the domain controller. However, these requirements may change in future releases of Windows Azure Connect.

- Windows Azure Connect is specifically intended to provide connectivity between roles running in the cloud and servers located on-premises. It is not suitable for connecting roles together; if you need to share resources between roles, a preferable solution is to use Windows Azure storage, Windows Azure queues, or databases tables deployed to the SQL Azure™ technology platform.

- You can only use Windows Azure Connect to establish network connectivity with servers running the Windows® operating system; the Windows Azure Connect endpoint software is not designed to operate with other operating systems. If you require cross-platform connectivity, you should consider using Windows Azure Service Bus.

Accessing On-Premises Services from Outside the Organization Using Windows Azure Service Bus Relay

Windows Azure Service Bus Relay provides the communication infrastructure that enables you to expose a service to the Internet from behind your firewall or NAT router. The Windows Azure Service Bus Relay service provides an easy to use mechanism for connecting applications and services running on either side of the corporate firewall, enabling them to communicate safely without requiring a complex security configuration or custom messaging infrastructure.

Guidelines for Using Windows Azure Service Bus Relay

Windows Azure Service Bus Relay is ideal for enabling secure communications with a service running on-premises, and for establishing peer-to-peer connectivity. Using Windows Azure Service Bus Relay brings a number of benefits:

- It is fully compatible with Windows Communication Foundation (WCF). You can leverage your existing WCF knowledge without needing to learn a new model for implementing services and client applications. Windows Azure Service Bus Relay provides WCF bindings that extend those commonly used by many existing services.

- It is interoperable with platforms and technologies for operating systems other than Windows. Windows Azure Service Bus Relay is based on common standards such as HTTPS and TCP/SSL for sending and receiving messages securely, and it exposes an HTTP REST interface. Any technology that can consume and produce HTTP REST requests can connect to a service that uses Windows Azure Service Bus Relay. You can also build services using the HTTP REST interface.

- It does not require you to open any inbound ports in your organization's firewall. Windows Azure Service Bus Relay uses only outbound connections.

- It supports naming and discovery. Windows Azure Service Bus Relay maintains a registry of active services within your organization's namespace. Client applications that have the appropriate authorization can query this registry to discover these services, and download the metadata necessary to connect to these services and exercise the operations that they expose. The registry is managed by Windows Azure and exploits the scalability and availability that Windows Azure provides.

- It supports federated security to authenticate requests. The identities of users and applications accessing an on-premises service through Windows Azure Service Bus Relay do not have to be members of your organization's security domain.
- It supports many common messaging patterns, including two-way request/response processing, one-way messaging, service remoting, and multicast eventing.
- It supports load balancing. You can open up to 25 listeners on the same endpoint. When the Service Bus receives requests directed towards an endpoint, it load balances the requests across these listeners.

You should note that Windows Azure Service Bus Relay is not suitable for implementing all communication solutions. For example, it imposes a temporal dependency between the services running on-premises and the client applications that connect to them; a service must be running before a client application can connect to it otherwise the client application will receive an **EndpointNotFoundException** exception (this limitation applies even with the **NetOnewayRelayBinding** and **NetEventRelayBinding** bindings described in the section "Selecting a Binding for a Service" later in this appendix.) Furthermore, Windows Azure Service Bus Relay is heavily dependent on the reliability of the network; a service may be running, but if a client application cannot reach it because of a network failure the client will again receive an **EndpointNotFoundException** exception. In these cases using Windows Azure Service Bus queues may provide a better alternative; see the section "Implementing a Reliable Communications Channel across Boundaries Using Service Bus Queues" later in this appendix for more information.

You should consider using Windows Azure Service Bus Relay in the following scenarios:

- **An application running in the cloud requires controlled access to your service hosted on-premises. Your service is built using WCF.**

    This is the primary scenario for using Windows Azure Service Bus Relay. Your service is built using WCF and the Windows Azure SDK. It uses the WCF bindings provided by the **Microsoft.ServiceBus** assembly to open an outbound communication channel through the firewall to the Windows Azure Service Bus Relay service and wait for incoming requests. Client applications in the cloud also use the same Windows Azure SDK and WCF bindings to connect to the service and send requests through the Windows Azure Service Bus Relay service. Responses from the on-premises services are routed back through the same communications channel through the Windows Azure Service Bus Relay service to the client application. On-premises services can be hosted in a custom application or by using Internet Information Services (IIS). When using IIS, an administrator can configure the on-premises service to start automatically so that it registers a connection with Windows Azure Service Bus Relay and is available to receive requests from client applications.

Appendix C

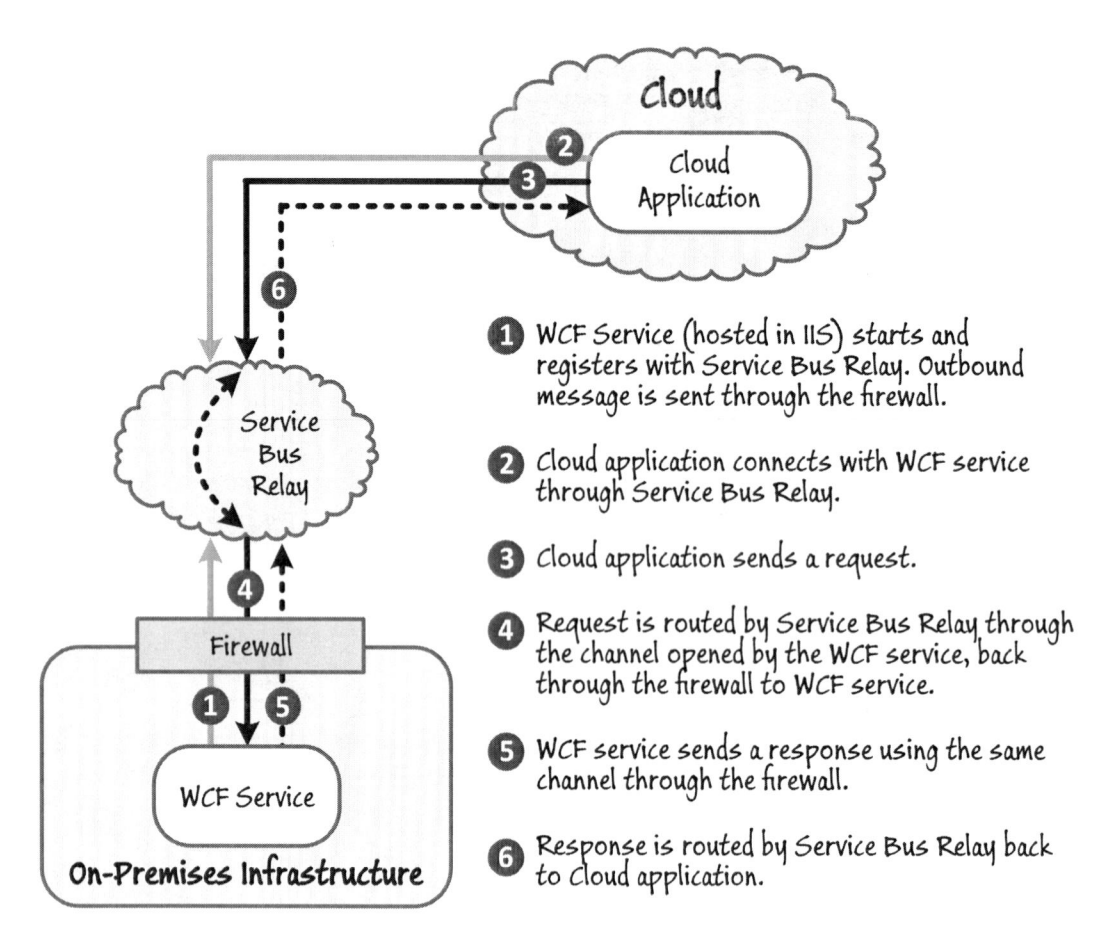

Figure 3
Routing requests and responses through Windows Azure Service Bus Relay

1. WCF Service (hosted in IIS) starts and registers with Service Bus Relay. Outbound message is sent through the firewall.
2. Cloud application connects with WCF service through Service Bus Relay.
3. Cloud application sends a request.
4. Request is routed by Service Bus Relay through the channel opened by the WCF service, back through the firewall to WCF service.
5. WCF service sends a response using the same channel through the firewall.
6. Response is routed by Service Bus Relay back to Cloud application.

This pattern is useful for providing remote access to existing on-premises services that were not originally accessible outside of your organization. You can build a façade around your services that publishes them in Windows Azure Service Bus Relay. Applications external to your organization and that have the appropriate security rights can then connect to your services through Windows Azure Service Bus Relay.

Initially all messages are routed through the Windows Azure Service Bus Relay service, but as an optimization mechanism a service exposing a TCP endpoint can use a direct connection, bypassing the Windows Azure Service Bus Relay service once the service and client have both been authenticated. The coordination of this direct connection is governed by the Windows Azure Service Bus Relay service. The direct socket connection algorithm uses only outbound connections for firewall traversal and relies on a mutual port prediction algorithm for NAT traversal. If a direct connection can be established, the relayed connection is automatically upgraded to the direct connection. If the direct connection cannot be established, the connection will revert back to passing messages through the Windows Azure Service Bus Relay service.

The NAT traversal algorithm is dependent on a very narrowly timed coordination and a best-guess prediction about the expected NAT behavior. Consequently the algorithm tends to have a very high success rate for home and small business scenarios with a small number of clients but degrades in its success rate as the size of the network increases.

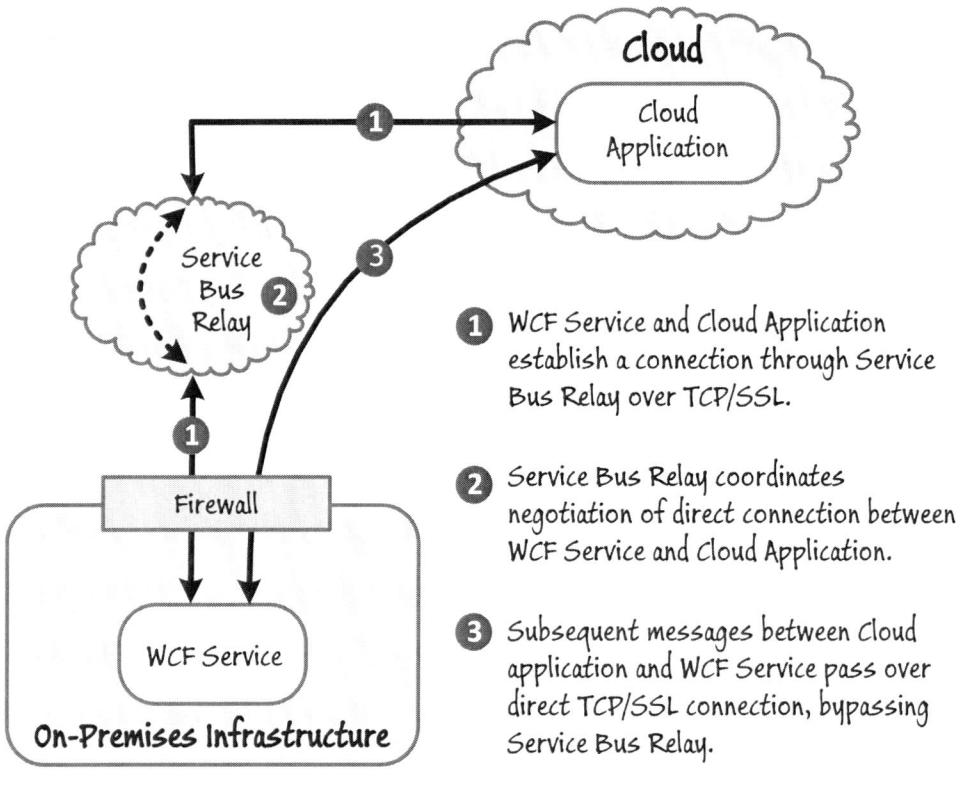

FIGURE 4
Establishing a direct connection over TCP/SSL

- **An application running at a partner organization requires controlled access to your service hosted on-premises. The client application is built by using a technology other than WCF.**

    In this scenario, a client application can use the HTTP REST interface exposed by the Windows Azure Service Bus Relay service to locate an on-premises service and send it requests.

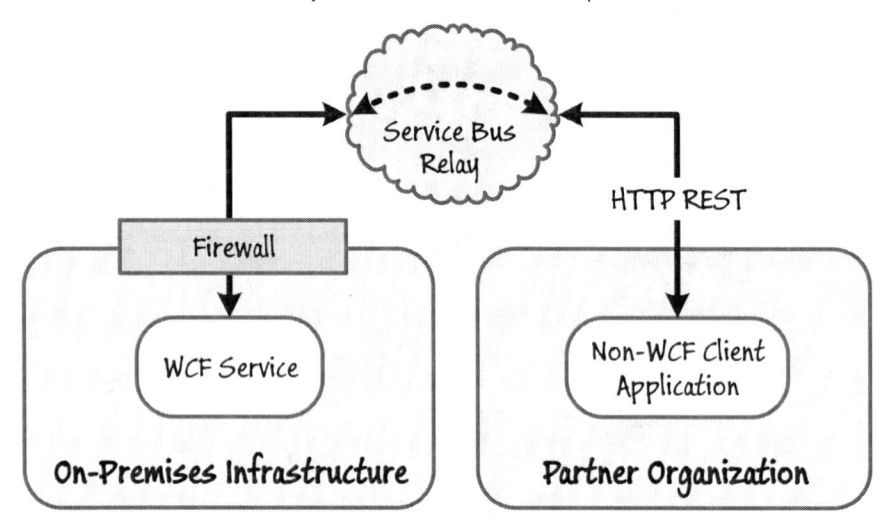

FIGURE 5
Connecting to an on-premises service by using HTTP REST requests

- **An application running in the cloud or at a partner organization requires controlled access to your service hosted on-premises. Your service is built by using a technology other than WCF.**

    You may have existing services that you have implemented using a technology such as Perl or Ruby. In this scenario, the service can use the HTTP REST interface to connect to the Windows Azure Service Bus Relay service and await requests.

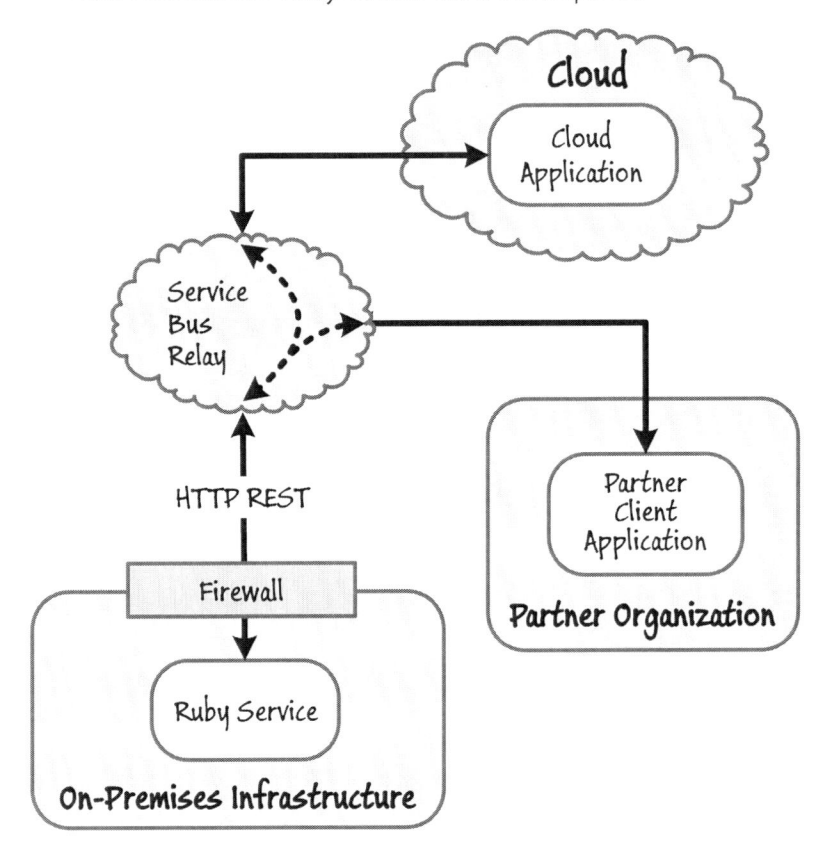

FIGURE 6
Connecting to an on-premises service built with Ruby using Windows Azure Service Bus Relay

- **An application running in the cloud or at a partner organization submits requests that can be sent to multiple services hosted on-premises.**

    In this scenario, a single request from a client application may be sent to and processed by more than one service running on-premises. Effectively, the message from the client application is multicast to all on-premises services registered at the same endpoint with the Windows Azure Service Bus Relay service. All messages sent by the client are one-way; the services do not send a response. This approach is useful for building event-based systems; each message sent by a client application constitutes an event, and services can transparently subscribe to this event by registering with the Windows Azure Service Bus Relay service.

> You can also implement an eventing system by using Service Bus topics and subscriptions. Windows Azure Service Bus Relay is very lightweight and efficient, but topics and subscriptions provide more flexibility. Guidelines for routing messages using Service Bus topics and subscriptions are provided in "Appendix D - Implementing Business Logic and Message Routing across Boundaries."

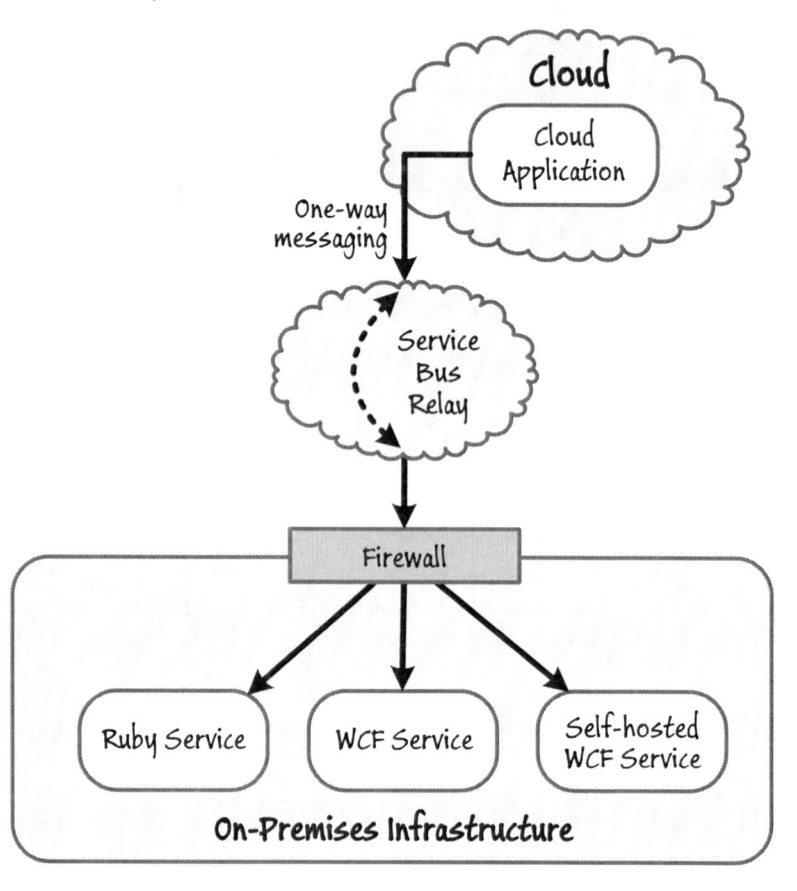

FIGURE 7
Multicasting using Windows Azure Service Bus Relay

- **An application running on-premises or at a partner organization requires controlled access to your service hosted in the cloud. Your service is implemented as a Windows Azure worker role and is built using WCF.**

    This scenario is the opposite of the situation described in the previous cases. In many situations, an application running on-premises or at a partner organization can access a WCF service implemented as a worker role directly, without the intervention of Windows Azure Service Bus Relay. However, this scenario is valuable if the WCF service was previously located on-premises and code running elsewhere connected via Windows Azure Service Bus Relay as described in the preceding examples, but the service has now been migrated to Windows Azure. Refactoring the service, relocating it to the cloud and publishing it through Windows Azure Service Bus Relay enables you to retain the same endpoint details, so client applications do not have to be modified or reconfigured in any way; they just carry on running as before. This architecture also enables you to more easily protect communications with the service by using the appropriate WCF bindings.

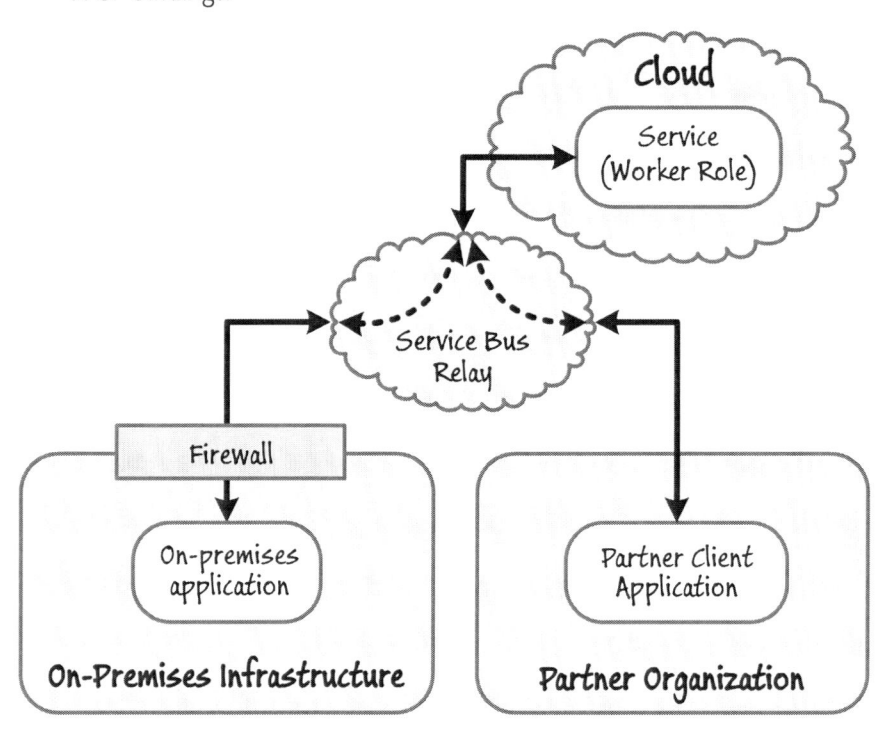

FIGURE 8
Routing requests to a worker role through Windows Azure Service Bus Relay

Guidelines for Securing Windows Azure Service Bus Relay

Windows Azure Service Bus Relay endpoints are organized by using Service Bus namespaces. When you create a new service that communicates with client applications by using Windows Azure Service Bus Relay you can use the Management Portal to generate a new service namespace. This namespace must be unique, and it determines the uniform resource identifier (URI) that your service exposes; client applications specify this URI to connect to your service through Windows Azure Service Bus Relay. For example, if you create a namespace with the value **TreyResearch** and you publish a service named **OrdersService** in this namespace, the full URI of the service is sb://treyresearch.servicebus.windows.net/OrdersService.

The services that you expose through Windows Azure Service Bus Relay can provide access to sensitive data, and are themselves valuable assets; therefore you should protect these services. There are several facets to this task:

- You should restrict access to your Service Bus namespace to authenticated services and client applications only. This requires that each service and client application runs with an associated identity that the Windows Azure Service Bus Relay service can verify. As described in "Appendix B - Authenticating Users and Authorizing Requests," Service Bus includes its own identity provider as part of the Windows Azure Access Control Service (ACS), and you can define identities and keys for each service and user running a client application. You can also implement federated security through ACS to authenticate requests against a security service running on-premises or at a partner organization.

    When you configure access to a service exposed through Windows Azure Service Bus Relay, the realm of the relying party application with which you associate authenticated identities is the URL of the service endpoint on which the service accepts requests.

- You should limit the way in which clients can interact with the endpoints published through your Service Bus namespace. For example, most client applications should only be able to send messages to a service (and obtain a response) while services should be only able to listen for incoming requests. Service Bus defines the claim type **net.windows.servicebus.action** which has the possible values **Send**, **Listen**, and **Manage**. Using ACS you can implement a rule group for each URI defined in your Service Bus namespace that maps an authenticated identity to one or more of these claim values.

> Remember that even though Service Bus is managed and maintained by one or more Microsoft datacenters, applications connect to Windows Azure Service Bus Relay across the Internet. Unauthorized applications that can connect to your Service Bus namespaces can implement common attacks, such as denial of service to disrupt your operations, or Man-in-the-Middle to steal data as it is passed to your services. Therefore, you should protect your Service Bus namespaces and the services that use it as carefully as you would defend your on-premises assets.

> Service Bus can also use third party identity providers, such as Windows Live ID, Google, and Yahoo!, but the default is to use the built-in identity provider included within ACS.

When a service starts running and attempts to advertise an endpoint, it provides its credentials to Windows Azure Service Bus Relay service. These credentials are validated, and are used to determine whether the service should be allowed to create the endpoint. A common approach used by many services is to define an endpoint behavior that references the **transportClientEndpointBehavior** element in the configuration file. This element has a **clientCredentials** element that enables a service to specify the name of an identity and the corresponding symmetric key to verify this identity. A client application can take a similar approach, except that it specifies the name and symmetric key for the identity running the application rather than that of the service.

For more information about protecting services through Windows Azure Service Bus Relay, see "Securing and Authenticating a Service Bus Connection" on MSDN.

Note that using the shared secret token provider is just one way of supplying the credentials for the service and the client application. When you specify this provider, ACS itself authenticates the name and key, and if they are valid it generates a Simple Web Token (SWT) containing the claims for this identity, as determined by the rules configured in ACS. These claims determine whether the service or client application has the appropriate rights to listen for, or send messages. Other authentication provider options are available, including using SAML tokens. You can also specify a different Security Token Service (STS) other than that provided by ACS to authenticate credentials and issue claims.

- When a client application has established a connection to a service through Windows Azure Service Bus Relay, you should carefully control the operations that the client application can invoke. The authorization process is governed by the way in which you implement the service and is outside the scope of ACS, although you can use ACS to generate the claims for an authenticated client, which your service can use for authorization purposes.
- All communications passing between your service and client applications is likely to pass across a public network or the Internet. You should protect these communications by using an appropriate level of data transfer security, such as SSL or HTTPS.

> If you are using WCF to implement your services, you should consider building a Windows Identity Foundation authorization provider to decouple the authorization rules from the business logic of your service.

> If you are using WCF to implement your services, implementing transport security is really just a matter of selecting the most appropriate WCF binding and then setting the relevant properties to specify how to encrypt and protect data.

APPENDIX C

Figure 9 illustrates the core recommendations for protecting services exposed through Windows Azure Service Bus Relay.

You can configure message authentication and encryption by configuring the WCF binding used by the service. For more information, see "Securing and Authenticating a Service Bus Connection" on MSDN.

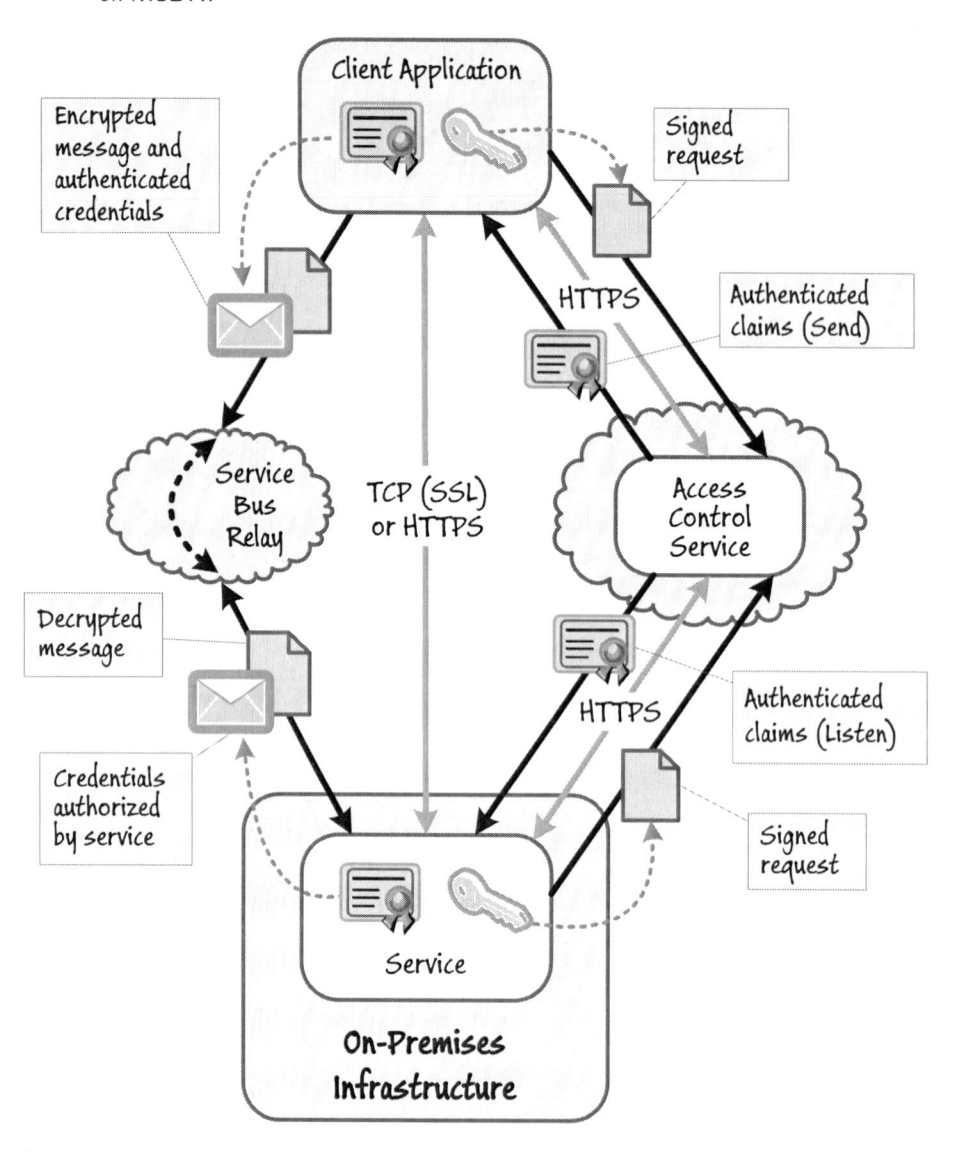

FIGURE 9
Recommendations for protecting services exposed through Windows Azure Service Bus Relay

Many organizations implement outbound firewall rules that are based on IP address allow-listing. In this configuration, to provide access to Service Bus or ACS you must add the addresses of the corresponding Windows Azure services to your firewall. These addresses vary according to the region hosting the services, and they may also change over time, but the following list shows the addresses for each region at the time of writing:

- **Asia (SouthEast):** 207.46.48.0/20, 111.221.16.0/21, 111.221.80.0/20
- **Asia (East):** 111.221.64.0/22, 65.52.160.0/19
- **Europe (West):** 94.245.97.0/24, 65.52.128.0/19
- **Europe (North):** 213.199.128.0/20, 213.199.160.0/20, 213.199.184.0/21, 94.245.112.0/20, 94.245.88.0/21, 94.245.104.0/21, 65.52.64.0/20, 65.52.224.0/19
- **US (North/Central):** 207.46.192.0/20, 65.52.0.0/19, 65.52.48.0/20, 65.52.192.0/19, 209.240.220.0/23
- **US (South/Central):** 65.55.80.0/20, 65.54.48.0/21, 65.55.64.0/20, 70.37.48.0/20, 70.37.64.0/18, 65.52.32.0/21, 70.37.160.0/21

> IP address allow-listing is not really a suitable security strategy for an organization when the target addresses identify a massively multi-tenant infrastructure such as Windows Azure (or any other public cloud platform, for that matter).

Guidelines for Naming Services in Windows Azure Service Bus Relay

If you have a large number of services, you should adopt a standardized convention for naming the endpoints for these services. This will help you manage, protect, and monitor services and the client applications that connect to them. Many organizations commonly adopt a hierarchical approach. For example, if Trey Research had sites in Chicago, New York, and Washington, each of which provided ordering and shipping services, an administrator might register URIs following the naming convention shown in this list:

- sb://treyresearch.servicebus.windows.net/chicago/ordersservice
- sb://treyresearch.servicebus.windows.net/chicago/shippingservice
- sb://treyresearch.servicebus.windows.net/newyork/ordersservice
- sb://treyresearch.servicebus.windows.net/newyork/shippingservice
- sb://treyresearch.servicebus.windows.net/washington/ordersservice
- sb://treyresearch.servicebus.windows.net/washington/shippingservice

However, when you register the URI for a service with Windows Azure Service Bus Relay, no other service can listen on any URI scoped at a lower level than your service. What this means that if in the future Trey Research decided to implement an additional orders service for exclusive customers, they could not register it by using a URI such as sb://treyresearch.servicebus.windows.net/chicago/ordersservice/exclusive.

To avoid problems such as this, you should ensure that the initial part of each URI is unique. You can generate a new GUID for each service, and prepend the city and service name elements of the URI with this GUID. In the Trey Research example, the URIs for the Chicago services, including the exclusive orders service, could be:

- sb://treyresearch.servicebus.windows.net/B3B4D086-BEB9-4773-97D3-064B0DD306EA/chicago/ordersservice
- sb://treyresearch.servicebus.windows.net/DD986578-EAB6-FC84-5490-075F34CD8B7A/chicago/ordersservice/exclusive
- sb://treyresearch.servicebus.windows.net/A8B3CC55-1256-5632-8A9F-FF0675438EEC/chicago/shippingservice

For more information about naming guidelines for Windows Azure Service Bus Relay services, see "*AppFabric Service Bus – Things You Should Know – Part 1 of 3 (Naming Your Endpoints).*"

Selecting a Binding for a Service

The purpose of Windows Azure Service Bus Relay is to provide a safe and reliable connection to your services running on-premises for client applications executing on the other side of your corporate firewall. Once a service has registered with the Windows Azure Service Bus Relay service, much of the complexity associated with protecting the service and authenticating and authorizing requests can be handled transparently outside the scope of the business logic of the service. If you are using WCF to implement your services, you can use the same types and APIs that you are familiar with in the **System.ServiceModel** assembly. The Windows Azure SDK includes transport bindings, behaviors, and other extensions in the **Microsoft.ServiceBus** assembly for integrating a WCF service with Windows Azure Service Bus Relay.

As with a regular WCF service, selecting an appropriate binding for a service that uses Windows Azure Service Bus Relay has an impact on the connectivity for client applications and the functionality and security that the transport provides. The **Microsoft.ServiceBus** assembly provides four sets of bindings:

> If you are familiar with building services and client applications using WCF, you should find Windows Azure Service Bus Relay quite straightforward.

- HTTP bindings; **BasicHttpRelayBinding**, **WSHttpRelayBinding**, **WS2007HttpRelayBinding** and **WebHttpRelayBinding**.

    These bindings are very similar to their standard WCF equivalents (**BasicHttpBinding**, **WSHttpBinding**, **WS2007-HttpBinding**, and **WebHttpBinding**) except that they are designed to extend the underlying WCF channel infrastructure and route messages through the Windows Azure Service Bus Relay service. They offer the same connectivity and feature set as their WCF counterparts, and they can operate over HTTP and HTTPS. For example, the **WS2007HttpRelayBinding** binding supports SOAP message-level security, reliable sessions, and transaction flow. These bindings open a channel to the Windows Azure Service Bus Relay service by using outbound connections only; you do not need to open any additional inbound ports in your corporate firewall.

- TCP binding; **NetTcpRelayBinding**.

    This binding is functionally equivalent to the **NetTcpBinding** binding of WCF. It supports duplex callbacks and offers better performance than the HTTP bindings although it is less portable. Client applications connecting to a service using this binding may be required to send requests and receive responses using the TCP binary encoding, depending on how the binding is configured by the service. Although this binding does not require you to open any additional inbound ports in your corporate firewall, it does necessitate that you open outbound TCP port 808, and port 828 if you are using SSL.

    This binding also supports the hybrid mode through the **ConnectionMode** property (the HTTP bindings do not support this type of connection). The default connection mode for this binding is **Relayed**, but you should consider setting it to **Hybrid** if you want to take advantage of the performance improvements that bypassing the Windows Azure Service Bus Relay service provides. However, the NAT prediction algorithm that establishes the direct connection between the service and client application requires that you also open outbound TCP ports 818 and 819 in your corporate firewall. Finally, note that the hybrid connection mode requires that the binding is configured to implement message-level security.

> The network scheme used for addresses advertised through the **NetTcpRelayBinding** binding is **sb** rather than the **net.tcp** scheme used by the WCF **NetTcpBinding** binding. For example, the address of the Orders service implemented by Trey Research could be **sb://treyresearch.servicebus.windows.net/OrdersService**

- One-way binding; **NetOnewayRelayBinding**.

    This binding implements one-way buffered messaging. A client application sends requests to a buffer managed by the Windows Azure Service Bus Relay service which delivers the message to the service. This binding is suitable for implementing a service that provides asynchronous operations as they can be queued and scheduled by the Windows Azure Service Bus Relay service, ensuring an orderly throughput without swamping the service. However message delivery is not guaranteed; if the service shuts down before the Windows Azure Service Bus Relay service has forwarded messages to it then these messages will be lost. Similarly, the order in which messages submitted by a client application are passed to the service is not guaranteed either.

    This binding uses a TCP connection for the service, so it requires outbound ports 808 and 828 (for SSL) to be open in your firewall.

- Multicasting binding; **NetEventRelayBinding**.

    This binding is a specialized version of the **NetOnewayRelayBinding** binding that enables multiple services to register the same endpoint with the Windows Azure Service Bus Relay service. Client applications can connect using either the **NetEventRelayBinding** binding or **NetOnewayRelayBinding** binding. All communications are one-way, and message delivery and order is not guaranteed.

    This binding is ideal for building an eventing system; N client applications can connect to M services, with the Windows Azure Service Bus Relay service effectively acting as the event hub. As with the **NetOnewayRelayBinding** binding, this binding uses a TCP connection for the service, so it requires outbound ports 808 and 828 (for SSL) to be open.

Windows Azure Service Bus Relay and Windows Azure Connect Compared

There is some overlap in the features provided by Windows Azure Service Bus Relay and Windows Azure Connect. However, when deciding which of these technologies you should use, consider the following points:

- Windows Azure Service Bus Relay can provide access to services that wrap on-premises resources. These services can act as façades that implement highly controlled and selective access to the wrapped resources. Client applications making requests can be authenticated with a service by using ACS and federated security; they do not have to provide an identity that is defined within your organization's corporate Windows domain.

    Windows Azure Connect is intended to provide direct access to resources that are not exposed publicly. You protect these resources by defining ACLs, but all client applications using these resources must be provisioned with an identity that is defined within your organization's corporate Windows domain.

- Windows Azure Service Bus Relay maintains a registry of publicly accessible services within a Windows Azure namespace. A client application with the appropriate security rights can query this registry and obtain a list of services and their metadata (if published), and use this information to connect to the service and invoke its operations. This mechanism supports dynamic client applications that discover services at runtime.

    Windows Azure Connect does not support enumeration of resources; a client application cannot easily discover resources at runtime.

- Client applications communicating with a service through Windows Azure Service Bus Relay can establish a direct connection, bypassing the Windows Azure Service Bus Relay service once an initial exchange has occurred.

All Windows Azure Connect requests pass through the Windows Azure Service Bus Relay service; you cannot make direct connections to resources (although the way in which Windows Azure Connect uses the Windows Azure Service Bus Relay service is transparent).

IMPLEMENTING A RELIABLE COMMUNICATIONS CHANNEL ACROSS BOUNDARIES USING SERVICE BUS QUEUES

Service Bus queues enable you to decouple services from the client applications that use them, both in terms of functionality (a client application does not have to implement any specific interface or proxy to send messages to a receiver) and time (a receiver does not have to be running when a client application posts it a message). Service Bus queues implement reliable, transactional messaging with guaranteed delivery, so messages are never inadvertently lost. Moreover, Service Bus queues are resilient to network failure; as long as a client application can post a message to a queue it will be delivered when the service is next able to connect to the queue.

When you are dealing with message queues, keep in mind that client applications and services can both send and receive messages. The descriptions in this section therefore refer to "senders" and "receivers" rather than client applications and services.

Service Bus Messages

A Service Bus message is an instance of the **BrokeredMessage** class. It consists of two elements; the message body which contains the information being sent, and a collection of message properties which can be used to add metadata to the message.

The message body is opaque to the Service Bus queue infrastructure and it can contain any application-defined information, as long as this data can be serialized. The message body may also be encrypted for additional security. The contents of the message are never visible outside of a sending or receiving application, not even in the Management Portal.

> The data in a message must be serializable. By default the **BrokeredMessage** class uses a **DataContractSerializer** object with a binary **XmlDictionaryWriter** to perform this task, although you can override this behavior and provide your own **XmlObjectSerializer** object if you need to customize the way the data is serialized. The body of a message can also be a stream.

In contrast, the Service Bus queue infrastructure can examine the metadata of a message. Some of the metadata items define standard messaging properties that an application can set; and are used by the Service Bus queues infrastructure for performing tasks such as uniquely identifying a message, specifying the session for a message, indicating the expiration time for a message if it is undelivered, and many other common operations. Messages also expose a number of system-managed read-only properties, such as the size of the message and the number of times a receiver has retrieved the message in **Peek-Lock** mode but not completed the operation successfully. Additionally, an application can define custom properties and add them to the metadata. These items are typically used to pass additional information describing the contents of the message, and they can also be used by Service Bus to filter and route messages to message subscribers.

Guidelines for Using Service Bus Queues

Service Bus queues are perfect for implementing a system based on asynchronous messaging. You can build applications and services that utilize Service Bus queues by using the Windows Azure SDK. This SDK includes APIs for interacting directly with the Service Bus queues object model, but it also provides bindings that enable WCF applications and services to connect to queues in a similar way to consuming Microsoft Windows Message Queuing queues in an enterprise environment.

Service Bus queues enable a variety of common patterns and can assist you in building highly elastic solutions as described in the following scenarios:

- A sender needs to post one or more messages to a receiver. Messages should be delivered in order and message delivery must be guaranteed, even if the receiver is not running when the sender posts the message.

    This is the most common scenario for using a Service Bus queue and is the typical model for implementing asynchronous processing. A sender posts a message to a queue and at some later point in time a receiver retrieves the message from the same queue. A Service Bus queue is an inherently first-in-first-out (FIFO) structure, and by default messages will be received in the order in which they are sent.

> Prior to the availability of Service Bus queues, Windows Azure provided message buffers. These are still available, but they are only included for backwards compatibility. If you are implementing a new system, you should use Service Bus queues instead.

> You should also note that Service Bus queues are different from Windows Azure storage queues, which are used primarily as a communication mechanism between web and worker roles running on the same site.

The sender and receiver are independent; they may be executing remotely from each other and, unlike the situation when using Windows Azure Service Bus Relay, they do not have to be running concurrently. For example, the receiver may be temporarily offline for maintenance. The queue effectively acts as a buffer, reliably storing messages for subsequent processing. An important point to note is that although Service Bus queues reside in the cloud, both the sender and the receiver can be located elsewhere. For example, a sender could be an on-premises application and a receiver could be a service running in a partner organization.

The Service Bus queue APIs in the Windows Azure SDK are actually wrappers around a series of HTTP REST interfaces. Applications and services built by using the Windows Azure SDK, and applications and services built using technologies not supported by the Windows Azure SDK, can all interact with Service Bus queues. Figure 10 shows an example architecture where the sender is an on-premises application and the receiver is a worker role running in the cloud. In this example, the on-premises application is built using the Java programming language and is running on Linux, so it performs HTTP REST requests to post messages. The worker role is a WCF service built using the Windows Azure SDK and the Service Bus queue APIs.

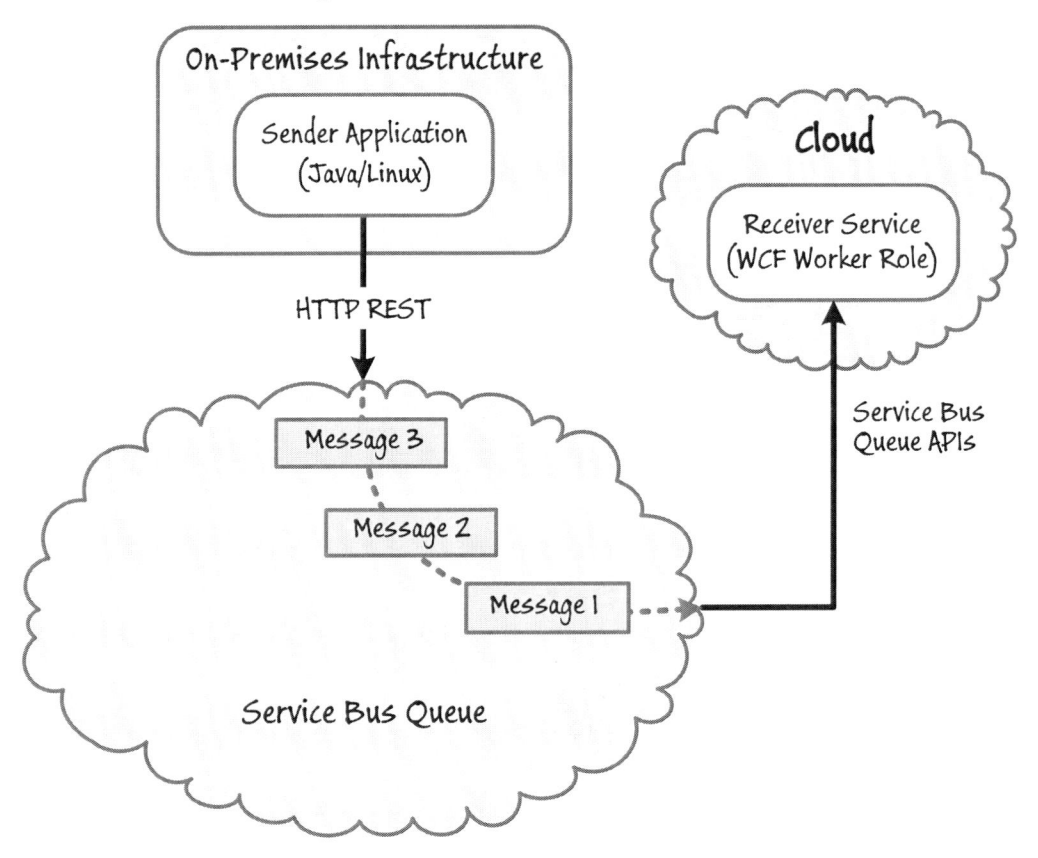

FIGURE 10
Sending and receiving messages, in order, using a Service Bus queue

- **Multiple senders running at partner organizations or in the cloud need to send messages to your system. These messages may require complex processing when they are received. The receiving application runs on-premises, and you need to optimize the time at which it executes so that you do not impact your core business operations.**

    Service Bus queues are highly suitable for batch-processing scenarios where the message-processing logic runs on-premises and may consume considerable resources. In this case, you may wish to perform message processing at off-peak hours so as to avoid a detrimental effect on the performance of your critical business components. To accomplish this style of processing, senders can post requests to a Service Bus queue while the receiver is offline. At some scheduled time you can start the receiver application running to retrieve and handle each message in turn. When the receiver has drained the queue it can shut down and release any resources it was using.

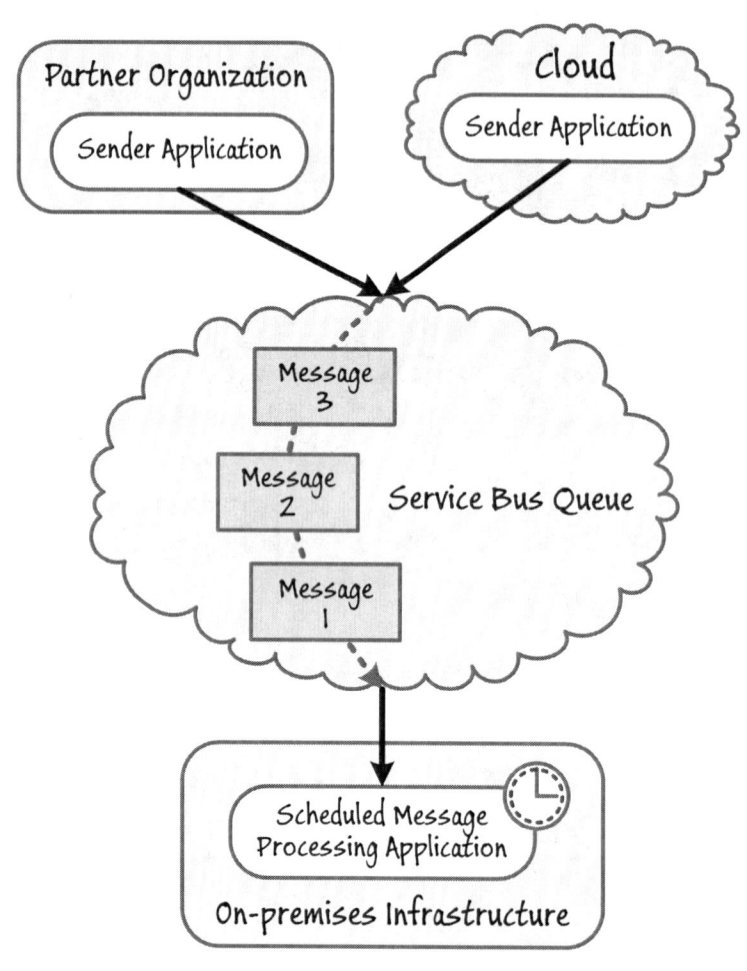

Figure 11
Implementing batch processing by using a Service Bus queue

You can use a similar solution to address the *fan-in* problem, where an unexpectedly large number of client applications suddenly post a high volume of requests to a service running on-premises. If the service attempts to process these requests synchronously it could easily be swamped, leading to poor performance and failures caused by client applications being unable to connect. In this case, you could restructure the service to use a Service Bus queue. Client applications post messages to this queue, and the service processes them in its own time. In this scenario, the Service Bus queue acts as a load-leveling technology, smoothing the workload of the receiver while not blocking the senders.

- **A sender posting request messages to a queue expects a response to these requests.**

    A message queue is an inherently asynchronous one-way communication mechanism. If a sender posting a request message expects a response, the receiver can post this response to a queue and the sender can receive the response from this queue. Although this is a straightforward mechanism in the simple case with a single sender posting a single request to a single receiver that replies with a single response, in a more realistic scenario there will be multiple senders, receivers, requests, and responses. In implementing a solution, you have to address two problems:
    - How can you prevent a response message being received by the wrong sender?
    - If a sender can post multiple request messages, how does the sender know which response message corresponds to which request?

    The answer to the first question is to create an additional queue that is specific to each sender. All senders post messages to the receiver using the same queue, but listen for the response on their own specific queues. All Service Bus messages have a collection of properties that you use to include metadata. A sender can populate the **ReplyTo** metadata property of a request message with a value that indicates which queue the receiver should use to post the response.

    All messages should have a unique **MessageId** value, set by the sender. The second problem can be handled by the sender setting the **CorrelationId** property of the response message to the value held in the **MessageId** of the original request. In this way, the sender can determine which response relates to which original request.

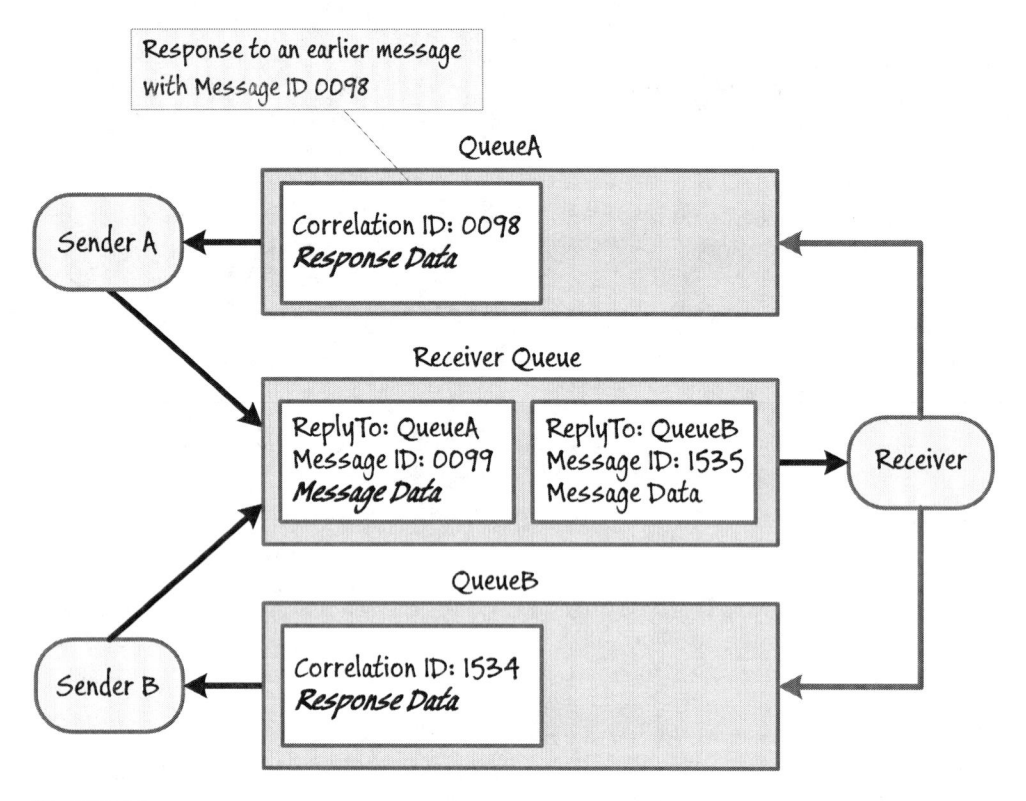

FIGURE 12
Implementing two-way messaging with response queues and message correlation

- **You require a reliable communications channel between a sender and a receiver.**

    You can extend the message-correlation approach if you need to implement a reliable communications channel based on Service Bus queues. Service Bus queues are themselves inherently reliable, but the connection to them across the network might not be, and neither might the applications sending or receiving messages. It may therefore be necessary not only to implement retry logic to handle the transient network errors that might occur when posting or receiving a message, but also to incorporate a simple end-to-end protocol between the sender and receiver in the form of acknowledgement messages.

    As a simple example, when a sender posts a message to a queue, it can wait (using an asynchronous task or background thread) for the receiver to respond with an acknowledgement. The **CorrelationId** property of the acknowledgement message should match the **MessageId** property of the original request. If no correlating response appears after a specified time interval, the sender can repost the message and wait again. This process can repeat until either the sender receives an acknowledgement, or a specified number of iterations have occurred; in which case the sender gives up and handles the situation as a failure to send the message.

However, it is possible that the receiver has retrieved the message posted by the sender and has acknowledged it, but this acknowledgement has failed to reach the sender. In this case, the sender may post a duplicate message that the receiver has already processed. To handle this situation, the receiver should maintain a list of message IDs for messages that it has handled. If it receives another message with the same ID, it should simply reply with an acknowledgement message but not attempt to repeat the processing associated with the message.

> *Do not use the duplicate detection feature of Service Bus queues to eliminate duplicate messages in this scenario. If you enable duplicate detection, repeated request or acknowledgement messages may be silently removed causing the end-to-end protocol to fail. For example, if a receiver reposts an acknowledgement message, duplicate detection may cause this reposted message to be removed and the sender will eventually abort, possibly causing the system to enter an inconsistent state; the sender assumes that the receiver has not received or processed the message while the receiver is not aware that the sender has aborted.*

- **Your system experiences spikes in the number of messages posted by senders and needs to handle a highly variable volume of messages in a timely manner.**

    Service Bus queues are a good solution for implementing load-leveling, preventing a receiver from being overwhelmed by a sudden influx of messages. However, this approach is only useful if the messages being sent are not time sensitive. In some situations, it may be important for a message to be processed within a short period of time. In this case, the solution is to add further receivers listening for messages on the same queue. This *fan-out* architecture naturally balances the load amongst the receivers as they compete for messages from the queue; the semantics of message queuing prevents the same message from being dequeued by two concurrent receivers. A monitor process can query the length of the queue, and dynamically start and stop receiver instances as the queue grows or drains.

    Senders do not have to be modified in any way as they continue to post messages to the queue in the same manner as before. This solution even works for implementing two-way messaging, as shown in Figure 13.

> You can use the Enterprise Library Autoscaling Application Block to monitor the length of a Service Bus queue and start or stop worker roles acting as receivers, as necessary.

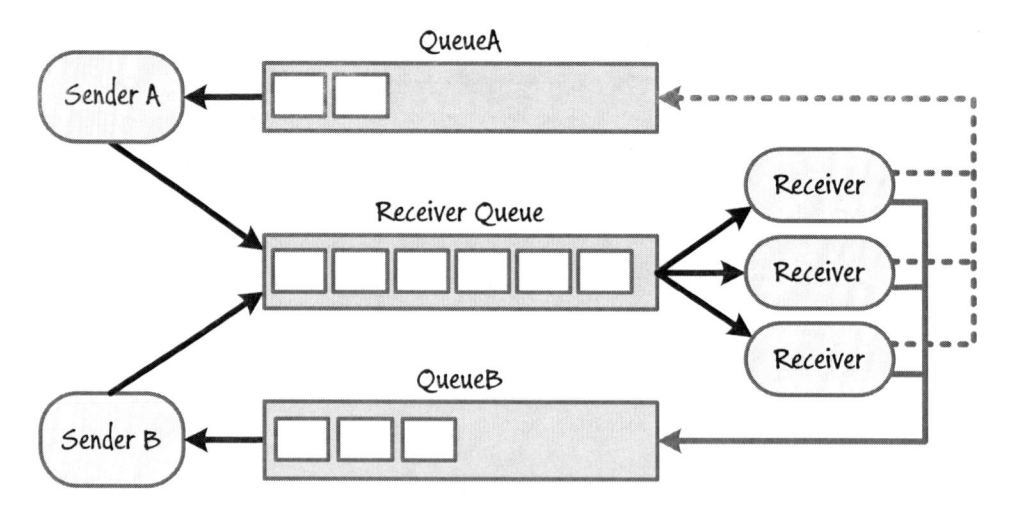

FIGURE 13
Implementing load-balancing with multiple receivers

- **A sender posts a series of messages to a queue. The messages have dependencies on each other and must be processed by the same receiver. If multiple receivers are listening to the queue, the same receiver must handle all the messages in the series.**

    In this scenario, a message might convey state information that sets the context for a set of following messages. The same message receiver that handled the first message may be required to process the subsequent messages in the series.

    In a related case, a sender might need to send a message that is bigger than the maximum message size (currently 256Kb). To address this problem, the sender can divide the data into multiple smaller messages. These messages will need to be retrieved by the same receiver and then reassembled into the original large message for processing.

    Service Bus queues can be configured to support sessions. A session consists of a set of messages that comprise a single conversation. All messages in a session must be handled by the same receiver. You indicate that a Service Bus queue supports sessions by setting the **RequiresSession** property of the queue to true when it is created. All messages posted to this queue must have their **SessionId** property set to a string value. The value stored in this string identifies the session, and all messages with the same **SessionId** value are considered part of the same session. Notice that it is possible for multiple senders to post messages with the same session ID, in which case all of these messages are treated as belonging to the same session.

> You should bear in mind that a sudden influx of a large number of requests might be the result of a denial of service attack. To help reduce the threat of such an attack, it is important to protect the Service Bus queues that your application uses to prevent unauthorized senders from posting messages. For more information, see the section "Guidelines for Securing Service Bus Queues" later in this appendix.

A receiver willing to handle the messages in the session calls the **AcceptMessageSession** method of a **QueueClient** object. This method establishes a session object that the receiver can use to retrieve the messages for the session, in much the same way as retrieving the messages from an ordinary queue. However, the **AcceptMessageSession** method effectively pins the messages in the queue that comprise the session to the receiver and hides them from all other receivers. Another receiver calling **AcceptMessageSession** will receive messages from the next available session. Figure 14 shows two senders posting messages using distinct session IDs; each sender generates its own session. The receivers for each session only handle the messages posted to that session.

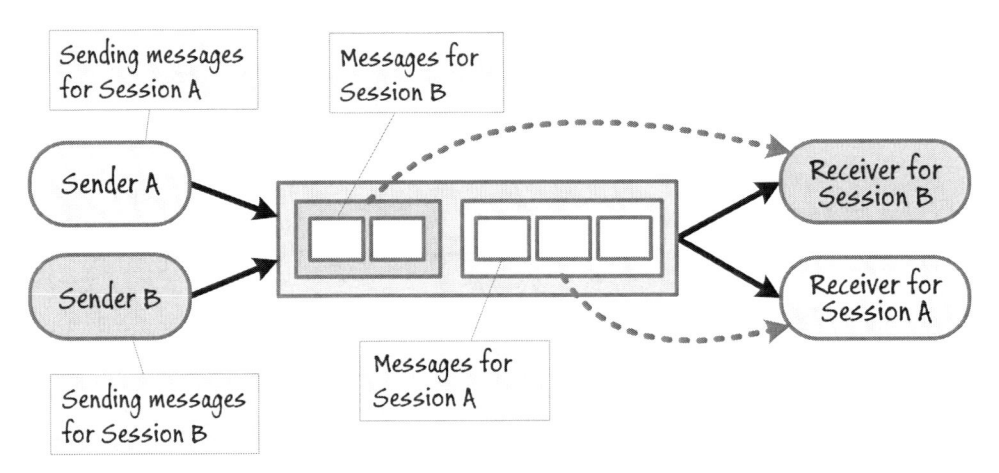

FIGURE 14
Using sessions to group messages

You can also establish duplex sessions if the receiver needs to send a set of messages as a reply. You achieve this by setting the **ReplyToSessionId** property of a response message with the value in the **SessionId** property of the received messages before replying to the sender.
The sender can then establish its own session and use the session ID to correlate the messages in the response session with the original requests.
A message session can include session state information, stored with the messages in the queue. You can use this information to track the work performed during the message session and implement a simple finite state machine in the message receiver. When a receiver retrieves a message from a session, it can store information about the work performed while processing the message in the session state and write this state information back to the session in the queue. If the receiver should fail or terminate unexpectedly, another instance can connect to the session, retrieve the session state information, and continue the necessary processing where the previous failed receiver left off. Figure 15 illustrates this scenario.

> Use the **GetState** and **SetState** methods of a **MessageSession** object to retrieve and update the state information for a message session.

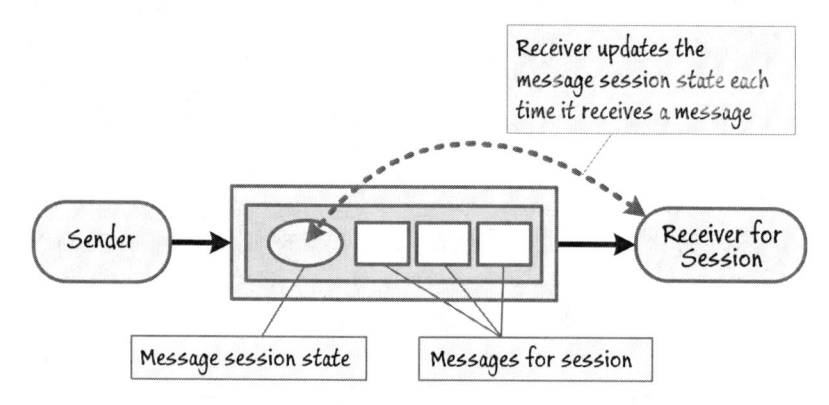

FIGURE 15
Retrieving and storing message session state information

It is possible for a session to be unbounded—there might be a continuous stream of messages posted to a session at varying and sometimes lengthy intervals. In this case, the message receiver should be prepared to hibernate itself if it is inactive for a predefined duration. When a new message appears for the session, another process monitoring the system can reawaken the hibernated receiver which can then resume processing.

- **A sender needs to post one or more messages to a queue as a singleton operation. If some part of the operation fails, then none of the messages should be sent and they must all be removed from the queue.**

  The simplest way to implement a singleton operation that posts multiple messages is by using a local transaction. You initiate a local transaction by creating a **TransactionScope** object. This is a programmatic construct that defines the scope for a set of tasks that comprise a single transaction.

  To post a batch of messages as part of the same transaction you should invoke the send operation for each message within the context of the same **TransactionScope** object. In effect, the messages are simply buffered and are not actually sent until the transaction completes. To ensure that all the messages are actually sent, you must complete the transaction successfully. If the transaction fails, none of the messages are sent but are instead removed from the queue. For more information about the **TransactionScope** class, see the topic *"TransactionScope Class"* on MSDN.

If you are sending messages asynchronously (the recommended approach), it may not be feasible to send messages within the context of a **TransactionScope** object. Note that if you are incorporating operations from other transactional sources, such as a SQL Server database, then these operations cannot be performed within the context of the same **TransactionScope** object; the resource manager that wraps Service Bus queues cannot share transactions with other resource managers.

In these scenarios, you can implement a custom-pseudo transactional mechanism based on manual failure detection, retry logic, and the duplicate message elimination (*dedupe*) of Service Bus queues.

To use dedupe, each message that a sender posts to a queue should have a unique message ID. If two messages are posted to the same queue with the same message ID, both messages are assumed to be identical and duplicate detection will remove the second message. Using this feature, in the event of failure in its business logic, a sender application can simply attempt to re-send a message as part of its failure/retry processing. If the message had been successfully posted previously the duplicate will be eliminated; the receiver will only see the first message. This approach guarantees that the message will always be sent at least once (assuming that the sender has a working connection to the queue) but it is not possible to easily withdraw the message if the failure processing in the business logic determines that the message should not be sent at all.

> If you attempt to use a **TransactionScope** object to perform local transactions that enlist a Service Bus queue and other resource managers, your code will throw an exception.

> You enable duplication detection by setting the **RequiresDuplicateDetection** property of the queue to true when you create it. It is not possible to change the value of this property on a queue that already exists. Additionally, you should set the **DuplicateDetectionHistoryTimeWindow** property to a **TimeSpan** value that indicates the period during which duplicate messages for a given message ID are discarded; if a new message with an identical message ID appears when this period has expired then it will be queued for delivery.

- **A receiver retrieves one or more messages from a queue, again as part of a transactional operation. If the transaction fails, then all messages must be replaced into the queue to enable them to be read again.**

    A message receiver can retrieve messages from a Service Bus queue by using one of two receive modes; **ReceiveAndDelete** and **PeekLock**. In **ReceiveAndDelete** mode, messages are removed from the queue as soon as they are read. In **PeekLock** mode, messages are not removed from the queue as they are read, but rather they are locked to prevent them from being retrieved by another concurrent receiver, which will instead retrieve the next available unlocked message. If the receiver successfully completes any processing associated with the message it can call the **Complete** method of the message, which removes the message from the queue. If the receiver is unable to handle the message successfully, it can call the **Abandon** method, which releases the lock but leaves the message on the queue. This approach is suitable for performing asynchronous receive operations.

    As with message send operations, a message receive operation performed using **PeekLock** mode can also be performed synchronously as part of a local transaction by defining a **TransactionScope** object, as long as the transaction does not attempt to enlist any additional resource managers. If the transaction does not complete successfully, all messages received and processed within the context of the **TransactionScope** object will be returned to the queue.

- **A receiver needs to examine the next message on the queue but should only dequeue the message if it is intended for the receiver.**

    In this scenario, the receiver can retrieve the message by using the **PeekLock** receive mode, copy the message into a local buffer and examine it. If the message is not intended for this receiver, it can quickly call the **Abandon** method of the message to make it available to another receiver.

    If a message contents are confidential and should only be read by a specific receiver, the sender can encrypt the message body with a key that only the correct receiver knows. The sender can also populate the **To** property of the message with an identifier that specifies the correct receiver. Message properties are not encrypted, so any receiver can retrieve the message, but if it does not recognize the address in the **To** property it will probably not have the appropriate key to decrypt the message contents, so it can abandon the message and leave it for the correct receiver.

> The **ReceiveAndDelete** receive mode provides better performance than the **PeekLock** receive mode, but **PeekLock** provides a greater degree of safety. In **ReceiveAndDelete** mode, if the receiver fails after reading the message then the message will be lost. In **PeekLock** mode, if the receive operation or message processing are not successfully completed, the message can be abandoned and returned to the queue from where it can be read again.
> The default receive mode for a Service Bus queue is **PeekLock**.

> Only **PeekLock** mode respects local transactions; **ReceiveAndDelete** mode does not.

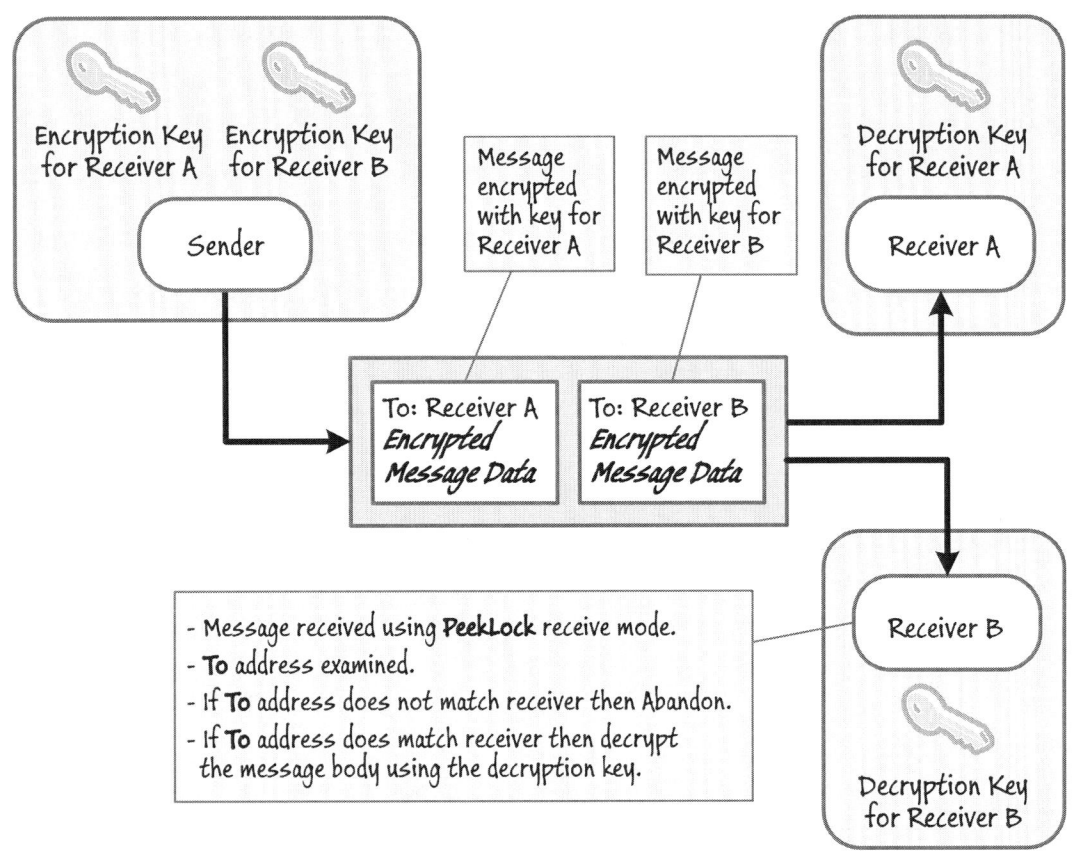

IMPLEMENTING CROSS-BOUNDARY COMMUNICATION 283

Figure 16
Using PeekLock with encryption to examine messages without dequeueing

Guidelines for Sending and Receiving Messages Using Service Bus Queues

You can implement the application logic that sends and receives messages using a variety of technologies:

- You can use the Service Bus queue APIs in the Windows Azure SDK. For more information and good practices for following this approach, see *"Best Practices for Leveraging Windows Azure Service Bus Brokered Messaging API"* on MSDN.

> As an alternative approach to implementing this scenario, you could consider using a Service Bus topic with a separate Service Bus subscription for each receiver. However, using subscriptions can become unwieldy and difficult to manage if there are a large or variable number of receivers.

You can also use the **Queue-Client** class in the **Microsoft.ServiceBus.Messaging** namespace to connect to a queue and send and receive messages. The **QueueClient** type is an abstract class that implements a superset of the functionality available in the **MessageSender** and **MessageReceiver** classes. The Windows Azure SDK provides additional types for sending messages to topics (**TopicClient**) and receiving messages from subscriptions (**SubscriptionClient**). However, the **MessageSender** and **MessageReceiver** classes abstract the differences between these types. For example, if you use a **MessageSender** to send and receive messages using queues, you can switch to using topics with minimal modifications to your code. Similarly, a **MessageReceiver** object enables you to retrieve messages from a queue and a subscription using the same code.

However, before you modify all of your existing code to use **MessageSender** and **MessageReceiver** objects, be aware that not all of the functionality implemented by the **QueueClient**, **TopicClient**, and **SubscriptionClient** types is available in the **MessageSender** and **MessageReceiver** classes. For example, the **MessageReceiver** class does not support sessions.

- You can use the Service Bus queue bindings to connect to a queue from WCF client applications and services. For more information, see *"How to: Build an Application with WCF and Service Bus Queues"* on MSDN.
- If you are building applications that connect to Service Bus queues by using a technology that does not support the Windows Azure SDK, you can use the HTTP REST interface exposed by Service Bus.

Sending and Receiving Messages Asynchronously
If you are using the Windows Azure SDK, you can implement applications that send and receive messages by using the **MessageSender** and **MessageReceiver** classes in the **Microsoft.ServiceBus.Messaging** namespace. These types expose the messaging operations described earlier in this appendix. The basic functionality for sending and receiving messages is available through the **Send** and **Receive** methods of these types. However, these operations are synchronous. For example, the **Send** method of the **MessageSender** class waits for the send operation to complete before continuing, and similarly the **Receive** method of the **MessageReceiver** class either waits for a message to be available or until a specified timeout period has expired. Remember that these methods are really just façades in front of a series of HTTP REST requests, and that the Service Bus queue is a remote service being accessed over the Internet. Therefore, your applications should assume that:

- **Send and receive operations may take an arbitrarily long time to complete, and your application should not block waiting for these operations to finish.**

    The **MessageSender** class exposes an asynchronous version of the **Send** method, and a **MessageReceiver** class provides an asynchronous implementation of the **Receive** method through the **BeginSend/EndSend** and **BeginReceive/EndReceive** method pairs respectively. You should use these methods in preference to their synchronous counterparts. These methods follow the standard asynchronous pattern implemented throughout the .NET Framework.

    The same issues arise with other operations, such as determining whether a queue exists, creating and deleting a queue, connecting to a queue, and querying the length of a queue. Therefore, you should perform these operations following the same robust, asynchronous approach.

- **A sender can post messages at any time, and a receiver may need to listen for messages on more than one queue.**

    There are many possible solutions to this problem, but the most common approaches involve using a thread or task to wait for messages on each queue, and triggering an event when a message is available. Application code that catches this event can then process the message. For example, you can define an **async** method that makes use of the **await** operator available in the Visual C#® language to create a series of tasks that wait for a message on each possible queue and raise an event. You can then use a framework such as the Microsoft Reactive Extensions to catch these events and process messages as they become available.

- **Send and receive operations could fail for a variety of reasons.**

    These can include a failure in connectivity between your application and the Service Bus in the cloud, a security violation caused by a change in the security implemented by the Service Bus queue (an administrator might decide to revoke or modify the rights of an identity for some reason), the queue being full (they have a finite size), and so on. Some of these failures might the result of transient errors, while others may be more permanent Asynchronous send and receive operations must incorporate the appropriate cancellation handling to enable any background threads to be tidied up appropriately and messaging resources released.

    When you retrieve a message from a queue, you can read it by using the **GetBody** method of the **BrokeredMessage** class. This method deserializes the message body. You can only deserialize this data once. This is an important factor to bear in mind when designing your fault-handling logic. If you attempt to call the **GetBody** method on the same message again (inside an exception handler, for example), it will fail. Therefore, if you anticipate requiring repeated access to the data in the message body, you should store this data in an appropriate object and use this object instead.

*If a receiver expects to handle multiple messages as part of a business operation, you can optimize the receiving process by using the prefetch functionality of the **QueueClient** class.*

*By default, when a **QueueClient** object performs the **Receive** method, only the next available message is taken from the queue. You can, however, set the **PrefetchCount** property of the **QueueClient** object to a positive integer value, and the **Receive** method will actually pull the specified number of messages from the queue (if they are available). The messages are buffered locally with the receiving application and are no longer available to other receivers. The **Receive** method then returns the first message from this buffer. Subsequent calls to **Receive** retrieve the remaining messages from the buffer until it is empty, when the next **Receive** operation will fetch the next batch of messages from the queue and buffer them. This approach makes more efficient use of the network bandwidth at the cost of lengthening the time taken to retrieve the first message. However, subsequent messages are returned much more quickly.*

Prefetched messages are subject to the same timeout semantics as unbuffered messages. If they are not processed within the timeout period starting from when they are fetched from the queue, then the receiver is assumed to have failed and the messages are returned to the queue. Therefore, if you implement prefetching, you should only buffer sufficient messages that the receiver can retrieve and process within this timeout period.

Scheduling, Expiring, and Deferring Messages

By default, when a sender posts a message to a queue, it is immediately available for a receiver to retrieve and process. However, you can arrange for a message to remain invisible when it is first sent and only appear on the queue at a later time. This technique is useful for scheduling messages that should only be processed after a particular point in time; for example, the data could be time sensitive and may not be released until after midnight. To specify the time when the message should appear on the queue and be available for processing, set the **ScheduledEnqueueTimeUtc property** of the **BrokeredMessage** object.

When a sender posts a message to a queue, that message might wait in the queue for some considerable time before a receiver picks it up. The message might have a lifetime after which it becomes stale and the information that it holds is no longer valid. In this case, if the message has not been received then it should be silently removed from the queue. You can achieve this by setting the **TimeToLive** property of the **BrokeredMessage** object when the sender posts the message.

In some situations, an application may not want to process the next available message but skip over it, retrieve subsequent messages, and only return to the skipped message later. You can achieve this by deferring the message, using the **Defer** method of the **BrokeredMessage** class. To implement this mechanism, an application must retrieve messages by using **PeekLock** mode. The **Defer** method leaves the message on the queue, but it is locked and unavailable to other receivers. At the appropriate juncture, the application can return to the message to process it, and then finish by calling the **Complete** or **Abandon** methods as described earlier in this appendix. In the event that a message is no longer useful or valid at the time that it is processed, the application can optionally dead letter it. Note that if the application fails, the lock eventually times out and the message becomes available in the queue. You can specify the lock duration by setting the **LockDuration** property of the queue when it is created.

Guidelines for Securing Service Bus Queues

Service Bus queues provide a messaging infrastructure for business applications. They are created and managed by Windows Azure, in the cloud. Consequently they are reliable and durable; once a sender has posted a message to a queue it will remain on the queue until it has been retrieved by a receiver or it has expired.

A Service Bus queue is held in a Service Bus namespace identified by a unique URI. You establish this URI when you create the namespace, and the URI structure is similar to that described in the section "Windows Azure Service Bus Relay Security Model" earlier in this appendix. An application instantiates a **MessagingFactory** object using this URI. The **MessagingFactory** object can then be used to create a **MessageSender** or **MessageReceiver** object that connects to the queue.

The Service Bus namespace provides the protection context for a queue, and the namespace holding your queues should only be made available to authenticated senders and receivers. You protect namespaces by using ACS, in a manner very similar to that described in the section "Guidelines for Securing Windows Azure Service Bus Relay" earlier in this appendix, except that the realm of the relying party application is the URI of the Service Bus namespace with the name of the Service Bus queue, topic, or subscription appended (such as http://treyresearch.servicebus.windows.net/orderstatusupdatequeue) rather than the address of a WCF service.

You can create an ACS rule group for this URI and assign the **net.windows.servicebus.action** claim type values **Send**, **Listen**, and **Manage** to authenticated identities, as appropriate. You should note that the **Send** and **Listen** claims each confer a very minimal set of privileges, enabling an application to post messages to a queue or retrieve messages from a queue respectively, but very little else. If your application needs to perform tasks such as creating a new queue, querying the number of messages currently posted to a queue, or even simply determining whether a queue with a given name exists, the application must run with an identity that has been granted the rights associated with the **Manage** claim.

All communications with a Service Bus queue occur over a TCP channel, encrypted by using SSL. If you need to implement additional security at the message level, you should encrypt the contents of messages and the receiver should be provided with the decryption key. In this way, if a message is somehow intercepted by a rogue receiver it will not be able to examine the contents of the message. Similarly, if the valid receiver of a message is not able to decrypt that message, it should be treated as a poison message from a rogue sender and moved to the dead letter queue.

> *You can also implement a mechanism to verify the identity of a sender posting a message to a Service Bus queue by adding an identity token to the header of the message. If this token is missing or unrecognized by the receiving application, the message should be treated as suspect. For an example of how to implement this approach, see the section "Securing Messages" in Chapter 4, "Implementing Reliable Messaging and Communications with the Cloud."*

More Information

All links in this book are accessible from the book's online bibliography available at: *http://msdn.microsoft.com/en-us/library/hh968447.aspx*.

- "Developing Applications for the Cloud on the Microsoft Windows Azure Platform" on CodePlex at *http://wag.codeplex.com* and "Developing Applications for the Cloud (2nd Edition)" on MSDN at *http://msdn.microsoft.com/en-us/library/ff966499.aspx*.
- Windows Azure Connect Team Blog at *http://blogs.msdn.com/b/windows_azure_connect_team_blog*.
- "Windows Azure Connect" at *http://msdn.microsoft.com/en-us/library/gg433122.aspx*.
- "Securing and Authenticating a Service Bus Connection" at *http://msdn.microsoft.com/en-us/library/dd582773.aspx*.
- "AppFabric Service Bus – Things You Should Know – Part 1 of 3 (Naming Your Endpoints)" at *http://windowsazurecat.com/2011/05/appfabric-service-bus-things-you-should-know-part-1-of-3-naming-your-endpoints*.
- "TransactionScope Class" at *http://msdn.microsoft.com/en-us/library/system.transactions.-transactionscope.aspx*.
- "Best Practices for Leveraging Windows Azure Service Bus Brokered Messaging API" at *http://msdn.microsoft.com/en-us/library/hh545245(v=VS.103).aspx*.
- "How to: Build an Application with WCF and Service Bus Queues" at *http://msdn.microsoft.com/en-us/library/windowsazure/hh243674.aspx*.

APPENDIX D Implementing Business Logic and Message Routing across Boundaries

A simple, reliable messaging strategy enables secure point-to-point communications between components participating in a distributed system. The implementation determines the degree of independence that message senders and receivers have from each other, but the system still needs some means for physically directing messages to a destination. In many solutions, this mechanism may be built into the application logic; a sender using RPC-style communications to communicate with a Windows Communication Foundation (WCF) service might specify the address of the destination, or an application implementing message-oriented communications might direct messages to a queue that a specific receiver listens on. This opaque approach can make it difficult to change the way in which messages are routed if destinations change their location, or new destinations are added, without reworking the underlying business logic.

Decoupling the data flow from the business logic of your applications brings many benefits. For example, by following this strategy you can transparently scale your solution to incorporate additional service instances to process messages during times of heavy load, you can monitor and audit messages without interrupting the natural progression from sender to receiver, or you can easily integrate additional services into your solution by extending the list of message destinations.

This appendix examines some of the common challenges associated with directing and controlling the flow of messages, and presents possible solutions and good practice for decoupling this data flow from the application logic when using the Windows Azure™ technology platform.

Use Cases and Challenges

Many hybrid applications must process business rules or workflows that contain conditional tests, and which result in different actions based on the results of the rules. For example, an application may need to update a stock database, send the order to the appropriate transport and warehouse partner, perform auditing operations on the content of the order (such as checking the customer's credit limit), and store the order in another database for accounting purposes. These operations may involve services and resources located both in the cloud and on-premises. Building an extensible solution based on these use cases typically requires that you address the specific challenges described in the following sections.

Separating the Business Logic from Message Routing

Description: As part of its business processing, an application needs to send and receive messages from other services which may be located anywhere.

> Designing the data flow to be independent from the implementation of the application logic can help to ensure that your solution is adaptable and extensible if business requirements should quickly change, and scalable as the traffic to your services increases.

The business operations performed by a distributed application are primarily concerned with gathering, examining, and processing data. The gathering and processing aspects may involve sending requests and passing messages to other services. However, the underlying business logic should not be tied to the location of these services; if a service migrates to a different address, you should not have to modify the way in which the application works if it is still performing the same business functions. Separating the business logic from the message routing helps to achieve this location independence.

Decoupling the message routing from the application logic also enables you to partition messages based on administrator-defined criteria, and then route these messages to the most appropriate instance of a service. These criteria are independent of the application. For example, to reduce response times an administrator might decide to run multiple instances of a service that processes mortgage requests in the cloud. An application running on a mortgage advisor's desktop can submit mortgage requests, and these requests can be transparently routed to a specific instance of the mortgage processing service based on some attribute of the mortgage advisor such as their assigned advisor number; requests from advisors 1 through 30 might be directed to one instance of the service, requests from advisors 31 to 60 might be directed to another instance, and so on. As the mortgage application is rolled out in new offices and more advisors come on line, the administrator can monitor how the workload of the services is divided and, if necessary, run more instances of the service and reconfigure the system to distribute mortgage processing requests more evenly.

You must decide where the logic that controls the routing of messages will reside. If you are aiming to completely decouple this flow from the business logic of the application it should not be managed by the code that sends messages, and equally it should not be incorporated into the elements that receive and process messages. This implies that the data flow logic must be contained within the middleware components that connect senders to receivers.

The middleware elements effectively act as a message broker, intelligently routing messages to destinations. The approach that you take to implement this broker should provide configurable routing logic that is independent from components sending and receiving messages, and be robust and compatible with whichever approach you have taken to implement reliable messaging.

Routing Messages to Multiple Destinations

Description: As part of its business processing, an application may need to send the same message to any number of services, and this list of services may vary over time.

It is frequently necessary to transmit the same message to different receivers, such as an order shipping service and a stock control service in an order processing system. Being able to transparently route messages to multiple destinations enables you to build extensible solutions. If your solution needs to incorporate new partners or services, such as an auditing service for example, you can easily include the additional destinations without modifying the applications that send messages to them.

This use case requires that the middleware components that route messages can transparently copy them and send them to the appropriate destinations. This in turn means that the middleware elements must be configurable to support this message duplication without introducing any dependencies on the sending or receiving logic.

Cross-Cutting Concerns

Message routing has a high dependency on the underlying messaging platform, and the cross-cutting concerns relating to message routing are essentially a superset of those that you must address when implementing cross-boundary communications. The following sections summarize these areas of concern.

Security

The messaging infrastructure must be robust and secure; it should prevent unauthorized applications and services from sending or receiving messages.

The messages must be protected and made available only to their intended recipients. This restriction also applies to the routing technology you use; the middleware implementing the router should be able to make decisions on where to send a message without needing to examine the data in the body of the message. If message confidentiality is paramount, then you should be able to encrypt the message body without losing the ability to route messages.

If your solution copies and dispatches messages to multiple destinations, message duplication must be performed in a controlled and secure manner, again without the middleware requiring direct access to the data held in message bodies.

Reliability

The underlying messaging technology must not lose messages as they are transmitted, and the routing mechanism must reliably arrange for messages to be sent to the correct destination(s).

Responsiveness and Availability

The messaging platform should not inhibit the flow of the business logic of your system. If the business logic sends messages that cannot be delivered immediately because of some transient problem in the infrastructure (such as a network failure between the messaging platform and the destination service), the business logic should be allowed to continue. The messages should be delivered transparently when the issue is resolved.

Interoperability

You may need to route messages between services built by using different technologies. The routing mechanism should be compatible with the technologies on which these services are based.

Windows Azure Technologies for Routing Messages

The primary Windows Azure technology that supports safe, reliable, responsive, and interoperable messaging between distributed services and applications is Service Bus queues.

A Service Bus queue is a simple first-in-first-out structure with additional features such as timeouts, transactional support, and dead-lettering. A Service Bus queue enables a variety of common messaging scenarios as described in "Appendix C - Implementing Cross-Boundary Communication." Windows Azure extends Service Bus queues with Service Bus topics and Service Bus subscriptions. These extensions enable you to incorporate a range of message routing options into your solutions, taking advantage of the security and reliability that Service Bus queues provide.

The following sections provide more detail on using Service Bus topics and subscriptions to implement message routing.

Separating the Business Logic from Message Routing Using Service Bus Topics and Subscriptions

Service Bus topics and subscriptions enable you to direct messages to different receivers based on application-defined criteria. They provide the advantages exhibited by Service Bus queues facilitating decoupling a message sender from a receiver but, in addition, they enable messages to be routed to one or more receivers by using information stored in the metadata of these messages.

A sender application posts a message to a Service Bus topic using much the same technique as when posting to a Service Bus queue. However, the sender typically adds one or more custom properties to the metadata of the message, and this information is used to route the message to the most appropriate receiver. While a Service Bus topic represents the sending end of a queue, a Service Bus subscription represents the receiving end; a receiver application waits for incoming messages by connecting to a Service Bus subscription. A Service Bus topic can be associated with multiple Service Bus subscriptions.

All subscriptions have an associated filter. If you don't specify how to filter data when you create a subscription, the default filter simply passes all messages from the topic through to the subscription.

A message is routed from a Service Bus topic to a Service Bus subscription by defining a filter that examines the metadata and custom properties attached the message. A filter is a predicate attached to a Service Bus subscription, and all messages that match the predicate are directed towards that Service Bus subscription. Filters enable you to define simple message routing rules that might otherwise require writing a substantial amount of code.

Guidelines for Using Service Bus Topics and Subscriptions to Route Messages

Service Bus topics and subscriptions are suitable for implementing simple, static routing of messages from a sender to a receiver. The filters that direct messages from a topic to a subscription are separate from the business logic of the sending and receiving applications. All a sender has to do is provide the metadata (in the form of message properties and values) that the filters can examine and use to make routing decisions. Figure 1 depicts the message flow from a sender though a topic and three subscriptions to the corresponding receivers. The data represents parcels being shipped, and messages are filtered by the **Weight** property added by the sender; Receiver A receives all messages where the value of the **Weight** property is less than 100, Receiver B receives messages for weights between 100 and 199, and Receiver C receives messages for all weights of 200 and over.

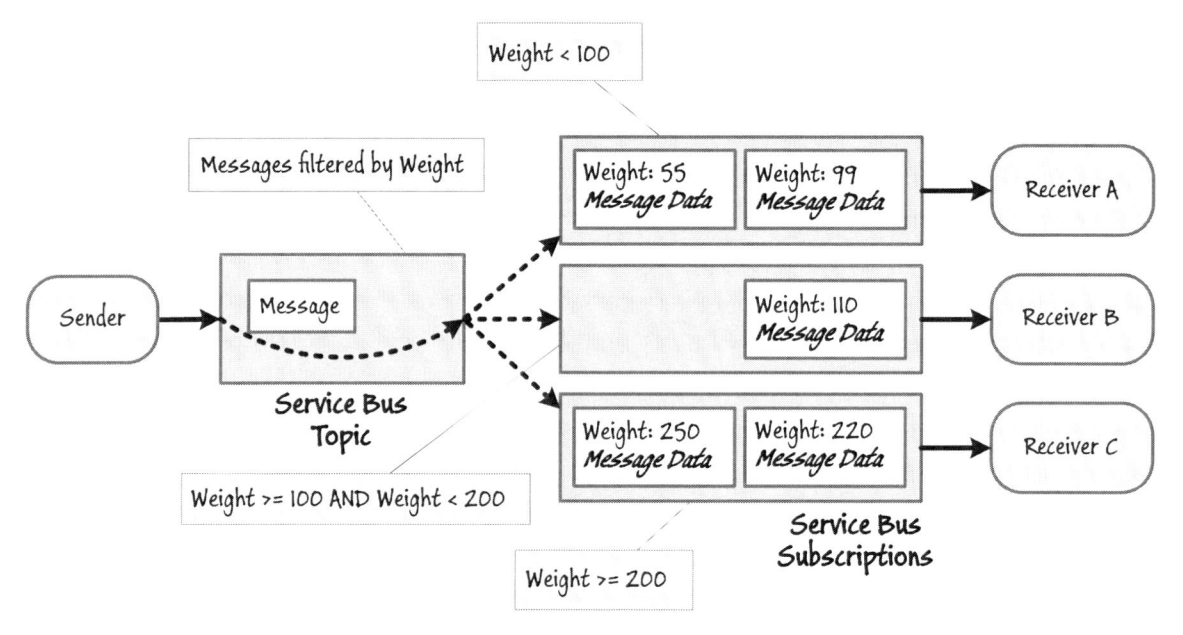

FIGURE 1
Routing messages to different receivers through a Service Bus topic and subscriptions

Service Bus topics and subscriptions expose a programmatic model through a series of APIs in the Windows Azure SDK. Like Service Bus queues, these APIs are wrappers around a series of REST interfaces, so you can utilize topics and subscriptions from technologies and platforms not directly supported by the Windows Azure SDK.

Service Bus topics and subscriptions enable you to address a number of scenarios with requirements beyond those that you can easily implement by using Service Bus queues, as follows:
- Your system generates a number of messages, each of which must be handled by a specific receiver. New receivers may be added over time, but for ease of maintenance you don't want to have to modify the business logic of the sender application when this occurs. This is the primary scenario for using Service Bus topics and subscriptions rather than queues.

> You should factor out the logic that determines the message routing from the main business logic of the application. In this way, if the algorithm that defines the routing changes, the business logic of the application does not have to be updated.

As an example, consider an order processing system where customers place orders using a web application. These orders must be fulfilled and shipped to customers. The company uses a number of transport partners to deliver goods, and the delivery partner selected depends on the location of the customer. The web application is not actually concerned with which transport partner is used, but simply posts the relevant details of each order to a Service Bus topic together with metadata that indicates the location of the customer. Occasionally new transport partners may be added or existing partners removed, but the business logic in the orders web application should not have to change when this happens. Each transport partner has its own Service Bus subscription to this topic, with a filter that examines the location metadata and directs the order message to the subscription as appropriate.

If your application communicates with third-party organizations, it must be able to interact with the systems used by that partner; these systems are unlikely to be based on Service Bus subscriptions. For example, international commercial transport providers typically have their own custom systems based on exposed web services that you must use when interacting with their systems. Therefore it may be necessary to construct a set of adapters that retrieve the messages for each partner from the appropriate Service Bus subscription, translate the messages into the format expected by the partner, and then communicate with the partner using its own web service interface. This is the *push* model for messaging, and implementing these adapters is your responsibility. The Service Bus subscriptions these adapters use do not have to be exposed to the world outside of your organization, so the authentication and authorization requirements can be handled directly by using Service Bus security and the Windows Azure Access Control Service (ACS).

Alternatively, if a partner does not publish a programmatic interface for interacting with its systems but is willing to host the logic for connecting to the Service Bus subscription on its own premises, it can connect directly to the Service Bus subscription, retrieve and reformat messages into a style compatible with its own internal processes, and then invoke these processes. This is the *pull* model for messaging. The logic for communicating with Service Bus can be factored out into a separate connector component to provide ease of maintenance should the mechanism used to communicate with your application change in the future.

You must make the Service Bus subscription endpoint accessible to the transport partner. This may necessitate implementing federated security across the Service Bus and the transport partner's own security domain.

Figure 2 shows the architecture of a possible solution based on communicating with three commercial transport providers. The partners for locations A and C expose functionality as a set of web services, so you must use adapters to communicate with them. The transport partner for location B does not publish a public interface for its services, but instead implements its own connector logic for pulling messages from the Service Bus subscription.

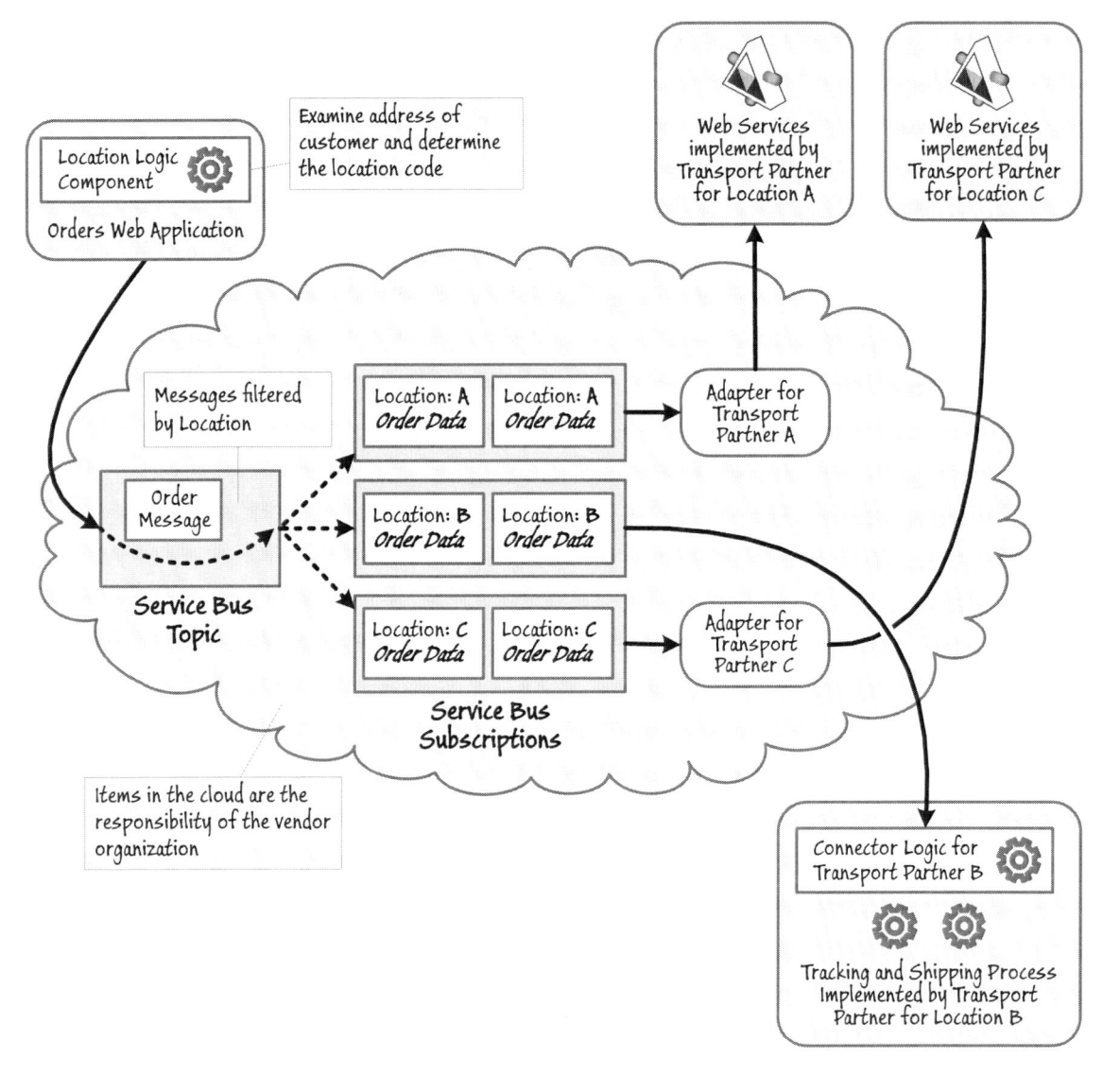

Figure 2
Decoupling a sender application from the message routing logic using a Service Bus topic and subscriptions

- Your system generates a variety of messages. Some of these message are high priority and must be processed as soon as possible, others are less urgent and can be handled at some later convenient time, while still further messages have a limited lifetime; if they are not handled within a specified timeframe they should simply be discarded.

Service Bus topics and subscriptions provide a mechanism to implement a *priority queue*. When a sender posts a message, it can tag it with a **Priority** property, and you can define subscriptions that filter messages by this property.

Urgent messages can be handled by a subscription with a pool of message receivers. You can monitor the queue length of this subscription, and if it exceeds some predefined length you can start new listeners. This technique is similar to the *fan-out* architecture for queues handling a sudden influx of messages described in the section "Guidelines for Using Service Bus Queues" in "Appendix C - Implementing Cross-Boundary Communication."

Lower priority messages can be handled by a subscription with a fixed number of receivers, implementing a *load-leveling* system as described in Appendix C.

Messages with a limited lifetime can be handled by using a subscription with a short value for the **DefaultMessageTimeToLive** property. Additionally, if no trace of the message is required after it has expired, you can set the **EnableDeadLetteringOnMessageExpiration** property of the subscription to false. In this configuration, if a message expires before it is received it will automatically be discarded.

Figure 3 shows the structure of this system. In this example, messages marked as **Critical** are high priority and must be processed immediately, messages marked as **Important** must be processed soon (ideally within the next 10 minutes), while messages marked as **Information** are non-urgent and if they are not handled within the next 20 minutes the data that they contain will be redundant and they can be discarded.

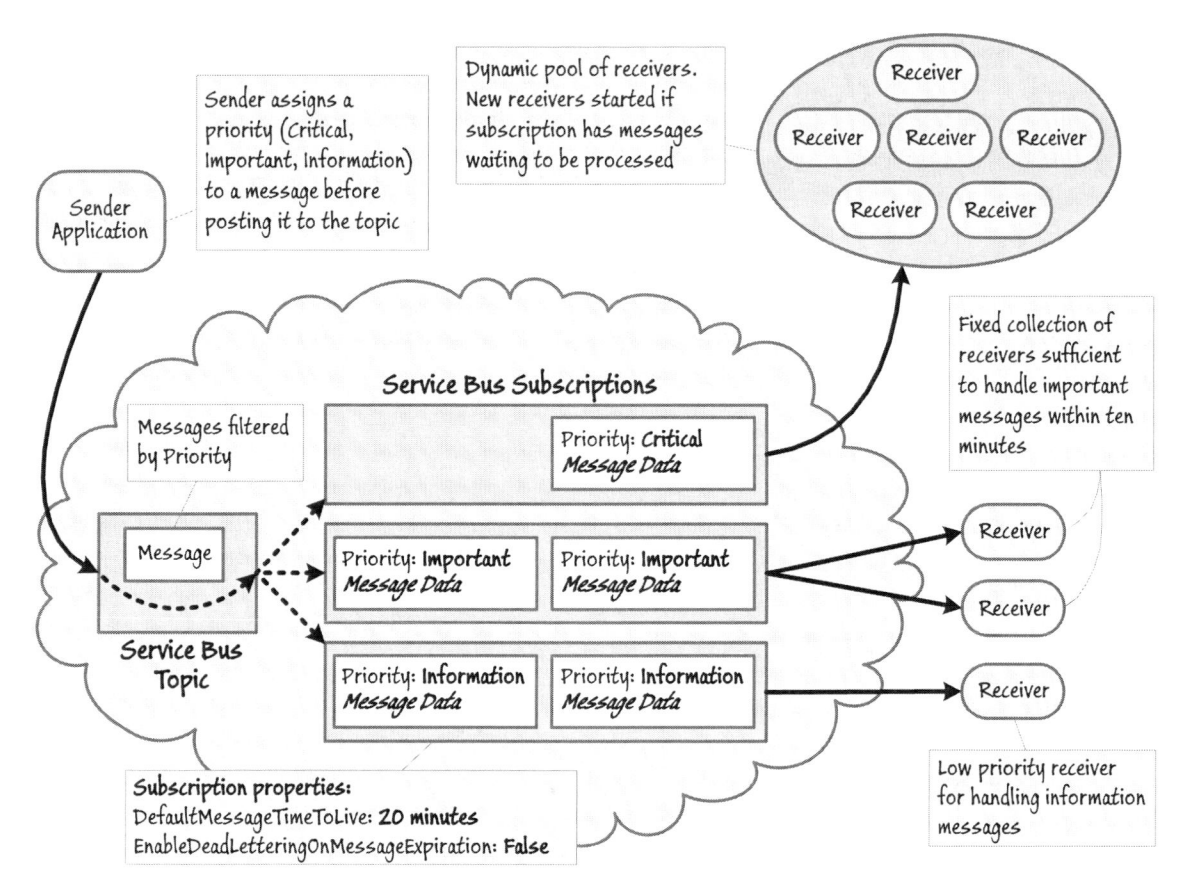

FIGURE 3
Prioritizing messages by using a Service Bus topic and subscriptions

- Senders posting messages expect a response to this message from the receiver. The number of senders can vary significantly over time.

    The section "Guidelines for Using Service Bus Queues" in "Appendix C - Implementing Cross-Boundary Communication" describes how a sender can post a message to a queue, receive a response on another queue, and correlate this response with the original message by using the **CorrelationId** property of the message. The sender specifies the queue on which to send the response in the **ReplyTo** property of the message, and the receiver populates the **CorrelationId** of the response message with a copy of the **MessageId** from the original request message.

    This approach is very straightforward and suitable for a reasonably small and static set of senders, but it does not scale well and can become unwieldy if the number of senders changes quickly. This is because each sender requires its own Service Bus queue, these queues take time to construct, and each queue has an associated monetary cost; ideally if a queue is no longer required it should be removed. Service Bus topics and subscriptions provide a better variation on this approach in a dynamic environment.

Service Bus subscriptions support the notion of *subscription correlation*. This mechanism enables an application to connect to a topic and create a subscription that filters messages based on the **CorrelationId** property. Service Bus subscriptions provide the **CorrelationFilter** filter specifically for this purpose. To implement subscription correlation, you perform the following tasks:

- The sender creates a message and populates the **MessageId** property with a unique value.
- The sender connects to the Service Bus subscription on which it expects to receive a response, and adds a **CorrelationFilter** to this subscription specifying the **MessageId** property of the original message. All senders share this same topic, but the filter ensures that each sender only receives the responses to the messages that it sent.

A single subscription can have more than one associated filter. A message passes through to a subscriber as long as one filter expression matches the message properties. However, if more than one expression matches then the same message will appear multiple times; once for each match.

- The sender posts the message to a Service Bus topic on which one or more receivers have subscriptions.
- A receiver retrieves the message, based on any filtering applied to the subscription, and then processes the message.
- The receiver creates a response message and populates the **CorrelationId** property with a copy of the value in the **MessageId** property of the original message.
- The receiver posts the response message to the Service Bus topic shared by all sender applications.
- When a message with a value in the **CorrelationId** property that matches the original **MessageId** appears in the topic, the **CorrelationFilter** for the appropriate subscription ensures that it is passed to the correct sender.

> The **CorrelationFilter** filter has been designed specifically for this scenario, and it provides an extremely efficient mechanism for filtering messages. In contrast, although a **SqlFilter** filter is more flexible, it has to undergo lexicographical analysis when it is created, and it has greater runtime costs.

Figure 4 shows the structure of this of this solution.

Receivers copy **MessageId** property of original message into **CorrelationId** property of response message

Senders populate **MessageID** property of message

Service Bus Subscriptions

Sender A → Service Bus Topic → Message

MessageId: **99** Message Data → Receiver

MessageId: **502** Message Data | MessageId: **342** Message Data → Receiver

Sender B

Senders filter response messages by **CorrelationID**

Response to earlier message **98** sent by Sender **A**

CorrelationId: **98** Message Data

Service Bus Topic → Response

CorrelationId: **341** Message Data

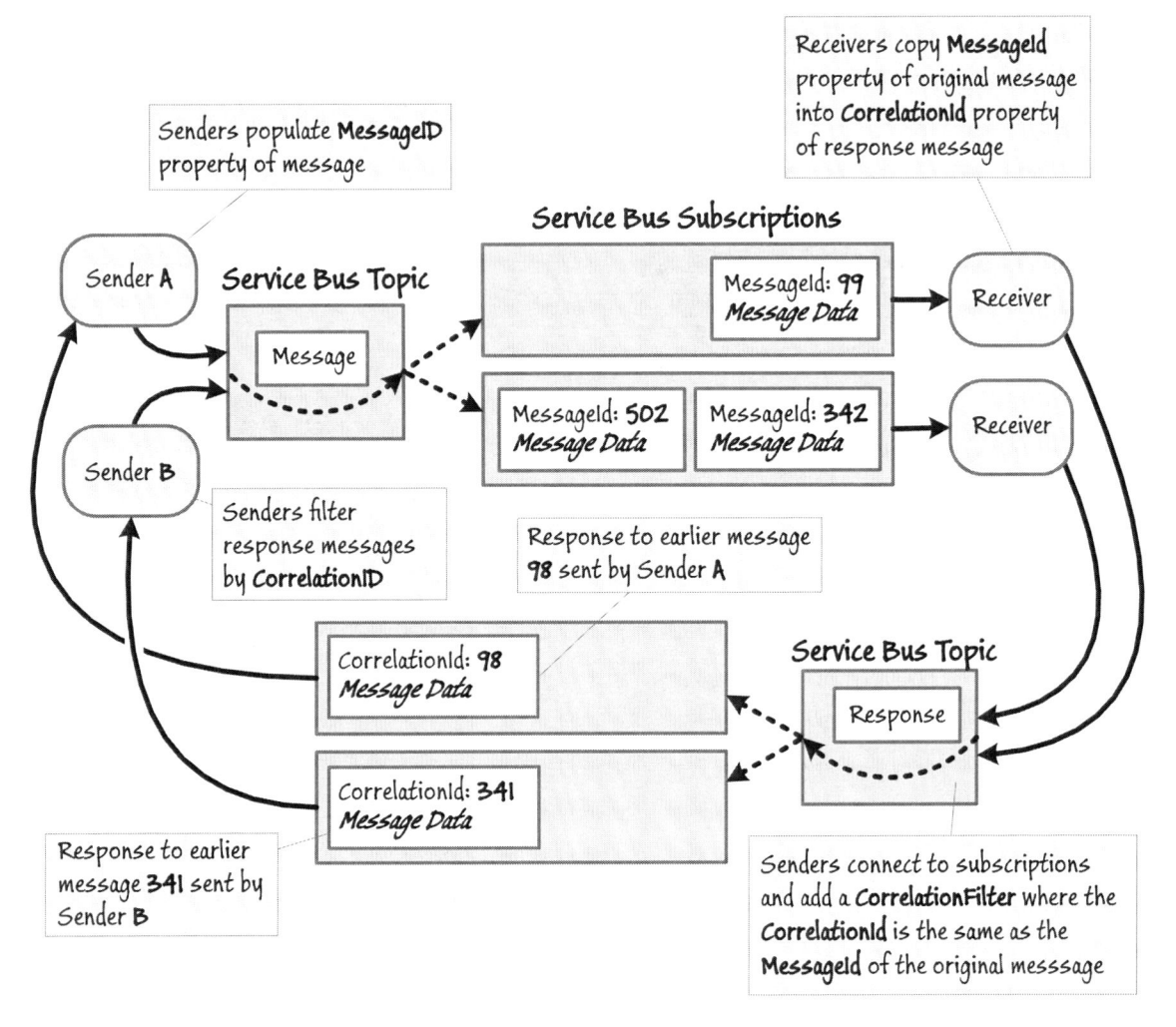

Response to earlier message **341** sent by Sender **B**

Senders connect to subscriptions and add a **CorrelationFilter** where the **CorrelationId** is the same as the **MessageId** of the original messsage

FIGURE 4
Using subscription correlation to deliver response messages to a sender

- Your system handles a continuous, large volume of messages. To maintain performance, you previously decided to implement a scale-out mechanism by using Service Bus queues, but you have found that you need more control over which receivers process which messages; you need to direct messages to receivers running on specific servers.

As described in the section "Guidelines for Using Service Bus Queues" in "Appendix C - Implementing Cross-Boundary Communication" Service Bus queues enable you to implement a simple load leveling mechanism for processing messages; multiple receivers can listen to the same queue, and the result is an approximate round robin distribution of messages. However, you may require more control over which receiver handles which messages that the round robin approach does not provide. For example, your system may require that all messages with the **Warehouse** property set to **A** are processed by a receiver running on a server physically close to warehouse **A**, messages marked as **B** are handled by a receiver running on a server close to warehouse **B**, and so on.

Service Bus topics and subscriptions provide a useful mechanism for partitioning the message-processing workload, based on one or more properties of each message. You can define a set of mutually exclusive filters that cover different ranges of the values in the message properties and direct each range to a different subscription. The various receivers listening to these subscriptions can run on specific on-premises servers, or they can be implemented as worker roles in the cloud. Additionally, each subscription can have multiple receivers. In this case, the receivers compete for messages from that subscription echoing the round robin load leveling technique used for a queue.

Figure 5 shows an example structure for the warehouse system. The receivers are all built using the Windows Azure SDK and connect directly to the various Service Bus subscriptions. Warehouse B expects more traffic than warehouse A, so messages for warehouse B are handled by multiple receivers, all running on hardware located locally to warehouse B.

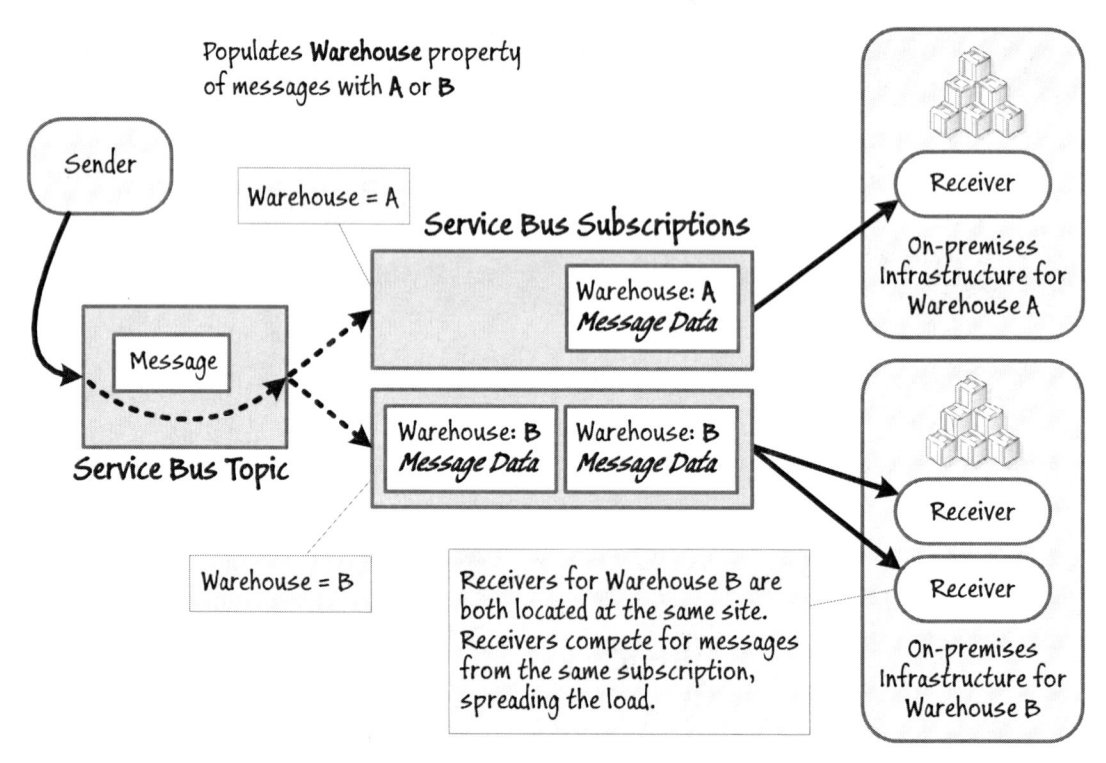

FIGURE 5
Scaling out by using a Service Bus topic and subscriptions

- Messages received from a subscription may be need to be forwarded to a final destination for additional processing, depending on system-defined criteria. This forwarding mechanism should be transparent to the sender as the forwarding criteria and final destinations may change. The receiver applications should also be decoupled from any changes to these criteria and final destinations.

    Consider the example of an ordering processing system where a web application sends orders to a receiving application through a Service Bus topic. The receiving application is responsible for arranging the packaging and dispatch of the order. All orders may be subjected to additional scrutiny and auditing depending on their value. This examination is performed by a separate set of processes, implementing auditing logic defined by the organization's order handling policy.

    For example if the value of the order is below 100 the order is simply logged, if the value is between 100 and 499 the order is logged and the details are printed for later scrutiny by an auditor, and if the value is 500 or more the order is logged and the details are emailed directly to an auditor for immediate examination. The auditor might choose to cancel the order if the customer does not meet certain requirements. However, these threshold values may change, and the business logic for the receiving application need to be insulated from this change.

    You can accomplish this level of decoupling by using a filter rule action. A filter rule action can change, add, or remove a message property when the message is retrieved from a Service Bus subscription. This action is performed transparently as the message is retrieved by the receiving application. The receiving application can create a copy of the message to perform its own processing, and repost the received message to another topic that routes the message on to the appropriate destination based on the updated property set.

> If you originally built the sender application by using the **Send** method of a **MessageSender** object to post messages to a Service Bus queue, you do not have to modify this part of the code because you can use the same method to post messages to a topic. All you need to do is create and populate the appropriate message properties required by the subscription filter before sending them to the topic. To receive messages from a Service Bus subscription, use a **SubscriptionClient** object.

Figure 6 shows a possible structure for the order processing example. The sender adds the total cost of the order as a property called **TotalCost** to the initial order message, together with other properties (not shown) that are used to route the message to the receiving application (labeled "Forwarding Receiver"). When the receiving application retrieves a message, a filter rule action is applied that automatically adds a property called **PriceRange** to each message. The value of the **PriceRange** property is set to **Low**, **Medium**, or **High** according to the cost; a cost below 100 is **Low**, a cost between 100 and 499 is **Medium**, and cost of 500 or more is **High**. The receiving application performs whatever processing is required. At the same time, it posts a copy of the received message, which now has a **PriceRange** property appended, to the Service Bus topic that the various Auditing Receivers subscribe to. The Auditing Receivers' subscriptions filter the message by the **PriceRange** property to route them to the receiver that performs the appropriate operations, as described earlier.

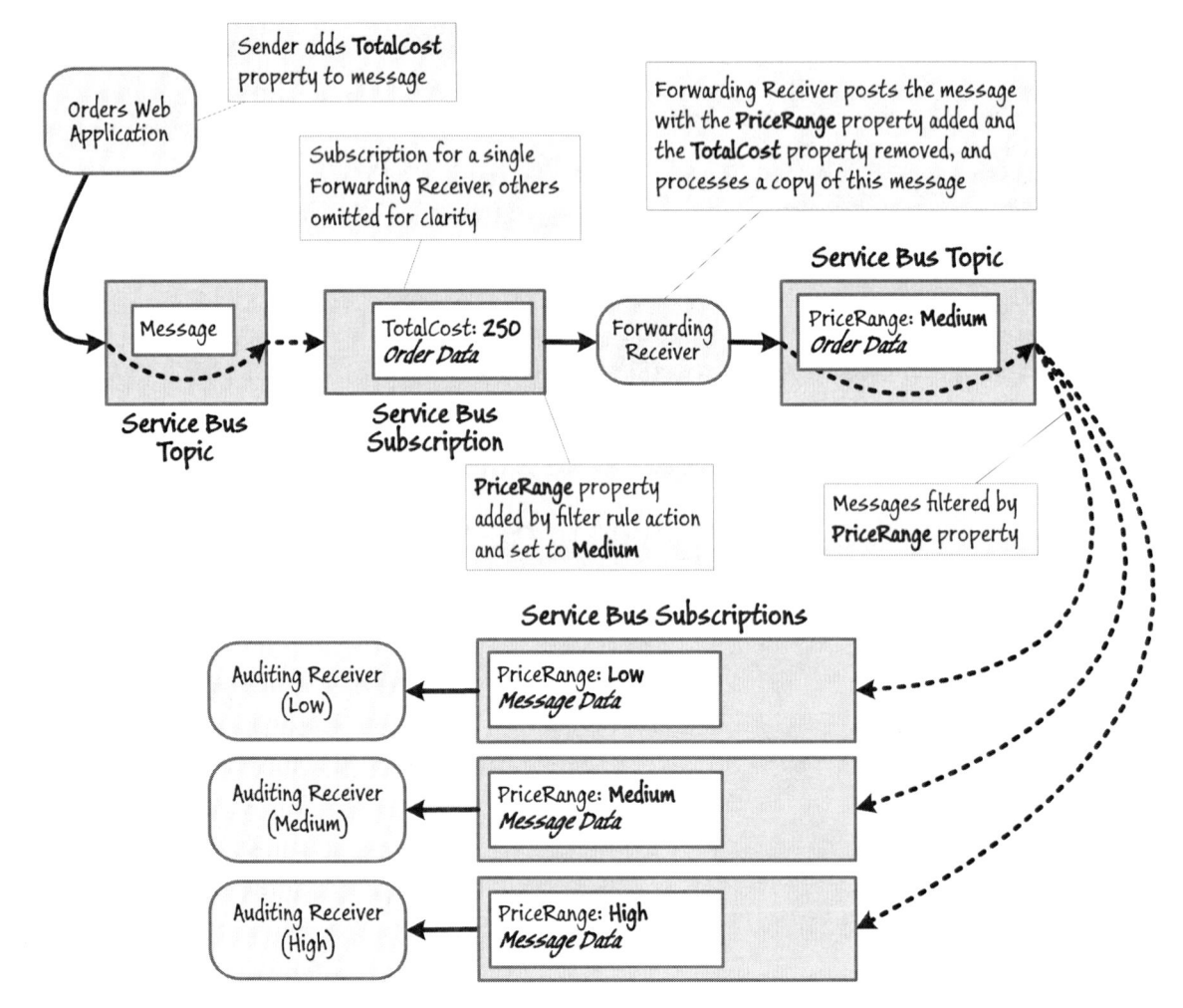

FIGURE 6
FORWARD-ROUTING MESSAGES BY USING A FILTER RULE ACTION

The following code example shows how to add the filter rule actions used by this example to a subscription on which the Forwarding Receiver application listens. Note that these filter rule actions also remove the **TotalCost** property from the message as it is not actually required to route the message to the auditing application; the forward routing is based solely on the **PriceRange** property. The full details of the order are still available to the Auditing Receiver in the body of the message, however.

```C#
...
// Define action rule filters
var ruleLowPrice = new RuleDescription()
{
  Action = new SqlRuleAction(
    "set PriceRange='Low';remove TotalCost"),
  Filter = new SqlFilter("TotalCost < 100"),
  Name = "LowPrice"
};

var ruleMediumPrice = new RuleDescription()
{
  Action = new SqlRuleAction(
    "set PriceRange='Medium';remove TotalCost"),
  Filter = new SqlFilter(
    "TotalCost >= 100 AND TotalCost < 500"),
  Name = "MediumPrice"
};

var ruleHighPrice = new RuleDescription()
{
  Action = new SqlRuleAction(
    "set PriceRange='High';remove TotalCost"),
  Filter = new SqlFilter("TotalCost >= 500"),
  Name = "HighPrice"
};

...
var subscriptionClient =
  messagingFactory.CreateSubscriptionClient(...);

// Add the rules to the subscription
subscriptionClient.AddRule(ruleLowPrice);
subscriptionClient.AddRule(ruleMediumPrice);
subscriptionClient.AddRule(ruleHighPrice);
```

Limitations of Using Service Bus Topics and Subscriptions to Route Messages

Service Bus topics and subscriptions only implement a simple routing mechanism. For security reasons, the filters that you define cannot access the body of a message, so they can only make decisions based on data exposed in message properties. Most commonly, you define filters by using the **SqlFilter** class. For optimization purposes, the conditions specified in these filters are limited to a subset of SQL92 syntax. You can perform direct comparisons of data and values by using common arithmetic and logical operators, but these filters do not support functions; for example, there is no **Substring** function. If you require routing based on more complex rules, you must implement this logic in your own code by creating a receiver that examines the data of a message and then reposting it to another queue or topic as necessary.

For more information about the types of expressions supported by the **SqlFilter** class, see the topic "*SqlFilter.SqlExpression Property*" on MSDN.

Routing Messages to Multiple Destinations Using Service Bus Topics and Subscriptions

The previous section described using filters that partition messages into distinct non-overlapping groups and direct each group to a Service Bus subscription, and each message is sent exclusively to a single subscription. However, it is also possible for different subscriptions to have filters with overlapping predicates. In this case, a copy of the same message is routed to each matching subscription. This mechanism provides a means for routing messages to multiple destinations.

The converse situation is also true; if all subscriptions have filters that fail to match the properties for a message it will remain queued on a Service Bus topic until it expires.

Guidelines for Using Service Bus Topics and Subscriptions to Route Messages to Multiple Destinations

Filters with overlapping predicates enable a number of powerful scenarios. The following list describes some common cases:

> "Appendix A - Replicating, Distributing, and Synchronizing Data" includes some additional patterns for using Service Bus topics and subscriptions to query and update data in a system that uses replicated data sources.

- Your system enables sender applications to post requests to services, but all of these requests must be logged for auditing or diagnostic purposes. This logging must be transparent to the sender applications.

This is an example of the most general pattern for posting messages to multiple destinations. Services can use subscriptions to retrieve their intended messages, but all messages must additionally be posted to a logging service so that they can be recorded and stored. The Windows Azure SDK provides the **TrueFilter** type specifically for this purpose. This filter matches all messages, and any subscription that utilizes this filter will automatically be fed with a replica of every message sent to the topic.

Figure 7 shows an example system that uses a **TrueFilter** to copy messages to an audit log for later examination.

> The **TrueFilter** is the default filter for a subscription; if you don't specify a filter when you create a subscription, the **TrueFilter** is applied automatically.

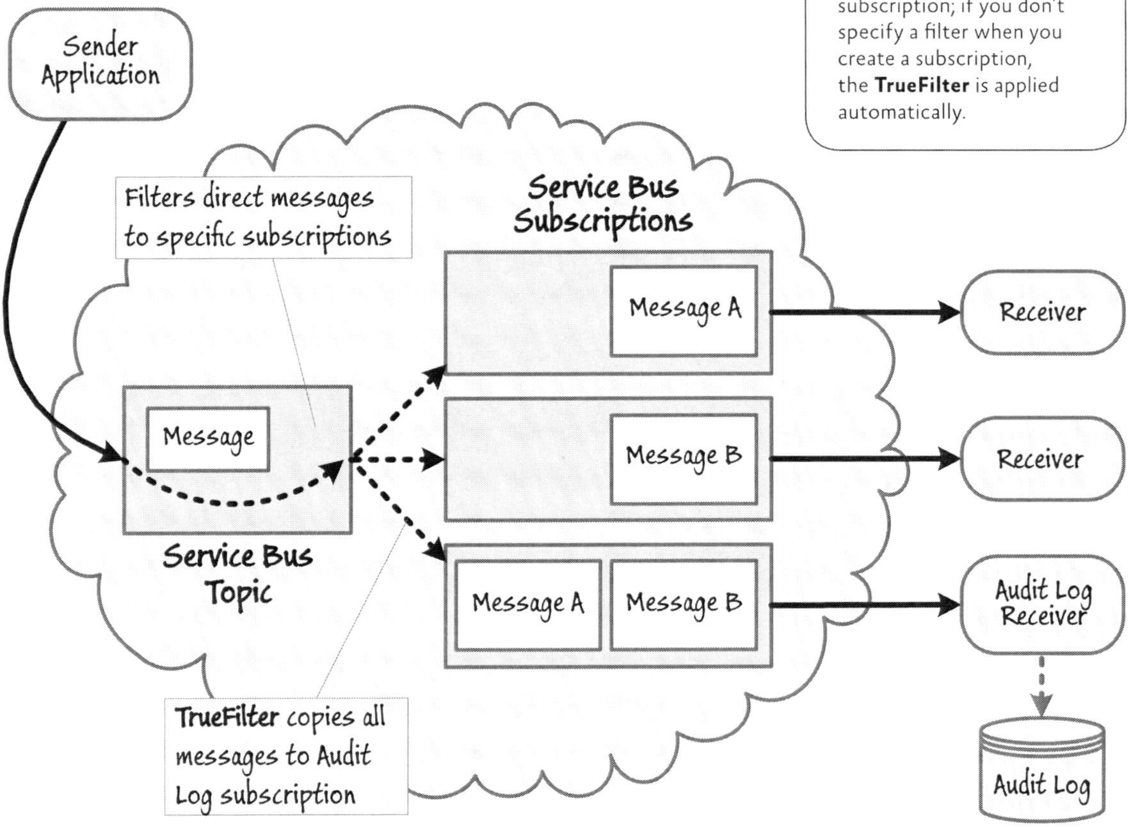

FIGURE 7
Logging messages by using a TrueFilter

The Audit Log Receiver is simply an example application that may benefit from this approach. Any functionality that requires a copy of messages that pass through your system can be implemented in a similar way. For example, you could implement an application that measures the number of messages flowing into your system over a given period of time and displays the results, giving an indication of the message throughput and performance of your solution.

Of course, you can also be more selective. Rather than using a **TrueFilter**, you can define an overlapping **SqlFilter** that captures messages based on property values, and these messages will be routed to the destination receivers expecting to process these message as well as the Audit Log Receiver application.

- You system raises a number of business events. Each event may be handled by zero or more processes, and your system must be able to add or remove processes that can handle these events without impacting the business logic of your system. The event handlers may be running remotely from the processes that raise the events.

  Processes that trigger events do so to inform interested parties that something significant has happened. The processes that listen for events are inherently asynchronous and are decoupled from the processes that raise events. Using Service Bus topics and subscriptions provides an excellent basis for building such a system, especially given the requirement that the event handlers may be located anywhere and events should be delivered reliably.

  In messaging terms, an application can notify interested parties of an event simply by posting a message that contains the event data to a Service Bus topic. Each application that is interested in an event can create its own subscription where the filter specifies the conditions for the messages that constitute the event. The topic on which a sender application posts event messages can have the **DefaultMessageTimeToLive** property set appropriately, so that if no applications subscribe to the event, then it will be discarded when this period expires.

> Do not attempt to share the same event subscription between two separate applications if they must both be notified of the event; they will compete for event messages routed to the subscription, and each message will only be passed to one of the applications.

Figure 8 shows an example from a basic system controlling the assembly line in a manufacturing plant. When production is due to start, the Controller application posts a "Start Machinery" message to a Service Bus topic. Each machine involved in the assembly process is driven by software that listens for this message, and when it occurs the software driver starts the machine. Similarly, when production is halted, the Controller application posts a "Stop Machinery" message, and the software drivers for each machine shut it down in a controlled manner. The Controller application has no knowledge of the machinery involved in the production line, and hardware can be added or removed without requiring the Controller application to be modified.

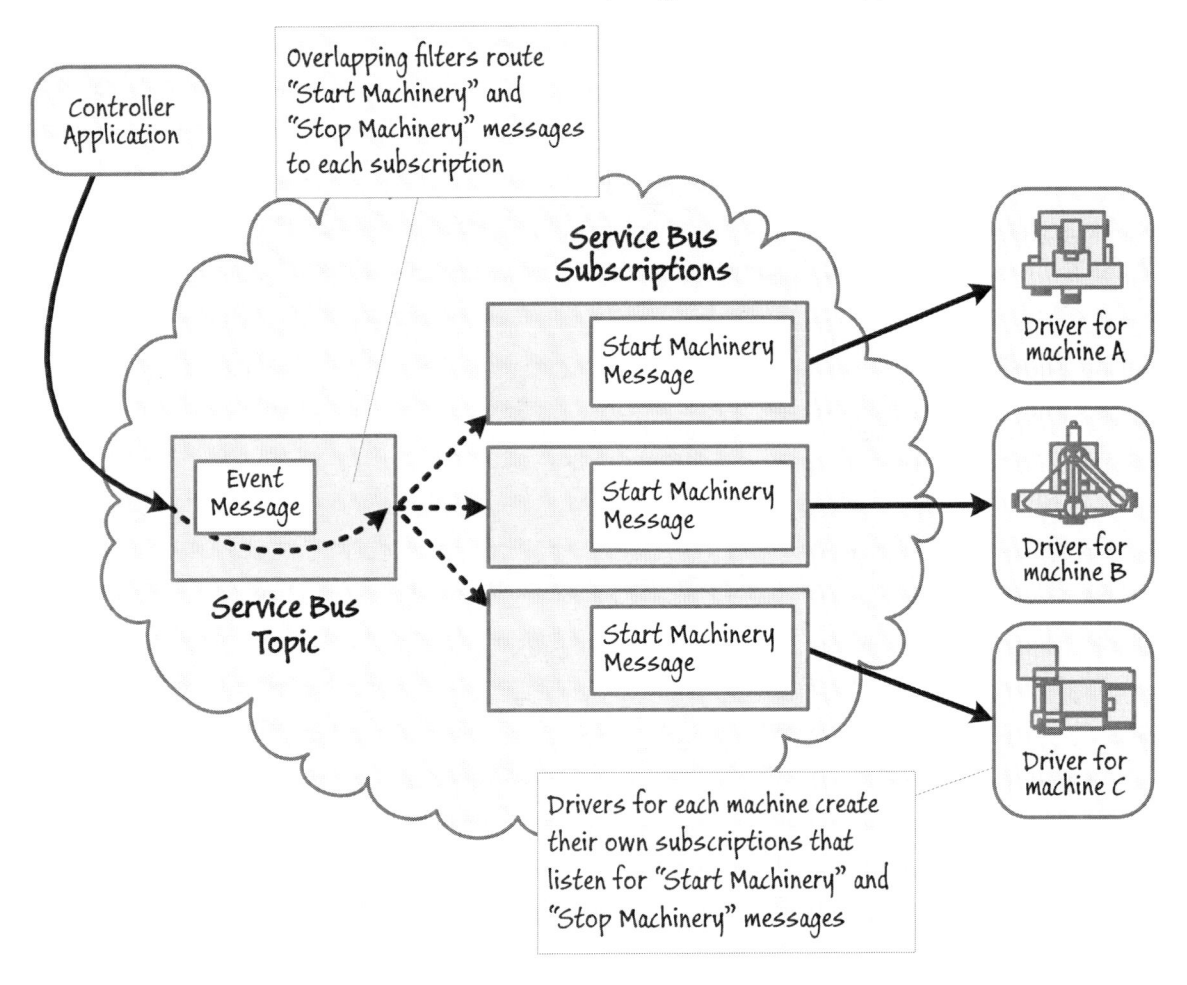

FIGURE 8
Controlling a production line by using events based on Service Bus subscriptions

For a detailed example and more information about using Service Bus to communicate between roles and applications, see "How to Simplify & Scale Inter-Role Communication Using Windows Azure Service Bus."

Limitations of Using Service Bus Topics and Subscriptions to Route Messages to Multiple Destinations

It is important to understand that, while Service Bus topics and subscriptions can provide reliable delivery of messages to one or more destinations, this delivery is not instantaneous. Topics and subscriptions reside in the cloud, and there will inevitably be some delay caused by the network latency associated with the Internet. Additionally, filters defined by using the **SqlFilter** type are subject to runtime evaluation, and the properties attached to each message must be examined and compared against every filter associated with each subscription. If a topic has a large number of subscriptions (a topic can have up to 2000 subscriptions in the current release of Service Bus), then this evaluation and examination may take some time to perform.

Security Guidelines for Using Service Bus Topics and Subscriptions

Service Bus topics and subscriptions are subject to the same security mechanism as Service Bus queues. You configure security, create identities, and associate privileges with these identities by using ACS. See "Appendix B - Authenticating Users and Authorizing Requests" for more information about using ACS. You can grant the privileges associated with the **Manage** and **Send** claims to a topic, and the privileges associated with the **Manage** and **Listen** claims to a subscription. When an application connects to a Service Bus namespace topic or subscription, it must authenticate with ACS and provide the identity used by the application. See "Appendix C - Implementing Cross-Boundary Communication" for further details about connecting to a Service Bus namespace and providing identity information.

As with Service Bus queues, all communications with Service Bus topics and subscriptions occur over a TCP channel and are automatically protected by using SSL.

More Information

All links in this book are accessible from the book's online bibliography available at: *http://msdn.microsoft.com/en-us/library/hh968447.aspx*.

- "SqlFilter.SqlExpression Property" at *http://msdn.microsoft.com/en-us/library/microsoft.servicebus.messaging.sqlfilter.sqlexpression.aspx*.
- "How to Simplify & Scale Inter-Role Communication Using Windows Azure Service Bus" at *http://windowsazurecat.com/2011/08/how-to-simplify-scale-inter-role-communication-using-windows-azure-service-bus/*.

APPENDIX E

Maximizing Scalability, Availability, and Performance

A key feature of the Windows Azure™ technology platform is the robustness that the platform provides. A typical Windows Azure solution is implemented as a collection of one or more roles, where each role is optimized for performing a specific category of tasks. For example, a web role is primarily useful for implementing the web front-end that provides the user interface of an application, while a worker role typically executes the underlying business logic such as performing any data processing required, interacting with a database, orchestrating requests to and from other services, and so on. If a role fails, Windows Azure can transparently start a new instance and the application can resume.

However, no matter how robust an application is, it must also perform and respond quickly. Windows Azure supports highly scalable services through the ability to dynamically start and stop instances of an application, enabling a Windows Azure solution to handle an influx of requests at peak times, while scaling back as the demand lowers, reducing the resources consumed and the associated costs.

However, scalability is not the only issue that affects performance and response times. If an application running in the cloud accesses resources and databases held in your on-premises servers, bear in mind that these items are no longer directly available over your local high-speed network. Instead the application must retrieve this data across the Internet with its lower bandwidth, higher latency, and inherent unpredictably concerning reliability and throughput. This can result in increased response times for users running your applications or reduced throughput for your services.

> If you are building a commercial system, you may have a contractual obligation to provide a certain level of performance to your customers. This obligation might be specified in a service level agreement (SLA) that guarantees the response time or throughput. In this environment, it is critical that you understand the architecture of your application, the resources that it utilizes, and the tools that Windows Azure provides for building and maintaining an efficient system.

Of course, if your application or service is now running remotely from your organization, it will also be running remotely from your users. This might not seem like much of an issue if you are building a public-facing website or service because the users would have been remote prior to you moving functionality to the cloud, but this change may impact the performance for users inside your organization who were previously accessing your solution over a local area network. Additionally, the location of an application or service can affect its perceived availability if the path from the user traverses network elements that are heavily congested, and network connectivity times out as a result. Finally, in the event of a catastrophic regional outage of the Internet or a failure at the datacenter hosting your applications and services, your users will be unable to connect.

This appendix considers issues associated with maintaining performance, reducing application response times, and ensuring that users can always access your application when you relocate functionality to the cloud. It describes solutions and good practice for addressing these concerns by using Windows Azure technologies.

Requirements and Challenges

The primary causes of extended response times and poor availability in a distributed environment are lack of resources for running applications, and network latency. Scaling can help to ensure that sufficient resources are available, but no matter how much effort you put into tuning and refining your applications, users will perceive that your system has poor performance if these applications cannot receive requests or send responses in a timely manner because the network is slow. A crucial task, therefore, is to organize your solution to minimize this network latency by making optimal use of the available bandwidth and utilizing resources as close as possible to the code and users that need them.

The following sections identify some common requirements concerning scalability, availability, and performance, summarizing many of the challenges you will face when you implement solutions to meet these requirements.

Managing Elasticity in the Cloud

Description: Your system must support a varying workload in a cost-effective manner.

Many commercial systems must support a workload that can vary considerably over time. For much of the time the load may be steady, with a regular volume of requests of a predictable nature. However, there may be occasions when the load dramatically and quickly increases. These peaks may arise at expected times; for example, an accounting system may receive a large number of requests as the end of each month approaches when users generate their month-end reports, and it may experience periods of increased usage towards the end of the financial year. In other types of application the load may surge unexpectedly; for example, requests to a news service may flood in if some dramatic event occurs.

The cloud is a highly scalable environment, and you can start new instances of a service to meet demand as the volume of requests increases. However, the more instances of a service you run, the more resources they occupy; and the costs associated with running your system rise accordingly. Therefore it makes economic sense to scale back the number of service instances and resources as demand for your system decreases.

How can you achieve this? One solution is to monitor the solution and start up more service instances as the number of requests arriving in a given period of time exceeds a specified threshold value. If the load increases further, you can define additional thresholds and start yet more instances. If the volume of requests later falls below these threshold values you can terminate the extra instances. In inactive periods, it might only be necessary to have a minimal number of service instances. However, there are a couple of challenges with this solution:

- You must automate the process that starts and stops service instances in response to changes in system load and the number of requests. It is unlikely to be possible to perform these tasks manually as peaks and troughs in the workload may occur at any time.
- The number of requests that occur in a given interval might not be the only measure of the workload; for example, a small number of requests that each incur intensive processing might also impact performance. Consequently the process that predicts performance and determines the necessary thresholds may need to perform calculations that measure the use of a complex mix of resources.

Reducing Network Latency for Accessing Cloud Applications

Description: Users should be connected to the closest available instance of your application running in the cloud to minimize network latency and reduce response times.

A cloud application may be hosted in a datacenter in one part of the world, while a user connecting to the application may be located in another, perhaps on a different continent. The distance between users and the applications and services they access can have a significant bearing on the response time of the system. You should adopt a strategy that minimizes this distance and reduces the associated network latency for users accessing your system.

If your users are geographically dispersed, you could consider replicating your cloud applications and hosting them in datacenters that are similarly dispersed. Users could then connect to the closest available instance of the application. The question that you need to address in this scenario is how do you direct a user to the most local instance of an application?

> Remember that starting and stopping service instances is not an instantaneous operation. It may take 10-15 minutes for Windows Azure to perform these tasks, so any performance measurements should include a predictive element based on trends over time, and initiate new service instances so that they are ready when required.

Maximizing Availability for Cloud Applications

Description: Users should always be able to connect to the application running in the cloud.

How do you ensure that your application is always running in the cloud and that users can connect to it? Replicating the application across datacenters may be part of the solution, but consider the following issues:

- What happens if the instance of an application closest to a user fails, or no network connection can be established?

- The instance of an application closest to a user may be heavily loaded compared to a more distant instance. For example, in the afternoon in Europe, traffic to datacenters in European locations may be a lot heavier than traffic in the Far East or West Coast America. How can you balance the cost of connecting to an instance of an application running on a heavily loaded server against that of connecting to an instance running more remotely but on a lightly-loaded server?

Optimizing the Response Time and Throughput for Cloud Applications

Description: The response time for services running in the cloud should be as low as possible, and the throughput should be maximized.

Windows Azure is a highly scalable platform that offers high performance for applications. However, available computing power alone does not guarantee that an application will be responsive. An application that is designed to function in a serial manner will not make best use of this platform and may spend a significant period blocked waiting for slower, dependent operations to complete. The solution is to perform these operations asynchronously, and this approach has been described throughout this guide.

Aside from the design and implementation of the application logic, the key factor that governs the response time and throughput of a service is the speed with which it can access the resources it needs. Some or all of these resources might be located remotely in other datacenters or on-premises servers. Operations that access remote resources may require a connection across the Internet. To mitigate the effects of network latency and unpredictability, you can cache these resources locally to the service, but this approach leads to two obvious questions:

- What happens if a resource is updated remotely? The cached copy used by the service will be out of date, so how should the service detect and handle this situation?

- What happens if the service itself needs to update a resource? In this case, the cached copy used by other instances of this or other services may now be out of date.

Caching is also a useful strategy for reducing contention to shared resources and can improve the response time for an application even if the resources that it utilizes are local. However, the issues associated with caching remain the same; specifically, if a local resource is modified the cached data is now out of date.

Windows Azure and Related Technologies

Windows Azure provides a number of technologies that can help you to address the challenges presented by each of the requirements in this appendix:

- **Enterprise Library Autoscaling Application Block**. You can use this application block to define performance indicators, measure performance against these indicators, and start and stop instances of services to maintain performance within acceptable parameters.

- **Windows Azure Traffic Manager**. You can use this service to reduce network latency by directing users to the nearest instance of an application running in the cloud. Windows Azure Traffic Manager can also detect whether an instance of a service has failed or is unreachable, automatically directing user requests to the next available service instance.

- **Windows Azure Caching**. You can use this service to cache data in the cloud and provide scalable, reliable, and shared access for multiple applications.

- **Content Delivery Network (CDN)**. You can use this service to improve the response time of web applications by caching frequently accessed data closer to the users that request it.

> The cloud is not a magic remedy for speeding up applications that are not designed with performance and scalability in mind.

Windows Azure Caching is primarily useful for improving the performance of web applications and services running in the cloud. However, users will frequently be invoking these web applications and services from their desktop, either by using a custom application that connects to them or by using a web browser. The data returned from a web application or service may be of a considerable size, and if the user is very distant it may take a significant time for this data to arrive at the user's desktop. CDN enables you to cache frequently queried data at a variety of locations around the world. When a user makes a request, the data can be served from the most optimal location based on the current volume of traffic at the various Internet nodes through which the requests are routed. Detailed information, samples, and exercises showing how to configure CDN are available on MSDN; see the topic "Windows Azure CDN." Additionally Chapter 3, "Accessing the Surveys Application" in the guide "Developing Applications for the Cloud, 2nd Edition" provides further implementation details.

The following sections describe the Enterprise Library Autoscaling Application Block, Windows Azure Traffic Manager, and Windows Azure Caching, and provide guidance on how to use them in a number of scenarios.

Managing Elasticity in the Cloud by Using the Microsoft Enterprise Library Autoscaling Application Block

It is possible to implement a custom solution that manages the number of deployed instances of the web and worker roles your application uses. However, this is far from a simple task and so it makes sense to consider using a prebuilt library that is sufficiently flexible and configurable to meet your requirements.

The Enterprise Library Autoscaling Application Block (also known as "Wasabi") provides such a solution. It is part of the Microsoft Enterprise Library 5.0 Integration Pack for Windows Azure, and can automatically scale your Windows Azure application or service based on rules that you define specifically for that application or service. You can use these rules to help your application or service maintain its throughput in response to changes in its workload, while at the same time minimize and control hosting costs.

> External services that can manage autoscaling do exist but you must provide these services with your management certificate so that they can access the role instances, which may not be an acceptable approach for your organization.

Scaling operations typically alter the number of role instances in your application, but the block also enables you to use other scaling actions such as throttling certain functionality within your application. This means that there are opportunities to achieve very subtle control of behavior based on a range of predefined and dynamically discovered conditions. The Autoscaling Application Block enables you to specify the following types of rules:

- **Constraint rules**, which enable you to set minimum and maximum values for the number of instances of a role or set of roles based on a timetable.

- **Reactive rules**, which allow you to adjust the number of instances of a role or set of roles based on aggregate values derived from data points collected from your Windows Azure environment or application. You can also use reactive rules to change configuration settings so that an application can modify its behavior and change its resource utilization by, for example, switching off nonessential features or gracefully degrading its UI as load and demand increases.

Rules are defined in XML format and can be stored in Windows Azure blob storage, in a file, or in a custom store that you create.

By applying a combination of these rules you can ensure that your application or service will meet demand and load requirements, even during the busiest periods, to conform to SLAs, minimize response times, and ensure availability while still minimizing operating costs.

How the Autoscaling Application Block Manages Role Instances

The Autoscaling Application Block can monitor key performance indicators in your application roles and automatically deploy or remove instances. For example, Figure 1 shows how the number of instances of a role may change over time within the boundaries defined for the minimum and maximum number of instances.

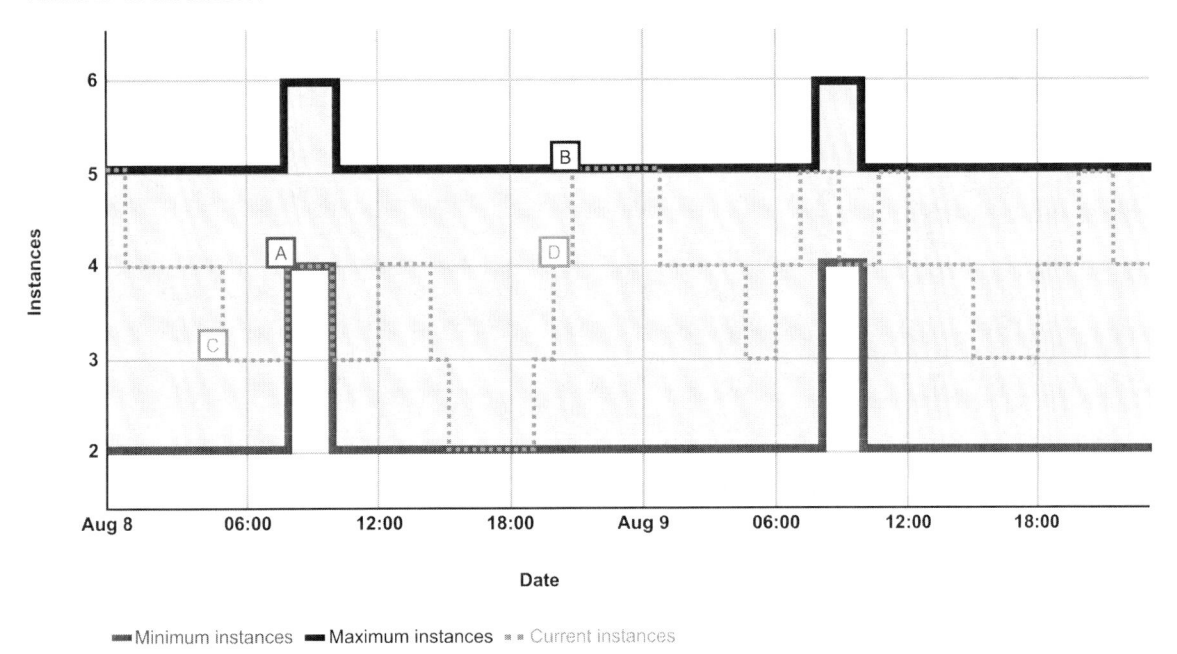

FIGURE 1
Data visualization of the scale boundaries and scale actions for a role

The behavior shown in Figure 1 was the result of the following configuration of the Autoscaling Application Block:

- A default **Constraint rule** that is always active, with the range set to a minimum of two and a maximum of five instances. At point **B** in the chart, this rule prevents the block from deploying any additional instances, even if the load on the application justifies it.
- A **Constraint rule** that is active every day from 08:00 for two hours, with the range set to a minimum of four and a maximum of six instances. The chart shows how, at point **A**, the block deploys a new instance of the role at 08:00.
- An **Operand** named **Avg_CPU_RoleA** bound to the average value over the previous 10 minutes of the Windows performance counter **\Processor(_Total)\% Processor Time**.
- A **Reactive rule** that increases the number of deployed role instances by one when the value of the **Avg_CPU_RoleA** operand is greater than 80. For example, at point **D** in the chart the block increases the number of roles to four and then to five as processor load increases.

- A Reactive rule that decreases the number of deployed role instances by one when the value of the **Avg_CPU_RoleA** operand falls below 20. For example, at point **C** in the chart the block has reduced the number of roles to three as processor load has decreased.

Constraint Rules

Constraint rules are used to proactively scale your application for the expected demand, and at the same time constrain the possible instance count, so that reactive rules do not change the instance count outside of that boundary. There is a comprehensive set of options for specifying the range of times for a constraint rule, including fixed periods and fixed durations, daily, weekly, monthly, and yearly recurrence, and relative recurring events such as the last Friday of each month.

Reactive Rules

Reactive rules specify the conditions and actions that change the number of deployed role instances or the behavior of the application. Each rule consists of one or more operands that define how the block matches the data from monitoring points with values you specify, and one or more actions that the block will execute when the operands match the monitored values.

Operands that define the data points for monitoring activity of a role can use any of the Windows® operating system performance counters, the length of a Windows Azure storage queue, and other built-in metrics. Alternatively you can create a custom operand that is specific to your own requirements, such as the number of unprocessed orders in your application.

Reactive rule conditions can use a wide range of comparison functions between operands to define the trigger for the related actions to occur. These functions include the typical greater than, greater than or equal, less than, less than or equal, and equal tests. You can also negate the tests using the **not** function, and build complex conditional expressions using **AND** and **OR** logical combinations.

Actions

The Autoscaling Application Block provides the following types of actions:
- The **setRange** action specifies the maximum and minimum number of role instances that should be available over a specified time period. This action is only applicable to Constraint rules.

By specifying the appropriate set of rules for the Autoscaling Application Block you can configure automatic scaling of the number of instances of the roles in your application to meet known demand peaks and to respond automatically to dynamic changes in load and demand.

The Autoscaling Application Block reads performance information collected by the Windows Azure diagnostics mechanism from Windows Azure storage. Windows Azure does not populate this with data from the Windows Azure diagnostics monitor by default; you must run code in your role when it starts or execute scripts while the application is running to configure the Windows Azure diagnostics to collect the required information and then start the diagnostics monitor.

- The **scale** action specifies that the block should increase or decrease the number of deployed role instances by an absolute or relative number. You specify the target role using the name, or you can define a scale group in the configuration of the block that includes the names of more than one role and then target the group so that the block scales all of the roles defined in the group.
- The **changeSetting** action is used for application throttling. It allows you to specify a new value for a setting in the application's service configuration file. The block changes this setting and the application responds by reading the new setting. Code in the application can use this setting to change its behavior. For example, it may switch off nonessential features or gracefully degrade its UI to better meet increased demand and load. This is usually referred to as application throttling.
- The capability to execute a custom action that you create and deploy as an assembly. The code in the assembly can perform any appropriate action, such as sending an email notification or running a script to modify a database deployed to the SQL Azure™ technology platform.

The Autoscaling Application Block logs events that relate to scaling actions and can send notification emails in response to the scaling of a role, or instead of scaling the role, if required. You can also configure several aspects of the way that the block works such as the scheduler that controls the monitoring and scaling activates, and the stabilizer that enforces "cool down" delays between actions to prevent repeated oscillation and optimize instance counts around the hourly boundary.

> You can use the Autoscaling Application Block to force your application to change its behavior automatically to meet changes in load and demand. The block can change the settings in the service configuration file, and the application can react to this to reduce its demand on the underlying infrastructure.

For more information, see "Microsoft Enterprise Library 5.0 Integration Pack for Windows Azure" on MSDN.

Guidelines for Using the Autoscaling Application Block

The following guidelines will help you understand how you can obtain the most benefit from using the Autoscaling Application Block:

- The Autoscaling Application Block can specify actions for multiple targets across multiple Windows Azure subscriptions. The service that hosts the target roles and the service that hosts the Autoscaling Application Block do not have to be in the same subscription. To allow the block to access applications, you must specify the ID of the Windows Azure subscription that hosts the target applications, and a management certificate that it uses to connect to the subscription.

You are charged by the hour for each Windows Azure role instance you deploy, even if you utilize only a few minutes of that hour. The stabilizer in the Autoscaling Application Block can help to reduce costs by forcing scale-out actions to take place only during the first few minutes of the hour, and scale-back actions to take place only during the last few minutes of the hour. You can specify these intervals so as to obtain maximum advantage from the hour for which you are charged.

- Consider using Windows Azure blob storage to hold your rules and service information. This makes it easy to update the rules and data when managing the application. Alternatively, if you want to implement special functionality for loading and updating rules, consider creating a custom rule store.
- You must define a constraint rule for each monitored role instance. Use the ranking for each constraint or reactive rule you define to control the priority where conditions overlap.
- Constraint rules do not take into account daylight saving times. They simply use the UTC offset that you specify at all times.
- Use scaling groups to define a set of roles that you target as one action to simplify the rules. This also makes it easy to add and remove roles from an action without needing to edit every rule.
- Consider using average times of half or one hour to even out the values returned by performance counters or other metrics to provide more consistent and reliable results. You can read the performance data for any hosted application or service; it does not have to be the one to which the rule action applies.
- Consider enabling and disabling rules instead of deleting them from the configuration when setting up the block and when temporary changes are made to the application.
- Remember that you must write code that initializes the Windows Azure Diagnostics mechanism when your role starts and copies the data to Windows Azure storage.
- Consider using the throttling behavior mechanism as well as scaling the number of roles. This can provide more fine-grained control of the way that the application responds to changes in load and demand. Remember that it can take 10-15 minutes for newly deployed role instances to start handling requests, whereas changes to throttling behavior occur much more quickly.
- Regularly analyze the information that the block logs about its activities to evaluate how well the rules are meeting your initial requirements, keeping the application running within the required constraints, and meeting any SLA commitments on availability and response times. Refine the rules based on this analysis.

Reducing Network Latency for Accessing Cloud Applications with Windows Azure Traffic Manager

Windows Azure Traffic Manager is a Windows Azure service that enables you to set up request routing and load balancing based on predefined policies and configurable rules. It provides a mechanism for routing requests to multiple deployments of your Windows Azure-hosted applications and services, irrespective of the datacenter location. The applications or services could be deployed in one or more datacenters.

Windows Azure Traffic Manager monitors the availability and network latency of each application you configure in a policy, on any HTTP or HTTPS port. If it detects that an application is offline it will not route any requests to it. However, it continues to monitor the application at 30 second intervals and will start to route requests to it, based on the configured load balancing policy, if it again becomes available.

Windows Azure Traffic Manager does not mark an application as offline until it has failed to respond three times in succession. This means that the total time between a failure and that application being marked as offline is three times the monitoring interval you specify.

In future releases of Windows Azure Traffic Manager you will be able to change the interval between the monitoring checks.

How Windows Azure Traffic Manager Routes Requests

Windows Azure Traffic Manager is effectively a DNS resolver. When you use Windows Azure Traffic Manager, web browsers and services accessing your application will perform a DNS query to Windows Azure Traffic Manager to resolve the IP address of the endpoint to which they will connect, just as they would when connecting to any other website or resource.

Windows Azure Traffic Manager uses the requested URL to identify the policy to apply, and returns an IP address resulting from evaluating the rules and configuration settings for that policy. The user's web browser or the requesting service then connects to that IP address, effectively routing them based on the policy you select and the rules you define.

This means that you can offer users a single URL that is aliased to the address of your Windows Azure Traffic Manager policy. For example, you could use a CNAME record to map the URL you want to expose to users of your application, such as **http://store.treyresearch.net**, in your own or your ISPs DNS to the entry point and policy of your Windows Azure Traffic Manager policy. If you have named your Windows Azure Traffic Manager namespace as **treyresearch** and have a policy for the **Orders** application named **ordersapp**, you would map the URL in your DNS to **http://ordersapp.treyresearch.trafficmanager.net**. All DNS queries for **store.treyresearch.net** will be passed to Windows Azure Traffic Manager, which will perform the required routing by returning the IP address of the appropriate deployed application. Figure 2 illustrates this scenario.

Windows Azure Traffic Manager does not perform HTTP redirection or use any other browser-based redirection technique because this would not work with other types of requests, such as from smart clients accessing web services exposed by your application. Instead, it acts as a DNS resolver that the client queries to obtain the IP address of the appropriate application endpoint. Windows Azure Traffic Manager returns the IP address of the deployed application that best satisfies the configured policy and rules.

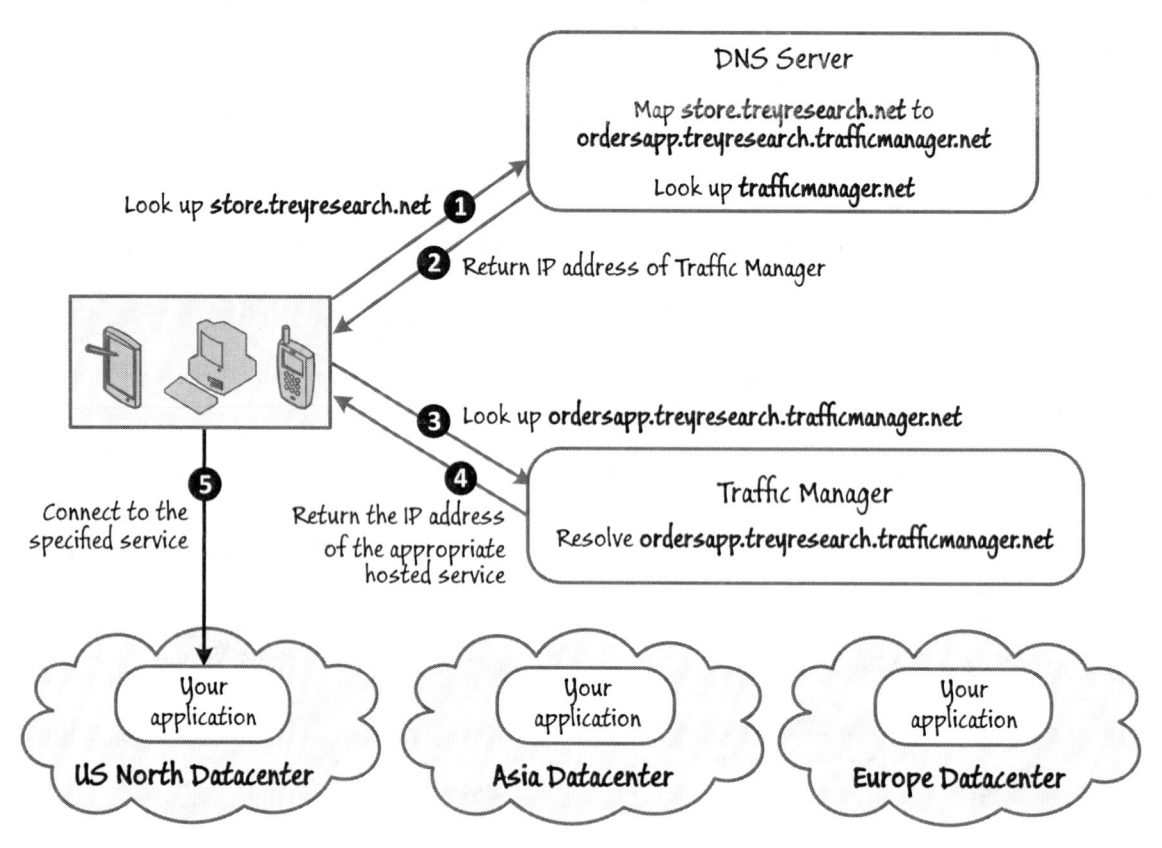

Figure 2
How Windows Azure Traffic Manager performs routing and redirection

Diagram flow:
1. Look up **store.treyresearch.net** → DNS Server (Map store.treyresearch.net to ordersapp.treyresearch.trafficmanager.net; Look up trafficmanager.net)
2. Return IP address of Traffic Manager
3. Look up **ordersapp.treyresearch.trafficmanager.net** → Traffic Manager (Resolve ordersapp.treyresearch.trafficmanager.net)
4. Return the IP address of the appropriate hosted service
5. Connect to the specified service

Datacenters: US North Datacenter, Asia Datacenter, Europe Datacenter (each hosting "Your application")

Global experiments undertaken by the team that develops Windows Azure Traffic Manager indicate that DNS updates typically propagate within the TTL specified in the records in 97% of cases. Changes to a policy will usually propagate to all of the Windows Azure Traffic Manager DNS resolvers within ten minutes. You can check the global propagation of DNS entries using a site such as http://www.just-dnslookup.com/.

The default time-to-live (TTL) value for the DNS responses that Windows Azure Traffic Manager will return to clients is 300 seconds (five minutes). When this interval expires, any requests made by a client application may need to be resolved again, and the new address that results can be used to connect to the service. For testing purposes you may want to reduce this value, but you should use the default or longer in a production scenario.

Remember that there may be intermediate DNS servers between clients and Windows Azure Traffic Manager that are likely to cache the DNS record for the time you specify. However, client applications and web browsers often cache the DNS entries they obtain, and so will not be redirected to a different application deployment until their cached entries expire.

Using Monitoring Endpoints

When you configure a policy in Windows Azure Traffic Manager you specify the port and relative path and name for the endpoint that Windows Azure Traffic Manager will access to test if the application is responding. By default this is port 80 and "/" so that Windows Azure Traffic Manager tests the root path of the application. As long as it receives an HTTP "200 OK" response within ten seconds, Windows Azure Traffic Manager will assume that the hosted service is online.

You can specify a different value for the relative path and name of the monitoring endpoint if required. For example, if you have a page that performs a test of all functions in the application you can specify this as the monitoring endpoint. Hosted applications and services can be included in more than one policy in Windows Azure Traffic Manager, so it is a good idea to have a consistent name and location for the monitoring endpoints in all your applications and services so that the relative path and name is the same and can be used in any policy.

If Windows Azure Traffic Manager detects that every service defined for a policy is offline, it will act as though they were all online, and continue to hand out IP addresses based on the type of policy you specify. This ensures that clients will still receive an IP address in response to a DNS query, even if the service is unreachable.

> If you implement special monitoring pages in your applications, ensure that they can always respond within ten seconds so that Windows Azure Traffic Manager does not mark them as being offline. Also consider the impact on the overall operation of the application of the processes you execute in the monitoring page.

Windows Azure Traffic Manager Policies

At the time of writing Windows Azure Traffic Manager offers the following three routing and load balancing policies, though more may be added in the future:

- The **Performance** policy redirects requests from users to the application in the closest data center. This may not be the application in the data center that is closest in purely geographical terms, but instead the one that provides the lowest network latency. This means that it takes into account the performance of the network routes between the customer and the data center. Windows Azure Traffic Manager also detects failed applications and does not route to these, instead choosing the next closest working application deployment.

Keep in mind that, when using the Performance policy, Windows Azure Traffic Manager bases its selection of target application on availability and average network latency, taking into account the geographical location of the originator of requests and the geographical location of each configured application in the policy (Windows Azure Traffic Manager periodically runs its own internal tests across the Internet between specific locations worldwide and each datacenter). This means that the closest one may always not be the geographically nearest, although this will usually be the case. However, if the application in the geographically nearest datacenter has failed to respond to requests, Windows Azure Traffic Manager may select a location that is not the geographically nearest.

- The **Failover** policy allows you to configure a prioritized list of applications, and Windows Azure Traffic Manager will route requests to the first one in the list that it detects is responding to requests. If that application fails, Windows Azure Traffic Manager will route requests to the next applications in the list, and so on. The Failover policy is useful if you want to provide backup for an application, but the backup application(s) are not designed or configured to be in use all of the time. You can deploy different versions of the application, such as restricted or alternative capability versions, for backup or failover use only when the main application(s) are unavailable. The Failover policy also provides an opportunity for staging and testing applications before release, during maintenance cycles, or when upgrading to a new version.

- The **Round Robin** policy routes requests to each application in turn; though it detects failed applications and does not route to these. This policy evens out the loading on each application, but may not provide users with the best possible response times as it ignores the relative locations of the user and data center.

To minimize network latency and maximize performance you will typically use the Performance policy to redirect all requests from all users to the application in the closest data center. The following sections describe the Performance policy. The other policies are described in the section *"Maximizing Availability for Cloud Applications with Windows Azure Traffic Manager"* later in this appendix.

Guidelines for Using Windows Azure Traffic Manager

The following list contains general guidelines for using Windows Azure Traffic Manager:

- When you name your hosted services and services, consider using a naming pattern that makes them easy to find and identify in the Windows Azure Traffic Manager list of services. Use a naming pattern makes it easier to search for related services using part of the name. Include the datacenter name in the service name so that it is easy to identify the datacenter in which the service is hosted.

- Ensure that Windows Azure Traffic Manager can correctly monitor your hosted applications or services. If you specify a monitoring page instead of the default "/" root path, ensure that the page always responds with an HTTP "200 OK" status, accurately detects the state of the application, and responds well within the ten seconds limit.

- To simplify management and administration, use the facility to enable and disable policies instead of adding and removing policies. Create as many policies as you need and enable only those that are currently applicable. Disable and enable individual services within a policy instead of adding and removing services.
- Consider using Windows Azure Traffic Manager as a rudimentary monitoring solution, even if you do not deploy your application in multiple datacenters or require routing to different deployments. You can set up a policy that includes all of your application deployments (including different applications) by using "/" as the monitoring endpoint. However, you do not direct client requests to Windows Azure Traffic Manager for DNS resolution. Instead, clients connect to the individual applications using the specific URLs you map for each one in your DNS. You can then use the Windows Azure Traffic Manager Web portal to see which deployments of all of the applications are online and offline.

Guidelines for Using Windows Azure Traffic Manager to Reduce Network Latency

The following list contains guidelines for using Windows Azure Traffic Manager to reduce network latency:

- Choose the Performance policy so that users are automatically redirected to the datacenter and application deployment that should provide best response times.
- Ensure that sufficient role instances are deployed in each application to ensure adequate performance, and consider using a mechanism such as that implemented by the Autoscaling Application Block (described earlier in this appendix) to automatically deploy additional instances when demand increases.
- Consider if the patterns of demand in each datacenter are cyclical or time dependent. You may be able to deploy fewer role instances at some times to minimize runtime cost (or even remove all instances so that users are redirected to another datacenter). Again, consider using a mechanism such as that described earlier in this appendix to automatically deploy and remove instances when demand changes.

If all of the hosted applications or services in a Performance policy are offline or unavailable (or availability cannot be tested due to a network or other failure), Windows Azure Traffic Manager will act as though all were online and route requests based on its internal measurements of global network latency based on the location of the client making the request. This means that clients will be able to access the application if it actually is online, or as soon as it comes back online, without the delay while Windows Azure Traffic Manager detects this and starts redirecting users based on measured latency.

Limitations of Using Windows Azure Traffic Manager

The following list identifies some of the limitations you should be aware of when using Windows Azure Traffic Manager:

- All of the hosted applications or services you add to a Windows Azure Traffic Manager policy must exist within the same Windows Azure subscription, although they can be in different namespaces.
- You cannot add hosted applications or services that are staged; they must be running in the production environment. However, you can perform a virtual IP address (VIP) swap to move hosted applications or services into production without affecting an existing Windows Azure Traffic Manager policy.

- All of the hosted applications or services must expose the same operations and use HTTP or HTTPS through the same ports so that Windows Azure Traffic Manager can route requests to any of them. If you expose a specific page as a monitoring endpoint, it must exist at the same location in every deployed application defined in the policy.
- Windows Azure Traffic Manager does not test the application for correct operation; it only tests for an HTTP "200 OK" response from the monitoring endpoint within ten seconds. If you want to perform more thorough tests to confirm correct operation, you should expose a specific monitoring endpoint and specify this in the Windows Azure Traffic Manager policy. However, ensure that the monitoring request (which occurs by default every 30 seconds) does not unduly affect the operation of your application or service.
- Take into account the effects of routing to different deployments of your application on data synchronization and caching. Users may be routed to a datacenter where the data the application uses may not be fully consistent with that in another datacenter.
- Take into account the effects of routing to different deployments of your application on the authentication approach you use. For example, if each deployment uses a separate instance of Windows Azure Access Control Service (ACS), users will need to sign in when rerouted to a different datacenter.

Maximizing Availability for Cloud Applications with Windows Azure Traffic Manager

Windows Azure Traffic Manager provides two policies that you can use to maximize availability of your applications. You can use the Round Robin policy to distribute requests to all application deployments that are currently responding to requests (applications that have not failed). Alternatively, you can use the Failover policy to ensure that a backup deployment of the application will receive requests should the primary one fail. These two policies provide opportunities for two very different approaches to maximizing availability:

- The Round Robin policy enables you to scale out your application across datacenters to achieve maximum availability. Requests will go to a deployment in a datacenter that is online, and the more role instances you configure the lower the average load on each one will be. However, you are charged for each role and application deployment in every datacenter, and you should consider carefully how many role instances to deploy in each application and datacenter.

> There is little reason to use the Round Robin policy if you only deploy your application to one datacenter. You can maximize availability and scale it out simply by adding more role instances. However, the Failover policy is useful if you only deploy to one datacenter because it allows you to define reserve or backup deployments of your application, which may be different from the main highest priority deployment.

- The Failover policy enables you to deploy reserve or backup versions of your application that only receive client requests when all of the higher deployments in the priority list are offline. Unlike the Performance and Round Robin policies, this policy is suitable for use when you deploy to only one datacenter as well as when deploying the application to multiple datacenters. However, you are charged for each application deployment in every datacenter, and you should consider carefully how many role instances to deploy in each datacenter.

    A typical scenario for using the Failover policy is to configure an appropriate priority order for one or more deployments of the same or different versions of the application so that the maximum number of features and the widest set of capabilities are always available, even if services and systems that the application depends on should fail. For example, you may deploy a backup version that can still accept customer orders when the order processing system is unavailable, but stores them securely and informs the customer of a delay.

    By arranging the priority order to use the appropriate reserve version in a different datacenter, or a reduced functionality backup version in the same or a different datacenter, you can offer the maximum availability and functionality at all times. Figure 3 shows an example of this approach.

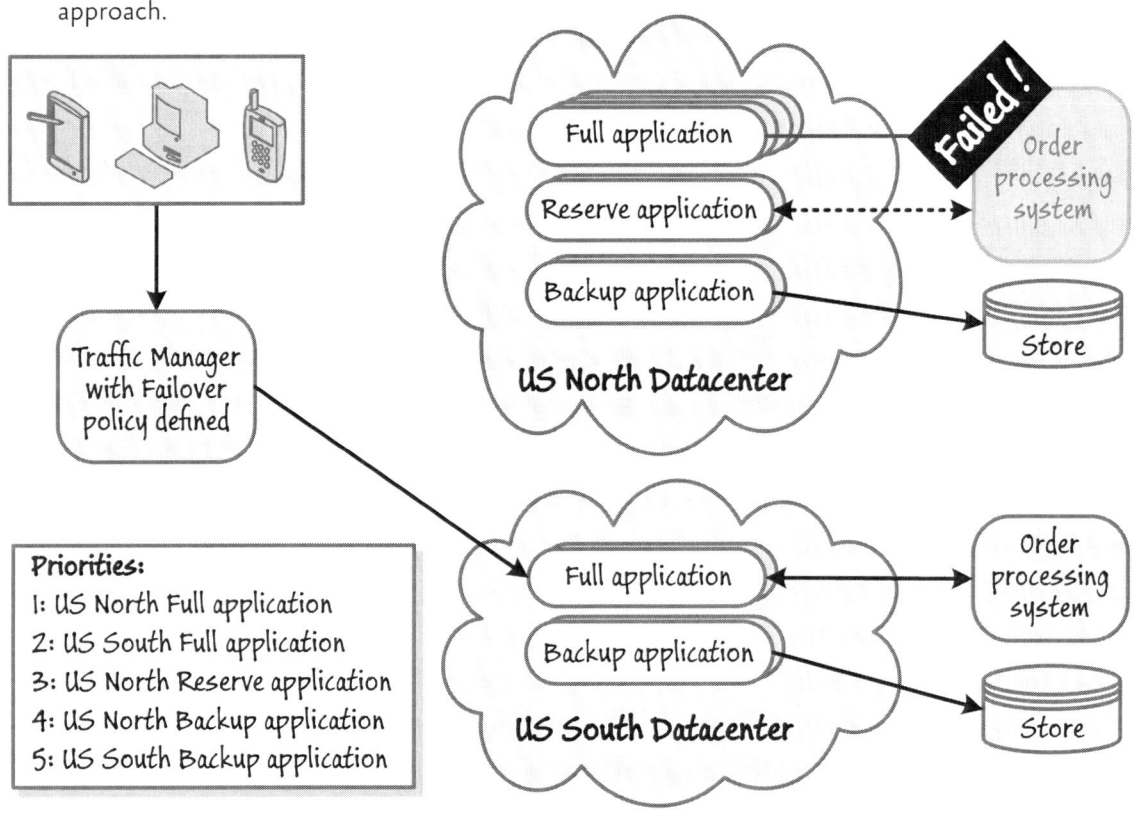

Figure 3
Using the Failover policy to achieve maximum availability and functionality

Guidelines for Using Windows Azure Traffic Manager to Maximize Availability

The following list contains guidelines for using Windows Azure Traffic Manager to maximize availability. Also see the sections "Guidelines for Using Windows Azure Traffic Manager" and "Limitations of Using Windows Azure Traffic Manager" earlier in this appendix.

- Choose the Round Robin policy if you want to distribute requests evenly between all deployments of the application. This policy is typically not suitable when you deploy the application in datacenters that are geographically widely separated as it will cause undue traffic across longer distances. It may also cause problems if you are synchronizing data between datacenters because the data in every datacenter may not be consistent between requests from the same client. However, it is useful for taking services offline during maintenance, testing, and upgrade periods.
- Choose the Failover policy if you want requests to go to one deployment of your application, and only change to another if the first one fails. Windows Azure Traffic Manager chooses the application nearest the top of the list you configured that is online. This policy is typically suited to scenarios where you want to provide backup applications or services.
- If you use the Round Robin policy, ensure that all of the deployed applications are identical so that users have the same experience regardless of the one to which they are routed.
- If you use the Failover policy, consider including application deployments that provide limited or different functionality, and will work when services or systems the application depends on are unavailable, in order to maximize the users' experience as far as possible.
- Consider using the Failover or Round Robin policy when you want to perform maintenance tasks, update applications, and perform testing of deployed applications. You can enable and disable individual applications within the policy as required so that requests are directed only to those that are enabled.
- Because a number of the application deployments will be lightly loaded or not servicing client requests (depending on the policy you choose), consider using a mechanism such as that provided by the Autoscaling Application Block, described earlier in this appendix, to manage the number of role instances for each application deployed in each datacenter to minimize runtime cost.

If all of the hosted applications or services in a Round Robin policy are offline or unavailable (or availability cannot be tested due to a network or other failure), Windows Azure Traffic Manager will act as though all were online and will continue to route requests to each configured application in turn. If all of the applications in a Failover policy are offline or unavailable, Windows Azure Traffic Manager will act as though the first one in the configured list is online and will route all requests to this one.

> *For more information about Windows Azure Traffic Manager, see "Windows Azure Traffic Manager."*

Optimizing the Response Time and Throughput for Cloud Applications by Using Windows Azure Caching

Windows Azure Caching service provides a scalable, reliable mechanism that enables you to retain frequently used data physically close to your applications and services. Windows Azure Caching runs in the cloud, and you can cache data in the same datacenter that hosts your code. If you deploy services to more than one datacenter, you should create a separate cache in each datacenter, and each service should access only the co-located cache. In this way, you can reduce the overhead associated with repeatedly accessing remote data, eliminate the network latency associated with remote data access, and improve the response times for applications referencing this data.

However, caching does not come without cost. Caching data means creating one or more copies of that data, and as soon as you make these copies you have concerns about what happens if you modify this data. Any updates have to be replicated across all copies, but it can take time for these updates to ripple through the system. This is especially true on the Internet where you also have to consider the possibility of network errors causing updates to fail to propagate quickly. So, although caching can improve the response time for many operations, it can also lead to issues of consistency if two instances of an item of data are not identical. Consequently, applications that use caching effectively should be designed to cope with data that may be stale but that eventually becomes consistent.

Do not use Windows Azure Caching for code that executes on-premises as it will not improve the performance of your applications in this environment. In fact, it will likely slow your system down due to the network latency involved in connecting to the cache in the cloud. If you need to implement caching for on-premises applications, you should consider using Windows Server AppFabric Caching instead. For more information, see *"Windows Server AppFabric Caching Features."*

Provisioning and Sizing a Windows Azure Cache

Windows Azure Caching is a service that is maintained and managed by Microsoft; you do not have to install any additional software or implement any infrastructure within your organization to use it. An administrator can easily provision an instance of the Caching service by using the Windows Azure Management Portal. The portal enables an administrator to select the location of the Caching service and specify the resources available to the cache. You indicate the resources to provision by selecting the size of the cache. Windows Azure Caching supports a number of predefined cache sizes, ranging from 128MB up to 4GB. Note that the bigger the cache size the higher the monthly charge.

> Windows Azure Caching is primarily intended for code running in the cloud, such as web and worker roles, and to gain the maximum benefit you implement Windows Azure Caching in the same datacenter that hosts your code.

The size of the cache also determines a number of other quotas. The purpose of these quotas is to ensure fair usage of resources, and imposes limits on the number of cache reads and writes per hour, the available bandwidth per hour, and the number of concurrent connections; the bigger the cache, the more of these resources are available. For example, if you select a 128MB cache, you can currently perform up to 40,000 cache reads and writes, occupying up to 1,400MB of bandwidth (MB per hour), spanning up to 10 concurrent connections, per hour. If you select a 4GB cache you can perform up to 12,800,000 reads and writes, occupying 44,800 MB of bandwidth, and supporting 160 concurrent users each hour.

> *The values specified here are correct at the time of writing, but these quotas are constantly under review and may be revised in the future. You can find information about the current production quota limits and prices at "Windows Azure Shared Caching FAQ."*

You can create as many caches as your applications require, and they can be of different sizes. However, for maximum cost effectiveness you should carefully estimate the amount of cache memory your applications will require and the volume of activity that they will generate. You should also consider the lifetime of objects in the cache. By default, objects expire after 48 hours and will then be removed. You cannot change this expiration period for the cache as a whole, although you can override it on an object by object basis when you store them in the cache. However, be aware that the longer an object resides in cache the more likely it is to become inconsistent with the original data source (referred to as the "authoritative" source) from which it was populated.

To assess the amount of memory needed, for each type of object that you will be storing:

1. Measure the size in bytes of a typical instance of the object (serialize objects by using the **NetDataContractSerializer** class and write them to a file),
2. Add a small overhead (approximately 1%) to allow for the metadata that the Caching service associates with each object,
3. Round this value up to the next nearest value of 1024 (the cache is allocated to objects in 1KB chunks),
4. Multiply this value by the maximum number of instances that you anticipate caching.

Sum the results for each type of object to obtain the required cache size. Note that the Management Portal enables you to monitor the current and peak sizes of the cache, and you can change the size of a cache after you have created it without stopping and restarting any of your services. However, the change is not immediate and you can only request to resize the cache once a day. Also, you can increase the size of a cache without losing objects from the cache, but if you reduce the cache size some objects may be evicted.

You should also carefully consider the other elements of the cache quota, and if necessary select a bigger cache size even if you do not require the volume of memory indicated. For example, if you exceed the number of cache reads and writes permitted in an hour, any subsequent read and write operations will fail with an exception. Similarly, if you exceed the bandwidth quota, applications will receive an exception the next time they attempt to access the cache. If you reach the connection limit, your applications will not be able to establish any new connections until one or more existing connections are closed.

You are not restricted to using a single cache in an application. Each instance of the Windows Azure Caching service belongs to a service namespace, and you can create multiple service namespaces each with its own cache in the same datacenter. Each cache can have a different size, so you can partition your data according to a cache profile; small objects that are accessed infrequently can be held in a 128MB cache, while larger objects that are accessed constantly by a large number of concurrent instances of your applications can be held in a 2GB or 4GB cache.

Implementing Services that Share Data by Using Windows Azure Caching

The Windows Azure Caching service implements an in-memory cache, located on a cache server in a Windows Azure datacenter, which can be shared by multiple concurrent services. It is ideal for holding immutable or slowly changing data, such as a product catalog or a list of customer addresses. Copying this data from a database into a shared cache can help to reduce the load on the database as well as improving the response time of the applications that use this data. It also assists you in building highly scalable and resilient services that exhibit reduced affinity with the applications that invoke them. For example, an application may call an operation in a service implemented as a Windows Azure web role to retrieve information about a specific customer. If this information is copied to a shared cache, the same application can make subsequent requests to query and maintain this customer information without depending on these requests being directed to the same instance of the Windows Azure web role. If the number of client requests increases over time, new instances of the web role can be started up to handle them, and the system scales easily. Figure 4 illustrates this architecture, where an on-premises applications employs the services exposed by instances of a web role. The on-premises application can be directed to any instance of the web role, and the same cached data is still available.

> Windows Azure Caching enables an application to pool connections. When connection pooling is configured, the same pool of connections is shared for a single application instance. Using connection pooling can improve the performance of applications that use the Caching service, but you should consider how this affects your total connection requirements based on the number of instances of your application that may be running concurrently. For more information, see *Understanding and Managing Connections in Windows Azure.*

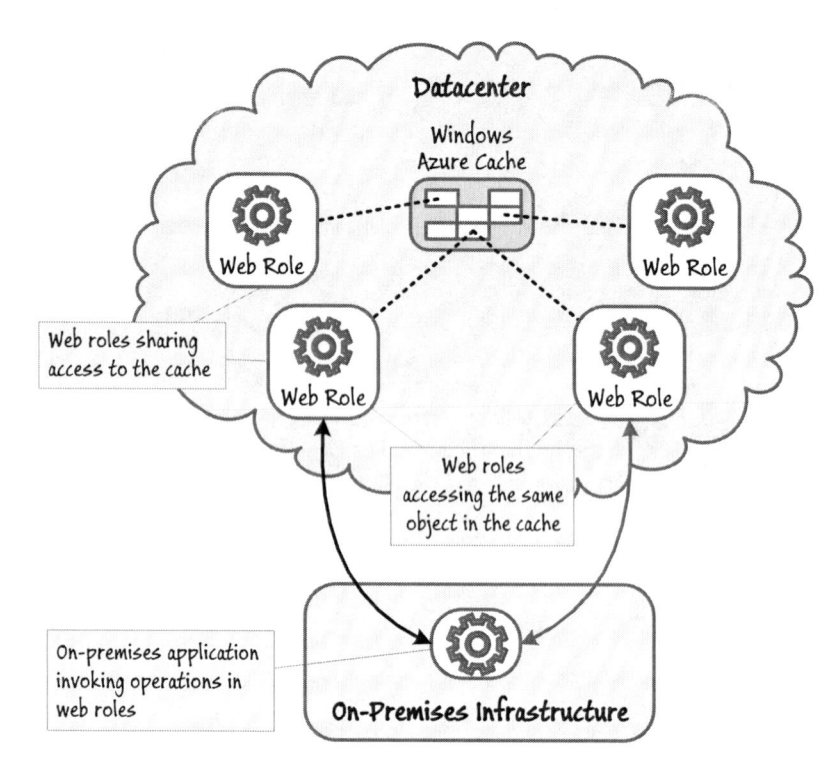

FIGURE 4
Using Windows Azure Caching to provide scalability

> Objects you store in the cache must be serializable.

Web applications access a shared cache by using the Windows Azure Caching APIs. These APIs are optimized to support the *cache-aside* programming pattern; a web application can query the cache to find an object, and if the object is present it can be retrieved. If the object is not currently stored in the cache, the web application can retrieve the data for the object from the authoritative store (such as a SQL Azure database), construct the object using this data, and then store it in the cache.

You can specify which cache to connect to either programmatically or by providing the connection information in a **dataCache-Client** section in the web application configuration file. You can generate the necessary client configuration information from the Management Portal, and then copy this information directly into the configuration file. For more information about configuring web applications to use Windows Azure Caching, see *"How to: Configure a Cache Client using the Application Configuration File for Windows Azure Caching."*

As described in the section "Provisioning and Sizing a Windows Azure Cache," an administrator specifies the resources available for caching data when the cache is created. If memory starts to run short, the Windows Azure Caching service will evict data on a least recently used basis. However, cached objects can also have their own independent lifetimes, and a developer can specify a period for caching an object when it is stored; when this time expires, the object is removed and its resources reclaimed.

For detailed information on using Windows Azure Caching APIs see *Developing for Windows Azure Shared Caching.*

> With the Windows Azure Caching service, your applications are not notified when an object is evicted from the cache or expires, so be warned.

Updating Cached Data

Web applications can modify the objects held in cache, but be aware that if the cache is being shared, more than one instance of an application might attempt to update the same information; this is identical to the update problem that you meet in any shared data scenario. To assist with this situation, the Windows Azure Caching APIs support two modes for updating cached data:

- **Optimistic, with versioning.**

    All cached objects can have an associated version number. When a web application updates the data for an object it has retrieved from the cache, it can check the version number of the object in the cache prior to storing the changes. If the version number is the same, it can store the data. Otherwise the web application should assume that another instance has already modified this object, fetch the new data, and resolve the conflict using whatever logic is appropriate to the business processing (maybe present the user with both versions of the data and ask which one to save). When an object is updated, it should be assigned a new unique version number when it is returned to the cache.

- **Pessimistic, with locking.**

    The optimistic approach is primarily useful if the chances of a collision are small, and although simple in theory the implementation inevitably involves a degree of complexity to handle the possible race conditions that can occur. The pessimistic approach takes the opposite view; it assumes that more than one instance of a web application is highly likely to try and simultaneously modify the same data, so it locks the data when it is retrieved from the cache to prevent this situation from occurring. When the object is updated and returned to the cache, the lock is released. If a web application attempts to retrieve and lock an object that is already locked by another instance, it will fail (it will not be blocked). The web application can then back off for a short period and try again. Although this approach guarantees the consistency of the cached data, ideally, any update operations should be very quick and the corresponding locks of a very short duration to minimize the possibility of collisions and to avoid web applications having to wait for extended periods as this can impact the response time and throughput of the application.

> An application specifies a duration for the lock when it retrieves data. If the application does not release the lock within this period, the lock is released by the Windows Azure Caching service. This feature is intended to prevent an application that has failed from locking data indefinitely. You should stipulate a period that will give your application sufficient time to perform the update operation, but not so long as to cause other instances to wait for access to this data for an excessive time.

If you are hosting multiple instances of the Windows Azure Caching service across different datacenters, the update problem becomes even more acute as you may need to synchronize a cache not only with the authoritative data source but also other caches located at different sites. Synchronization necessarily generates network traffic, which in turn is subject to the latency and occasionally unreliable nature of the Internet. In many cases, it may be preferable to update the authoritative data source directly, remove the data from the cache in the same datacenter as the web application, and let the cached data at each remaining site expire naturally, when it can be repopulated from the authoritative data source.

The logic that updates the authoritative data source should be composed in such a way as to minimize the chances of overwriting a modification made by another instance of the application, perhaps by including version information in the data and verifying that this version number has not changed when the update is performed.

The purpose of removing the data from the cache rather than simply updating it is to reduce the chance of losing changes made by other instances of the web application at other sites and to minimize the chances of introducing inconsistencies if the update to the authoritative data store is unsuccessful. The next time this data is required, a consistent version of the data will be read from the authoritative data store and copied to the cache.

If you require a more immediate update across sites, you can implement a custom solution by using Service Bus topics implementing a variation on the patterns described in the section "Replicating and Synchronizing Data Using Service Bus Topics and Subscriptions" in "Appendix A - Replicating, Distributing, and Synchronizing Data."

Both approaches are illustrated later in this appendix, in the section "Guidelines for Using Azure Caching."

The nature of the Windows Azure Caching service means that it is essential you incorporate comprehensive exception-handling and recovery logic into your web applications. For example:

- A race-condition exists in the simple implementation of the cache-aside pattern, which can cause two instances of a web application to attempt to add the same data to the cache. Depending on how you implement the logic that stores data in the cache, this can cause one instance to overwrite the data previously added by another (if you use the **Put** method of the cache), or it can cause the instance to fail with a **DataCacheException** exception (if you use the **Add** method of the cache). For more information, see the topic *"Add an Object to a Cache."*

- Be prepared to catch exceptions when attempting to retrieve locked data and implement an appropriate mechanism to retry the read operation after an appropriate interval, perhaps by using the Transient Fault Handling Application Block.

- You should treat a failure to retrieve data from the Windows Azure Caching service as a cache miss and allow the web application to retrieve the item from the authoritative data source instead.

- If your application exceeds the quotas associated with the cache size, your application may no longer be able to connect to the cache. You should log these exceptions, and if they occur frequently an administrator should consider increasing the size of the cache.

> Incorporating Windows Azure Caching into a web application must be a conscious design decision as it directly affects the update logic of the application. To some extent you can hide this complexity and aid reusability by building the caching layer as a library and abstracting the code that retrieves and updates cached data, but you must still implement this logic somewhere.

Implementing a Local Cache

As well as the shared cache, you can configure a web application to create its own local cache. The purpose of a local cache is to optimize repeated read requests to cached data. A local cache resides in the memory of the application, and as such it is faster to access. It operates in tandem with the shared cache. If a local cache is enabled, when an application requests an object, the caching client library first checks to see whether this object is available locally. If it is, a reference to this object is returned immediately without contacting the shared cache. If the object is not found in the local cache, the caching client library fetches the data from the shared cache and then stores a copy of this object in the local cache. The application then references the object from the local cache. Of course, if the object is not found in the shared cache, then the application must retrieve the object from the authoritative data source instead.

Once an item has been cached locally, the local version of this item will continue to be used until it expires or is evicted from the cache. However, it is possible that another application may modify the data in the shared cache. In this case the application using the local cache will not see these changes until the local version of the item is removed from the local cache. Therefore, although using a local cache can dramatically improve the response time for an application, the local cache can very quickly become inconsistent if the information in the shared cache changes. For this reason you should configure the local cache to only store objects for a short time before refreshing them. If the data held in a shared cache is highly dynamic and consistency is important, you may find it preferable to use the shared cache rather than a local cache.

After an item has been copied to the local cache, the application can then access it by using the same Windows Azure Caching APIs and programming model that operate on a shared cache; the interactions with the local cache are completely transparent. For example, if the application modifies an item and puts the updated item back into the cache, the Windows Azure Caching APIs update the local cache and also the copy in the shared cache.

A local cache is not subject to the same resource quotas as the shared cache managed by the Windows Azure Caching service. You specify the maximum number of objects that the cache can hold when it is created, and the storage for the cache is allocated directly from the memory available to the application.

> You enable local caching by populating the **LocalCacheProperties** member of the **DataCacheFactoryConfiguration** object that you use to manage your cache client configuration. You can perform this task programmatically or declaratively in the application configuration file. You can specify the size of the cache and the default expiration period for cached items. For more information, see the topic "Enable Windows Server AppFabric Local Cache (XML)."

Caching Web Application Session State

The Windows Azure Caching service enables you to use the **DistributedCacheSessionStateStoreProvider** session state provider for ASP.NET web applications and services. With this provider, you can store session state in a Windows Azure cache. Using a Windows Azure cache to hold session state gives you several advantages:

- It can share session state among different instances of ASP.NET web applications providing improved scalability,
- It supports concurrent access to same session state data for multiple readers and a single writer, and
- It can use compression to save memory and bandwidth.

You can configure this provider either through code or by using the application configuration file; you can generate the configuration information by using the Management Portal and copy this information directly into the configuration file. For more information, see *"How to: Configure the ASP.NET Session State Provider for Windows Azure Caching."*

Once the provider is configured, you access it programmatically through the **Session** object, employing the same code as an ordinary ASP.NET web application; you do not need to invoke the Windows Azure Caching APIs.

Caching HTML Output

The **DistributedCacheOutputCacheProvider** class available for the Windows Azure Caching service implements output caching for web applications. Using this provider, you can build scalable web applications that take advantage of the Windows Azure Caching service for caching the HTTP responses that they generate for web pages returned to client applications, and this cache can be shared by multiple instances of an application. This provider has several advantages over the regular per process output cache, including:

- You can cache larger amounts of output data.
- The output cache is stored externally from the worker process running the web application and it is not lost if the web application is restarted.
- It can use compression to save memory and bandwidth.

Again, you can generate the information for configuring this provider by using the Management Portal and copy this information directly into the application configuration file. For more information, see *"How to: Configure the ASP.NET Output Cache Provider for Windows Azure Caching."*

Like the **DistributedCacheSessionStateStoreProvider** class, the **DistributedCacheOutputCacheProvider** class is completely transparent; if your application previously employed output caching, you do not have to make any changes to your code.

Guidelines for Using Windows Azure Caching

The following scenarios describe some common scenarios for using Windows Azure Caching:

- **Web applications and services running in the cloud require fast access to data. This data is queried frequently, but rarely modified. The same data may be required by all instances of the web applications and services.**

    This is the ideal case for using Windows Azure Caching. In this simple scenario, you can configure the Windows Azure Caching service running in the same datacenter that hosts the web applications and services (implemented as web or worker roles). Each web application or service can implement the cache-aside pattern when it needs a data item; it can attempt to retrieve the item from cache, and if it is not found then it can be retrieved from the authoritative data store and copied to cache. If the data is static, and the cache is configured with sufficient memory, you can specify a long expiration period for each item as it is cached. Objects representing data that might change in the authoritative data store should be cached with a shorter expiration time; the period should reflect the frequency with which the data may be modified and the urgency of the application to access the most recently updated information.

    Figure 5 shows a possible structure for this solution. In this example, a series of web applications implemented as web roles, hosted in different datacenters, require access to customer addresses held in a SQL Server database located on-premises within an organization. To reduce the network latency associated with making repeated requests for the same data across the Internet, the information used by the web applications is cached by using the Windows Azure Caching service. Each datacenter contains a separate instance of the Caching service, and web applications only access the cache located in the same datacenter. The web applications only query customer addresses, although other applications running on-premises may make the occasional modification. The expiration period for each item in the cache is set to 24 hours, so any changes made to this data will eventually be visible to the web applications.

> To take best advantage of Windows Azure Caching, only cache data that is unlikely to change frequently.

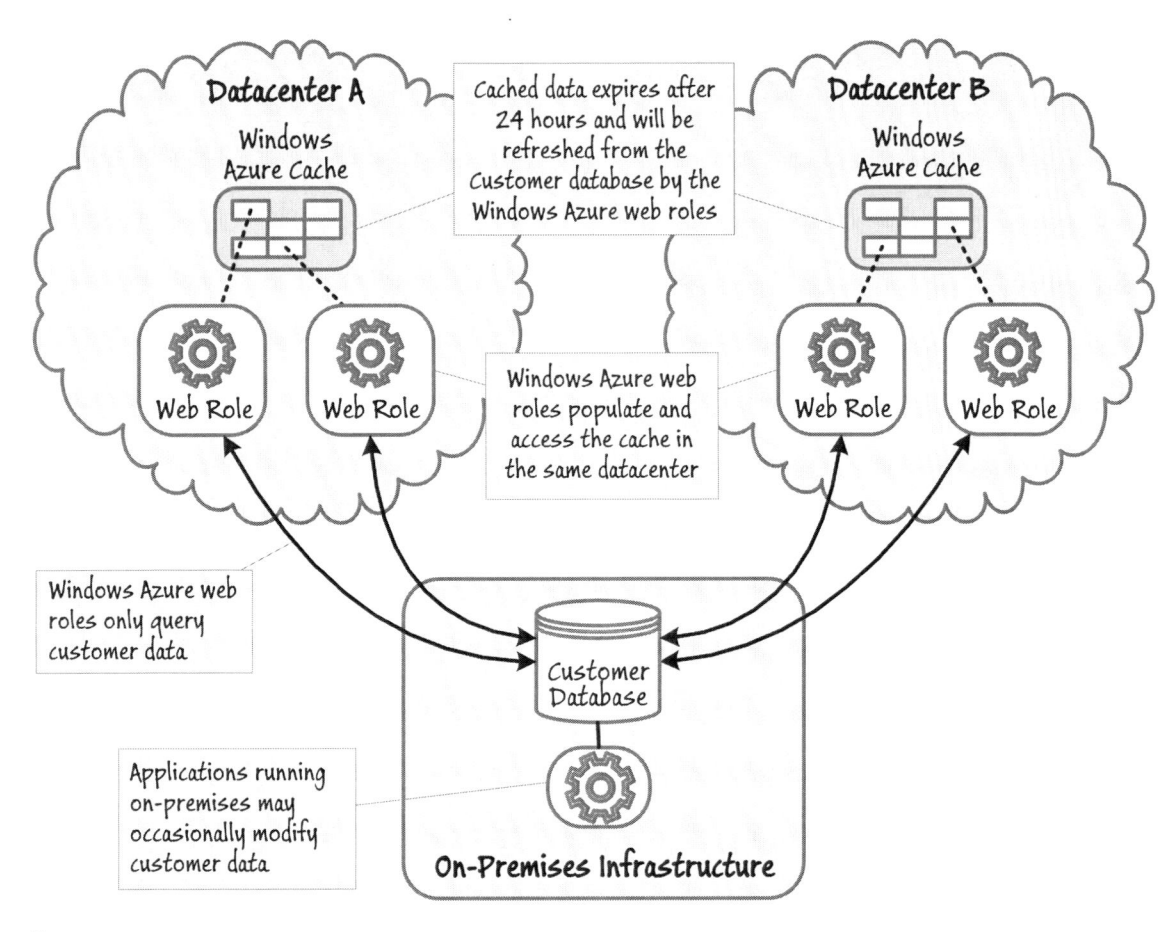

FIGURE 5
Caching static data to reduce network latency in web applications

- **Web applications and services running in the cloud require fast access to shared data, and they may frequently modify this data.**

    This scenario is a potentially complex extension of the previous case, depending on the location of the data, the frequency of the updates, the distribution of the web applications and services, and the urgency with which the updates must be visible to these web applications and services.

    In the most straightforward case, when a web application needs to update an object, it retrieves the item from cache (first fetching it from the authoritative data store if necessary), modifies this item in cache, and makes the corresponding change to the authoritative data store. However, this is a two-step process, and to minimize the chances of a race condition occurring all updates must follow the same order in which they perform these steps. Depending on the likelihood of a conflicting update being made by a concurrent instance of the application, you can implement either the optimistic or pessimistic strategy for updating the cache as described in the earlier section "Updating Cached Data." Figure 6 depicts this process. In this example, the on-premises Customer database is the authoritative data store.

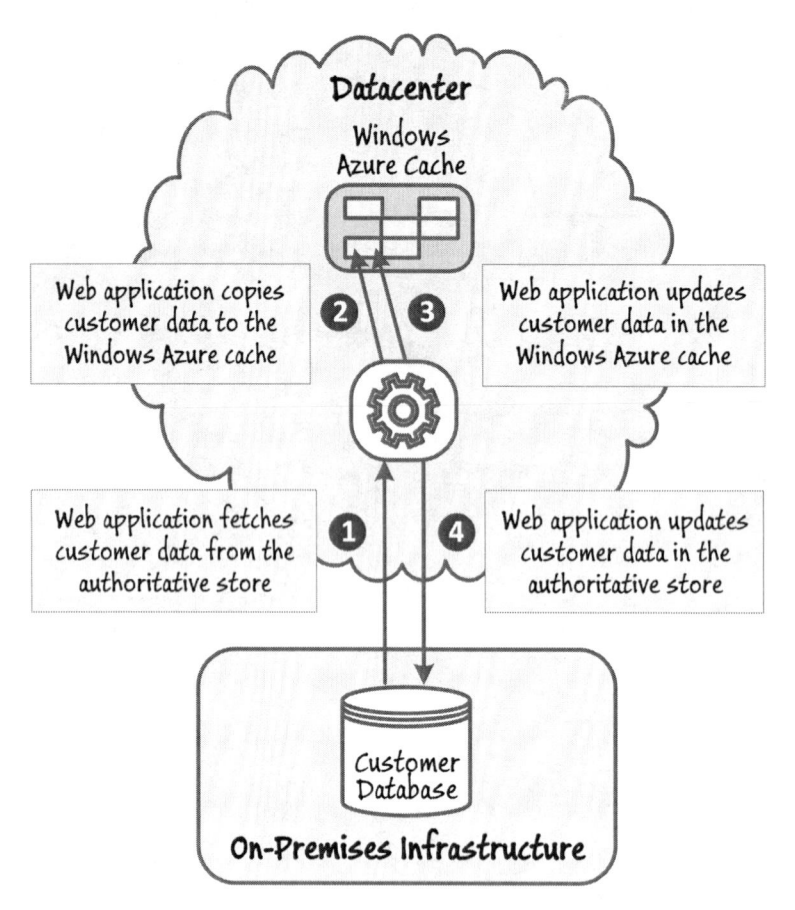

FIGURE 6
Updating data in the cache and the authoritative data store

The approach just described is suitable for a solution contained within a single datacenter. However, if your web applications and services span multiple sites, you should implement a cache at each datacenter. Now updates have to be carefully synchronized and coordinated across datacenters and all copies of the cached data modified. As described in the section "Updating Cached Data," you have at least two options available for tackling this problem:
- Only update the authoritative data store and remove the item from the cache in the datacenter hosting the web application. The data cached at each other datacenter will eventually expire and be removed from cache. The next time this data is required, it will be retrieved from the authoritative store and used to repopulate the cache.
- Implement a custom solution by using Service Bus topics similar to that described in the section "Replicating and Synchronizing Data Using Service Bus Topics and Subscriptions" in "Appendix A - Replicating, Distributing, and Synchronizing Data."

The first option is clearly the simpler of the two, but the various caches may be inconsistent with each other and the authoritative data source for some time, depending on the expiration period applied to the cached data. Additionally, the web applications and services may employ a local SQL Azure database rather than accessing an on-premises installation of SQL Server. These SQL Azure databases can be replicated and synchronized in each datacenter as described in "Appendix A - Replicating, Distributing, and Synchronizing Data." This strategy reduces the network latency associated with retrieving the data when populating the cache at the cost of yet more complexity if web applications modify this data; they update the local SQL Azure database, and these updates must be synchronized with the SQL Azure databases at the other datacenters.

Depending on how frequently this synchronization occurs, cached data at the other datacenters could be out of date for some considerable time; not only does the data have to expire in the cache, it also has to wait for the database synchronization to occur. In this scenario, tuning the interval between database synchronization events as well as setting the expiration period of cached data is crucial if a web application must minimize the amount of time it is prepared to handle stale information. Figure 7 shows an example of this solution with replicated instances of SQL Azure acting as the authoritative data store.

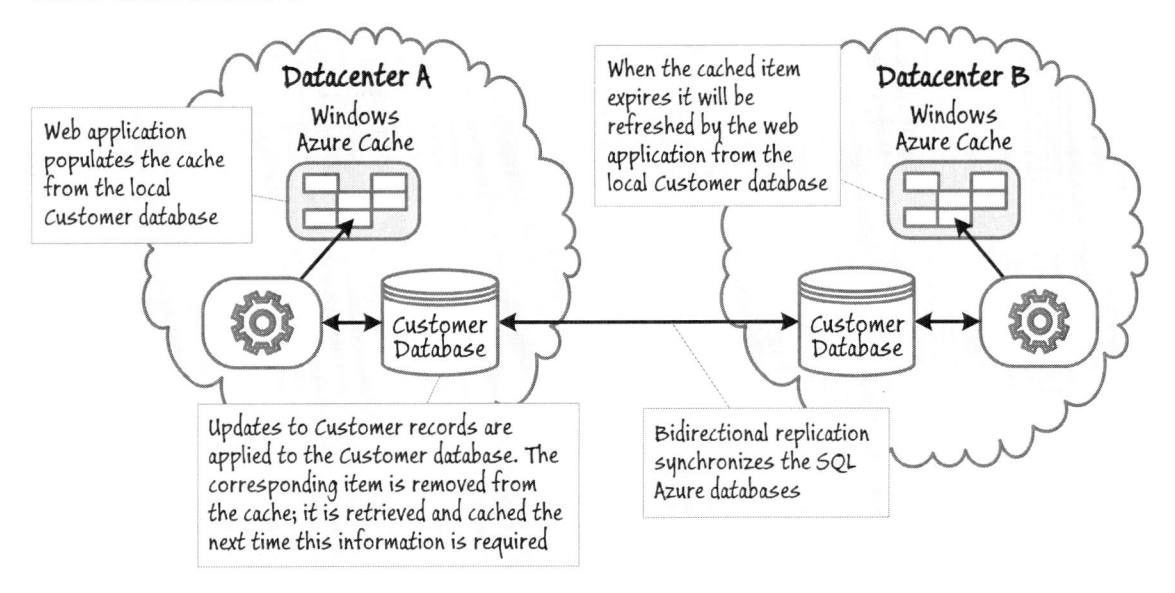

FIGURE 7
Propagating updates between Windows Azure caches and replicated data stores

Implementing a custom solution based on Service Bus topics and subscriptions is more complex, but results in the updates being synchronized more quickly across datacenters. Figure 8 illustrates one possible implementation of this approach. In this example, a web application retrieves and caches data in the Windows Azure cache hosted in the same datacenter. Performing a data update involves the following sequence of tasks:
- The web application updates the authoritative data store (the on-premises database).
- If the database update was successful, the web application duplicates this modification to the data held in the cache in the same datacenter.
- The web application posts an update message to a Service Bus topic.
- Receiver applications running at each datacenter subscribe to this topic and retrieve the update messages.
- The receiver application applies the update to the cache at this datacenter if the data is currently cached locally.

If the data is not currently cached at this datacenter the update message can simply be discarded.

The receiver at the datacenter hosting the web application that initiated the update will also receive the update message. You might include additional metadata in the update message with the details of the instance of the web application that posted the message; the receiver can then include logic to prevent it updating the cache unnecessarily (when the web application instance that posted the message is the same as the current instance).

Note that, in this example, the authoritative data source is located on-premises, but this model can be extended to use replicated instances of SQL Azure at each datacenter. In this case, each receiver application could update the local instance of SQL Azure as well as modifying the data in-cache.

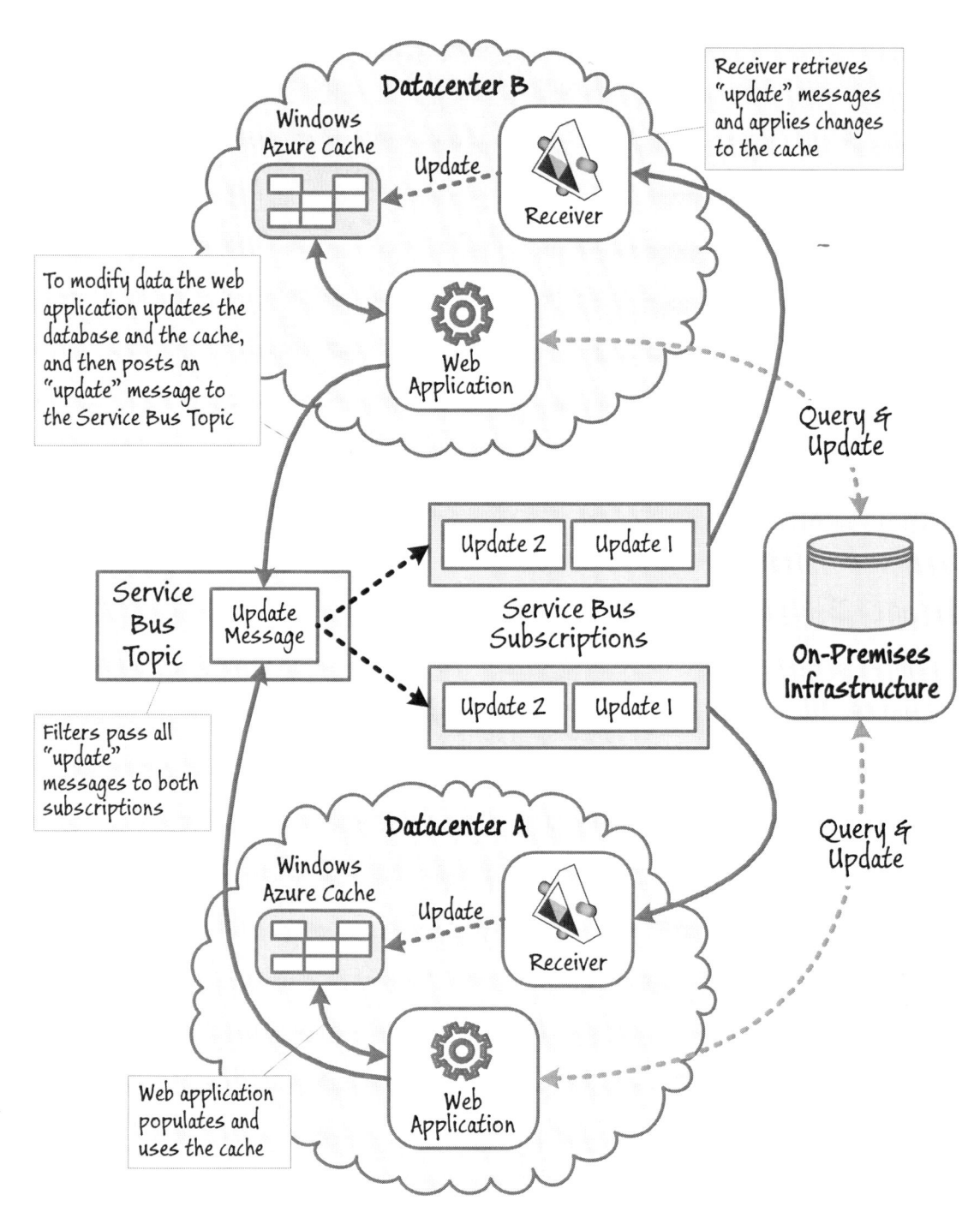

Figure 8
Propagating updates between Windows Azure caches and an authoritative data store

It is also possible that there is no permanent data store and the caches themselves act as the authoritative store. Examples of this scenario include online gaming, where the current game score is constantly updated but needs to be available to all instances of the game application. In this case, the cache at each datacenter holds a copy of all of the data, but the same general solution depicted by Figure 8, without the on-premises database, can still be applied.

- **A web application requires fast access to the data that it uses. This data is not referenced by other instances of the web application.**

  In this scenario, the data is effectively private to an instance of the web application and can be cached in-memory in the application itself. You can implement this solution in many ways, but the most convenient and extensible approach is probably to use the Windows Azure Caching APIs, and to configure the application as a Windows Azure cache client and enable the local cache properties. This configuration was described in the section "Implementing a Local Cache" earlier in this appendix. This approach also enables you to quickly switch to using a shared cache without modifying your code; you simply reconfigure the data cache client settings.

  As the data is not shared, updates are straightforward; the application can simply modify the data in the authoritative data source and, if successful, apply the same changes to the cached data in-memory (this will also update data in the shared cache from which the local cache is initially populated, as described in the in the section "Implementing a Local Cache."

  In a variant on this scenario, two or more instances of a web application cache data locally, but they access overlapping data from the authoritative data store. In this case, if one instance modifies the data and writes the changes to the authoritative data store, the cached data at the other instance is now out of date. This is essentially the same problem addressed earlier with multiple shared caches. If immediate synchronization between instances of the web application is important, then caching data in-memory is not the most suitable approach and it is best to use a shared cache. However, data in the local cache expires in a manner similar to that of a shared cache except the default expiration period is much shorter—5 minutes. If applications can handle stale data for a short while, then using a local cache configured with a suitable lifetime for cached objects may be appropriate.

> Unlike a shared cache, you can modify the default expiration time for a local cache. You can still override this period as you cache each object, but beware of attempting to retain objects in a local cache for a lengthy period as they might become stale very quickly.

Caching data in-memory in the web application speeds access to this data, but as described earlier it can reduce the consistency of your solution. You should also be aware of the increased memory requirements of your applications and the potential charges associated with hosting applications with an increased memory footprint, especially if they attempt to cache large amounts of data. You should carefully balance these benefits and concerns against the requirements of your application.

Figure 9 shows an example of this scenario with several instances of a web application using a local cache to optimize access to data held in an on-premises database. It does not matter whether the web application instances are located in the same or different datacenters, caching the data in-memory in each instance makes them independent from each other for query purposes. Some updates may occur, and in this example the data referenced by each instance overlaps. Therefore the cached objects are configured with a suitable expiration period to enable them to be refreshed appropriately and to prevent them from becoming too stale.

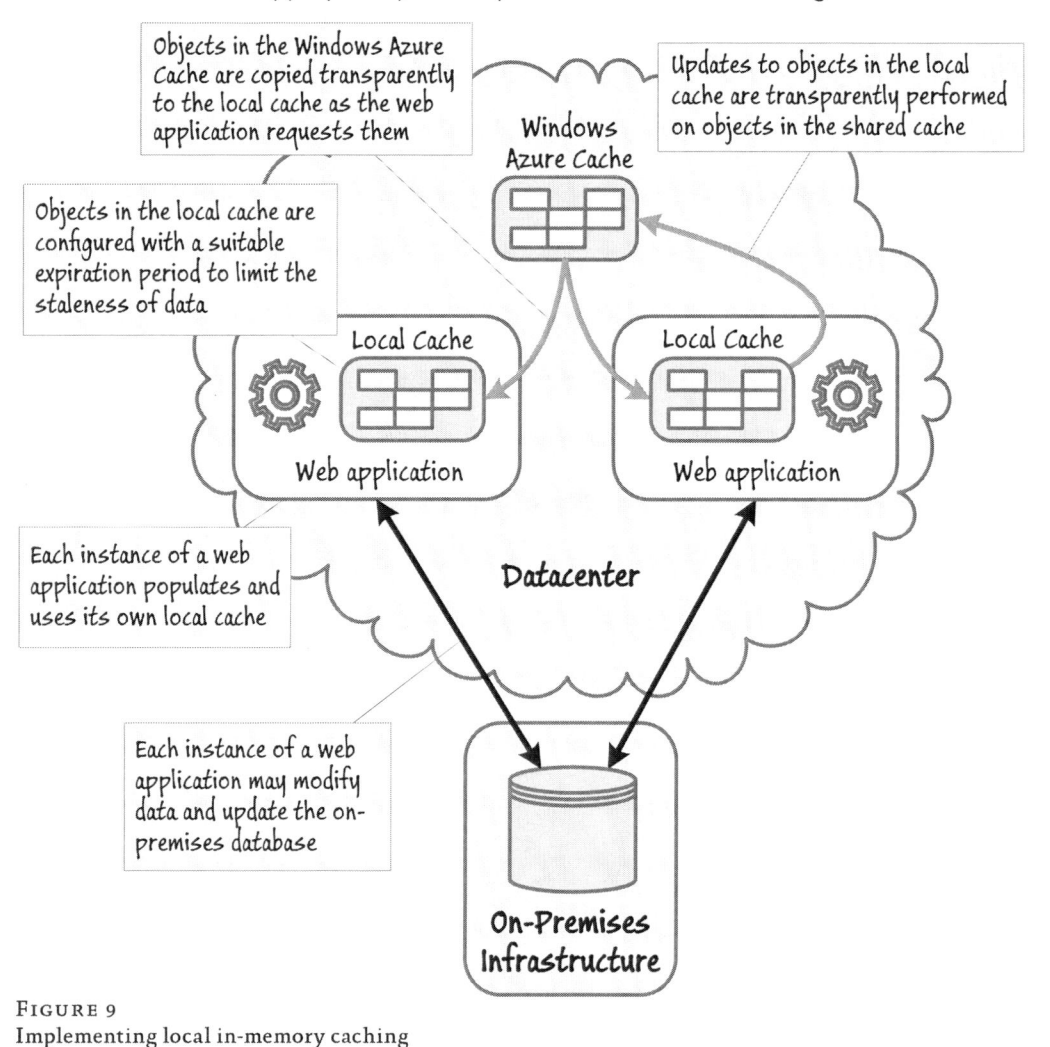

FIGURE 9
Implementing local in-memory caching

- **You have built a web application hosted by using a Windows Azure web role. The web application needs to cache session state information, but this information must not be pinned to a particular instance of the web role; if the web application fails and the web role is restarted, the session state information must not be lost.**

    One of the primary reasons for using Windows Azure to host web applications is the scalability that this platform provides. As the load on a web application increases, you can use a technology such as the Enterprise Library Autoscaling Application Block to automatically start new instances and distribute the work more evenly (for more information, see the section "Managing Elasticity in the Cloud by Using the Enterprise Library Autoscaling Application Block" earlier in this appendix.) Additionally, the reliability features of Windows Azure ensure that an application can be restarted if it should fail for some reason.

    However, these scalability and reliability features assume that a client using the web application can connect to any instance of the web application. If the web application uses sessions and stores session state information, then you must avoid tying this state information to a specific instance of the application. For example, if you are using ASP.NET to build a web application, session state is stored in-memory within the web application by default. In this model, a client connecting to different instances of the web application at different times may see different session state information each time it connects. This phenomenon is undesirable in a scalable web application.

    The **DistributedCacheSessionStateStoreProvider** session state provider enables you to configure a web application to store session state out-of-process, using the Windows Azure Caching service as the storage mechanism. Different instances of the web application can then access the same session state information. This provider is transparent to the web application, which can continue to use the same ASP.NET syntax to access session state information. For more information, refer to the section "Caching Web Application Session State" earlier in this appendix.

    Note that while the **DistributedCacheSessionStateStoreProvider** session state provider enables instances of web applications running in the same datacenter to share session data, each datacenter should be configured with its own shared cache. This may have an impact on your solution if you are using a technology such as Windows Azure Traffic Manager to route client requests to web applications. For example, the Windows Azure Traffic Manager Round Robin policy and some edge cases of the Performance policy may redirect a client to a different datacenter holding different session state for some requests, as shown in Figure 10.

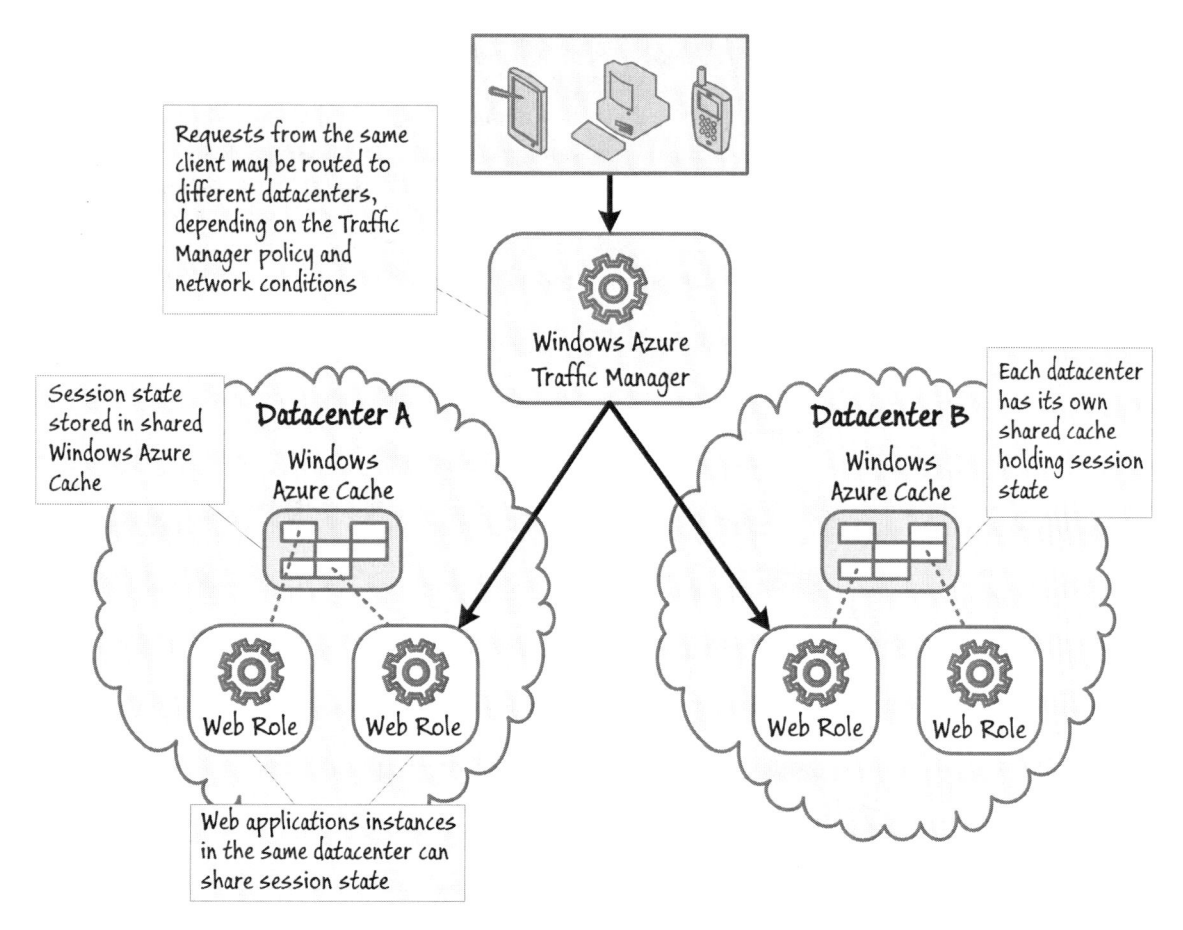

FIGURE 10
Client requests obtaining different session state from different datacenters

- **You have built a web application that performs complex processing and rendering of results based on a series of query parameters. You need to improve the response time of various pages served by this application, and avoid repeatedly performing the same processing when different clients request pages.**

    This is the classic scenario for implementing output caching. The output generated by an ASP.NET web page can be cached at the server hosting the web application, and subsequent requests to access the same page with the same query parameters can be satisfied by responding with the cached version of the page rather than generating a new response. For more information about how ASP.NET output caching works and how to use it, see *"Caching ASP.NET Pages."*

However, the default output cache provider supplied with ASP.NET operates on a per server basis. In the Windows Azure environment a web server equates to an instance of a web role, so using the default output cache provider causes each web role instance to generate its own cached output. If the volume of cached output is large and each cached page is the result of heavy, intensive processing, then each web role instance may end up duplicating this processing and storing a copy of the same results. The **DistributedCacheOutputCacheProvider** class enables web roles to store the output cache in a shared Windows Azure cache, removing this duplication of space and effort. For more information, see the section "Caching HTML Output" earlier in this appendix.

As with the session cache, you should create and use a separate shared cache for caching output data at each datacenter.

Limitations of Windows Azure Caching

The features provided by the Windows Azure Caching service are very similar to those of Windows Server AppFabric Caching; they share the same programmatic APIs and configuration methods. However the Windows Azure implementation provides only a subset of the features available to the Windows Server version. Currently, the Windows Azure Caching service has the following limitations compared to Windows Server AppFabric Caching:

- It does not support notifications. Your applications are not informed if an object expires or is evicted from cache.
- You cannot change the default expiration period for a shared cache. Objects expire in the shared cache after 48 hours, and you cannot modify this setting for the cache as a whole. However, you can override this value on an object by object basis as you store them in the cache. In contrast, you can modify the default expiration period for a local cache (the default duration is 5 minutes).
- You cannot disable the eviction policy. If there is insufficient space in the cache for a new object, older objects will be evicted following the least recently used principle.
- You cannot explicitly remove an item from the cache.
- You cannot partition cached data. A Windows Azure cache cannot contain user-defined named regions.
- You cannot add tags to cached data to assist with object searches.

Windows Azure Caching may remove some of these limitations in future releases.

You should also note that a Windows Azure cache automatically benefits from the reliability and scalability features of Windows Azure; you do not have to manage these aspects yourself. Consequently, many of the high availability features of Windows Server AppFabric Caching are not available because they are not required in the Windows Azure environment.

For more information about the differences between Windows Azure Caching and Windows Server AppFabric Caching, see the topic *"Differences Between Caching On-Premises and in the Cloud."*

Guidelines for Securing Windows Azure Caching

You access a Windows Azure cache through an instance of the Windows Azure Caching service. You generate an instance of the Windows Azure Caching service by using the Management Portal and specifying a new service namespace for the Caching service. The Caching service is deployed to a datacenter in the cloud, and has endpoints with URLs that are based on the name of the service namespace with the suffix ".cache.windows.net". Your applications connect to the Caching service using these URLs. The Caching service exposes endpoints that support basic HTTP connectivity (via port 22233) as well as SSL (via port 22243).

All connection requests from an application to the Windows Azure Caching service are authenticated and authorized by using ACS. To connect to the Caching service, an application must provide the appropriate authentication token.

More Information

All links in this book are accessible from the book's online bibliography available at:
http://msdn.microsoft.com/en-us/library/hh968447.aspx.

- "Windows Azure CDN" at *http://msdn.microsoft.com/en-us/gg405416*.
- Chapter 3, "Accessing the Surveys Application" in the guide "Developing Applications for the Cloud, 2nd Edition" at *http://msdn.microsoft.com/en-us/library/hh534477.aspx*.
- "Microsoft Enterprise Library 5.0 Integration Pack for Windows Azure" at *http://msdn.microsoft.com/en-us/library/hh680918(v=pandp.50).aspx*.
- "Windows Azure Traffic Manager" at *http://msdn.microsoft.com/en-us/gg197529*.
- "Windows Server AppFabric Caching Features" at *http://msdn.microsoft.com/en-us/library/ff383731.aspx*.
- "Windows Azure Shared Caching FAQ" at *http://msdn.microsoft.com/en-us/library/hh697522.aspx*.
- "Understanding and Managing Connections in Windows Azure" at *http://msdn.microsoft.com/en-us/library/hh552970.aspx*.
- "How to: Configure a Cache Client using the Application Configuration File for Windows Azure Caching" at *http://msdn.microsoft.com/en-us/library/windowsazure/gg278346.aspx*.
- "Developing for Windows Azure Shared Caching" at *http://msdn.microsoft.com/en-us/library/windowsazure/gg278342.aspx*.
- "Add an Object to a Cache" at *http://msdn.microsoft.com/en-us/library/ee790846.aspx*.

> Only web applications and services running in the cloud need to be provided with the authentication token for connecting to the Windows Azure Caching service as these are the only items that should connect to the cache. Utilizing a Windows Azure cache from code running externally to the datacenter provides little benefit other than for testing when using the Windows Azure compute emulator, and is not a supported scenario for production purposes.

- "Enable Windows Server AppFabric Local Cache (XML)" at *http://msdn.microsoft.com/en-us/library/ee790880.aspx*.
- "How to: Configure the ASP.NET Session State Provider for Windows Azure Caching" at *http://msdn.microsoft.com/en-us/library/windowsazure/gg278339.aspx*.
- "How to: Configure the ASP.NET Output Cache Provider for Windows Azure Caching" at *http://msdn.microsoft.com/en-us/library/windowsazure/gg185676.aspx*.
- "Caching ASP.NET Pages" at *http://msdn.microsoft.com/en-us/library/06bh14hk(v=VS.100).aspx*.
- "Differences Between Caching On-Premises and in the Cloud" at *http://msdn.microsoft.com/en-us/library/windowsazure/gg185678.aspx*.

APPENDIX F

Monitoring and Managing Hybrid Applications

A typical hybrid application comprises a number of components, built using a variety of technologies, distributed across a range of sites and connected by networks of varying bandwidth and reliability. With this complexity, it is very important to be able to monitor how well the system is functioning, and quickly take any necessary restorative action in the event of failure. However, monitoring a complex system is itself a complex task, requiring tools that can quickly gather performance data to help you analyze throughput and pinpoint the causes of any errors, failures, or other shortcomings in the system.

The range of problems can vary significantly, from simple failures caused by application errors in a service running in the cloud, through issues with the environment hosting individual elements, to complete systemic failure and loss of connectivity between components whether they are running on-premises or in the cloud.

Once you have been alerted to a problem, you must be able to take the appropriate steps to correct it and keep the system functioning. The earlier you can detect issues and the more quickly you can fix them, the less impact they will have on the operations that your business depends on, and on the customers using your system.

It is important to follow a systematic approach, not only when managing hybrid applications but also when deploying and updating the components of your systems. You should try to minimize the performance impact of the monitoring and management process, and you should avoid making the entire system unavailable if you need to update specific elements.

Collecting diagnostic information about the way in which your system operates is also a fundamentally important part in determining the capacity of your solution, and this in turn can affect any service level agreement (SLA) that you offer users of your system. By monitoring your system you can determine how it is consuming resources as the volume of requests increases or decreases, and this in turn can assist in assessing the resources required and the running costs associated with maintaining an agreed level of performance.

This appendix explores the challenges encountered in keeping your applications running well and fulfilling your obligations to your customers. It also describes the technologies and tools that the Windows Azure™ technology platform provides to help you monitor and manage your solutions in a proactive manner, as well as assisting you in determining the capacity and running costs of your systems.

Use Cases and Challenges

Monitoring and managing a hybrid application is a nontrivial task due to the number, location, and variety of the various moving parts that comprise a typical business solution. Gathering accurate and timely metrics is key to measuring the capacity of your solution and monitoring the health of the system. Additionally, well defined procedures for recovering elements in the event of failure are of paramount importance. You may also be required to collect routine auditing information about how your system is used, and by whom.

The following sections describe some common use cases for monitoring and maintaining a hybrid solution, and summarize many of the challenges you will encounter while performing the associated tasks.

Measuring and Adjusting the Capacity of Your System

Description: You need to determine the capacity of your system so that you can identify where additional resources may be required, and the running costs associated with these resources.

In a commercial environment, customers paying to use your service expect to receive a quality of service and level of performance defined by their SLA. Even nonpaying visitors to your web sites and services will anticipate that their experience will be trouble-free; nothing is more annoying to potential customers than running up against poor performance or errors caused by a lack of resources.

One way to measure the capacity of your system is to monitor its performance under real world conditions. Many of the issues associated with effectively monitoring and managing a system also arise when you are hosting the solution on-premises using your organization's servers. However, when you relocate services and functionality to the cloud, the situation can become much more complicated for a variety of reasons, including:

- The servers are no longer running locally and may be starting up and shutting down automatically as your solution scales elastically.
- There may be many instances of your services running across these servers.
- Applications in the cloud may be multi-tenanted.
- Communications may be brittle and sporadic.
- Operations may run asynchronously.

> Customers care about the quality of service that your system provides, not how well the network or the cloud environment is functioning. You should ensure that your system is designed and optimized for the cloud as this will help you set (and fulfill) realistic expectations for your users.

This is clearly a much more challenging environment than the on-premises scenario. You must consider not just how to gather the statistics and performance data from each instance of each service running on each server, but also how to consolidate this information into a meaningful view that an administrator can quickly use to determine the health of your system and determine the cause of any performance problems or failures. In turn, this requires you to establish an infrastructure that can unobtrusively collect the necessary information from your services and servers running in the cloud, and persist and analyze this data to identify any trends that can highlight scope for potential failure; such as excessive request queue lengths, processing bottlenecks, response times, and so on.

You can then take steps to address these trends, perhaps by starting additional service instances, deploying services to more datacenters in the cloud, or modifying the configuration of the system. In some cases you may also determine that elements of your system need to be redesigned to better handle the throughput required. For example, a service processing requests synchronously may need to be reworked to perform its work asynchronously, or a different communications mechanism might be required to send requests to the service more reliably.

Monitoring Services to Detect Performance Problems and Failures Early

Description: You need to maintain a guaranteed level of service.

In an ideal situation, software never fails and everything always works. This is unrealistic in the real world, but you should aim to give users the impression that your system is always running perfectly; they should not be aware of any problems that might occur.

However, no matter how well tested a system is there will be factors outside your control that can affect how your system functions; the network being a prime example. Additionally, unless you have spent considerable time and money performing a complete and formal mathematical verification of the components in your solution, you cannot guarantee that they are free of bugs. The key to maintaining a good quality of service is to detect problems before your customers do, diagnose their cause, and either repair these failures quickly or reconfigure the system to transparently redirect customer requests around them.

Remember that testing can only prove the presence and not the absence of bugs.

If designed carefully, the techniques and infrastructure you employ to monitor and measure the capacity of your system can also be used to detect failures. It is important that the infrastructure flags such errors early so that operations staff can take the proper corrective actions rapidly and efficiently. The information gathered should also be sufficient to enable operations staff to spot any trends, and if necessary dynamically reconfigure the system to add or remove resources as appropriate to accommodate any change in demand.

Recovering from Failure Quickly
Description: You need to handle failure systematically and restore functionality quickly.

Once you have determined the cause of a failure, the next task is to recover the failed components or reconfigure the system. In a live environment spanning many computers and supporting many thousands of users, you must perform this task in a thoroughly systematic, documented, and repeatable manner, and you should seek to minimize any disruption to service. Ideally, the steps that you take should be scripted so that you can automate them, and they must be robust enough to record and handle any errors that occur while these steps are being performed.

Logging Activity and Auditing Operations
Description: You need to record all operations performed by every instance of a service in every datacenter.

You may be required to maintain logs of all the operations performed by users accessing each instance of your service, performance variations, errors, and other runtime occurrences. These logs should be a complete, permanent, and secure record of events. Logging may be a regulatory requirement of your system, but even if is not, you may still need to track the resources accessed by each user for billing purposes.

An audit log should include a record of all operations performed by the operations staff, such as service shutdowns and restarts, reconfigurations, deployments, and so on. If you are charging customers for accessing your system, the audit log should also contain information about the operations requested by your customers and the resources consumed to perform these operations.

An error log should provide a date and time-stamped list of all the errors and other significant events that occur inside your system, such as exceptions raised by failing components and infrastructure, and warnings concerning unusual activity such as failed logins.

A performance log should provide sufficient data to help monitor and measure the health of the elements that comprise your system. Analytical tools should be available to identify trends that may cause a subsequent failure, such as a SQL Azure™ technology platform database nearing its configured capacity, enabling the operations staff to perform the actions necessary to prevent such a failure from actually occurring.

Deploying and Updating Components
Description: You need to deploy and update components in a controlled, repeatable, and reliable manner whilst maintaining availability of the system.

As in the use case for recovering from system failure, all component deployment and update operations should be performed in a controlled and documented manner, enabling any changes to be quickly rolled back in the event of deployment failure; the system should never be left in an indeterminate state. You should implement procedures that apply updates in a manner that minimizes any possible disruption for your customers; the system should remain available while any updates occur. In addition, all updates must be thoroughly tested in the cloud environment before they are made available to live customers.

Cross-Cutting Concerns

This section summarizes the major cross-cutting concerns that you may need to address when implementing a strategy for monitoring and managing a hybrid solution.

Performance
Monitoring a system and gathering diagnostic and auditing information will have an effect on the performance of the system. The solution you use should minimize this impact so as not to adversely affect your customers.

For diagnostic information, in a stable configuration, it might not be necessary to gather extensive statistics. However, during critical periods collecting more information might help you to detect and correct any problems more quickly. Therefore any solution should be flexible enough to allowing tuning and reconfiguration of the monitoring process as the situation dictates.

The policy for gathering auditing information is unlikely to be as flexible, so you should determine an efficient mechanism for collecting this data, and a compact mechanism for transporting and storing it.

Security
There are several security aspects to consider concerning the diagnostic and auditing data that you collect:
- Diagnostic data is sensitive as it may contain information about the configuration of your system and the operations being performed. If intercepted, an attacker may be able to use this information to infiltrate your system. Therefore you should protect this data as it traverses the network. You should also store this data securely.

- Diagnostic data may also include information about operations being performed by your customers. You should avoid capturing any personally identifiable information about these customers, storing it with the diagnostic data, or making it available to the operators monitoring your system.

- Audit information forms a permanent record of the tasks performed by your system. Depending on the nature of your system and the jurisdiction in which you organization operates, regulatory requirements may dictate that you must not delete this data or modify it in any way. It must be stored safely and protected resolutely.

Additionally, the monitoring, management, deployment, and maintenance tasks associated with the components that comprise your system are sensitive tasks that must only be performed by specified personnel. You should take appropriate steps to prevent unauthorized staff from being able to access monitoring data, deploy components, or change the configuration of the system.

Windows Azure and Related Technologies

Windows Azure provides a number of useful tools and APIs that you can employ to supervise and manage hybrid applications. Specifically you can use:

- Windows Azure Diagnostics to capture diagnostic data for monitoring the performance of your system. Windows Azure Diagnostics can operate in conjunction with the Enterprise Library Logging Application Block. Microsoft Systems Center Operations Manager also provides a management pack for Windows Azure, again based on Windows Azure Diagnostics.
- The Windows Azure Management Portal, which enables administrators to provision the resources and websites required by your applications. It also provides a means for implementing the various security roles required to protect these resources and websites. For more information, log in to the Management Portal at *http://windows.azure.com*.
- The Windows Azure Service Management API, which enables you to create your own custom administration tools, as well as carrying out scripted management tasks from the Windows PowerShell® command-line interface.

The following sections provide more information about the Windows Azure Service Management API and Windows Azure Diagnostics, and summarize good practice for utilizing them to support a hybrid application.

> You can configure Remote Desktop access for the roles running your applications and services in the cloud. This feature enables you to access the Windows logs and other diagnostic information on these machines directly, by means of the same procedures and tools that you use to obtain data from computers hosted on-premises.

Monitoring Services, Logging Activity, and Measuring Performance in a Hybrid Application by Using Windows Azure Diagnostics

An on-premises application typically consists of a series of well-defined elements running on a fixed set of computers, accessing a series of well-known resources. Monitoring such an application requires being able to transparently trap and record the various requests that flow between the components, and noting any significant events that occur. In this environment, you have total control over the software deployed and configuration of each computer. You can install tools to capture data locally on each machine, and you combine this data to give you an overall view of how well the system is functioning.

In a hybrid application, the situation is much more complicated; you have a dynamic environment where the number of compute nodes (implementing web and worker roles) running instances of your services and components might vary over time, and the work being performed is distributed across these nodes. You have much less control of the configuration of these nodes as they are managed and maintained by the datacenters in which they are hosted. You cannot easily install your own monitoring software to assess the performance and well-being of the elements located at each node. This is where Windows Azure Diagnostics is useful.

Windows Azure Diagnostics provides an infrastructure running on each node that enables you to gather performance and other diagnostic data about the components running on these nodes. It is highly configurable; you specify the information that you are interested in, whether it is data from event logs, trace logs, performance counters, IIS logs, crash dumps, or other arbitrary log files. For detailed information about how to implement Windows Azure Diagnostics and configure your applications to control the type of information that Windows Azure Diagnostics records, see *"Collecting Logging Data by Using Windows Azure Diagnostics"* on MSDN.

Windows Azure Diagnostics is designed specifically to operate in the cloud. As a result, it is highly scalable while attempting to minimize the performance impact that it has on the roles that are configured to use it. However, the diagnostic data itself is held locally in each compute node being monitored. This data is lost if the compute node is reset. Also, the Windows Azure diagnostic monitor applies a 4GB quota to the data that it logs; if this quota is exceeded, information is deleted on an age basis. You can modify this quota, but you cannot exceed the storage capacity specified for the web or worker role. In many cases you will likely wish to retain this data, and you can arrange to transfer the diagnostic information to Windows Azure storage, either periodically or on demand. The topics *"How to Schedule a Transfer"* and *"How to Perform an On-Demand Transfer"* provide information on how to perform these tasks.

> Transferring diagnostic data to Windows Azure storage on demand may take some time, depending on the volume of information being copied and the location of the Windows Azure storage. To ensure best performance, use a storage account hosted in the same datacenter as the compute node running the web or worker role being monitored. Additionally, you should perform the transfer asynchronously so that it minimizes the impact on the response time of the code.

Windows Azure storage is independent of any specific compute node and the information that it contains will not be lost if any compute node is restarted. You must create a storage account for holding this data, and you must configure the Windows Azure Diagnostics Monitor with the address of this storage account and the appropriate access key. For more information, see *"How to Specify a Storage Account for Transfers."* Event-based logs are transferred to Windows Azure table storage and file-based logs are copied to blob storage. The appropriate tables and blobs are created by the Windows Azure Diagnostics Monitor; for example, information from the Windows Event Logs is transferred to a table named **WADWindowsEventLogsTable**, data gathered from performance counters is copied to a table name **WADPerformanceCountersTable**. Crash dumps are transferred to a blob storage container under the path **wad-crash-dumps** and IIS 7.0 logs are copied to another blob storage container under the path **wad-iis-logfiles**.

Guidelines for Using Windows Azure Diagnostics

From a technical perspective, Windows Azure Diagnostics is implemented as a component within the Windows Azure SDK that supports that standard diagnostic APIs. This component is called the *Windows Azure diagnostic monitor*, and it runs in the cloud alongside each web role or worker role that you wish to gather information about.

You can configure the diagnostic monitor to determine the data that it should collect by using the Windows Azure Diagnostics configuration file, diagnostics.wadcfg. For more information, see *"How to Use the Windows Azure Diagnostics Configuration File."* Additionally, an application can record custom trace information by using a trace log. Windows Azure Diagnostics provides the **DiagnosticMonitorTraceListener** class to perform this task, and you can configure this type of tracing by adding the appropriate **<system.diagnostics>** section to the application configuration file of your web or worker role. See *"How to Configure the TraceListener in a Windows Azure Application"* for more information.

If you are building a distributed, hybrid solution, using Windows Azure Diagnostics enables you to gather the data from multiple role instances located on distributed compute nodes, and combine this data to give you an overall view of your system. You can use System Center Operations Manager, or alternatively there are an increasing number of third-party applications available that can work with the raw data available through Windows Azure Diagnostics to present the information in a variety of easy to digest ways. These tools enable you to determine the throughput, performance, and responsiveness of your system and the resources that it consumes. By analyzing this information, you can pinpoint areas of concern that may impact the operations that your system implements, and also evaluate the costs associated with performing these operations.

The following list suggests opportunities for incorporating Windows Azure Diagnostics into your solutions:

- **You need to provide a centralized view of your system to help ensure that you meet the SLA requirements of your customers and to maintain an accurate record of resource use for billing purposes. Your organization currently uses System Center Operations Manager to monitor and maintain services running on-premises.**

    If you have deployed Systems Center Operations Manager on-premises, you can also install the Monitoring Management Pack for Windows Azure. This pack operates in conjunction with Windows Azure Diagnostics on each compute node, enabling you to record and observe the diagnostics for your applications and services hosted in the cloud and integrate it with the data for the other elements of your solution that are running on-premises. This tool is also invaluable for assessing how the services that comprise your system are using resources, helping you to determine the costs that you should be recharging to your customers, if appropriate.

    Using Systems Center Operations Manager, you can configure alerts that are raised when various measurements exceed specified thresholds. You can associate tasks with these alerts, and automate procedures for taking any necessary corrective action. For example, you can arrange for additional role instances to be started if the response time for handling client requests is too long, or you can send an alert to an operator who can investigate the issue.

    For more information, see *"System Center Monitoring Pack for Windows Azure Applications."*

- **You need to provide a centralized view of your system to help monitor your application and maintain an audit record of selected operations. Your organization does not use System Center Operations Manager.**

    You can periodically transfer the diagnostic data to Windows Azure storage and examine it by using a utility such as Windows Azure Storage Explorer from Neudesic (see *http://azurestorageexplorer.codeplex.com/*). Additionally, if you have the Visual Studio® development system and the Windows Azure SDK, you can connect to Windows Azure storage and view the contents of tables and blobs by using Server Explorer. For more information, see *"How to View Diagnostic Data Stored in Windows Azure Storage."*

By default, the System Center Monitoring Pack for Windows Azure Applications monitors the deployment state of roles, the state of each hosted service and role, and the performance counters measuring ASP.NET performance, disk capacity, physical memory utilization, network adapter utilization, and processor performance.

However, these tools simply provide very generalized access to Windows Azure storage. If you need to analyze the data in detail it may be necessary to build a custom dashboard application that connects to the tables and blobs in the storage account, aggregates the information gathered for all the nodes, and generates reports that show how throughput varies over time. This enables you to identify any trends that may require you to allocate additional resources. You can also download the diagnostic data from Windows Azure storage to your servers located on-premises if you require a permanent historical record of this information, such as an audit log of selected operations, or you wish to analyze the data offline.

Figure 1 depicts the overall structure of a solution that gathers diagnostics data from multiple nodes and analyzes it on-premises. The diagnostics data is reformatted and copied into tables in a SQL Server database, and SQL Server Reporting Services outputs a series of reports that provide a graphical summary showing the performance of the system.

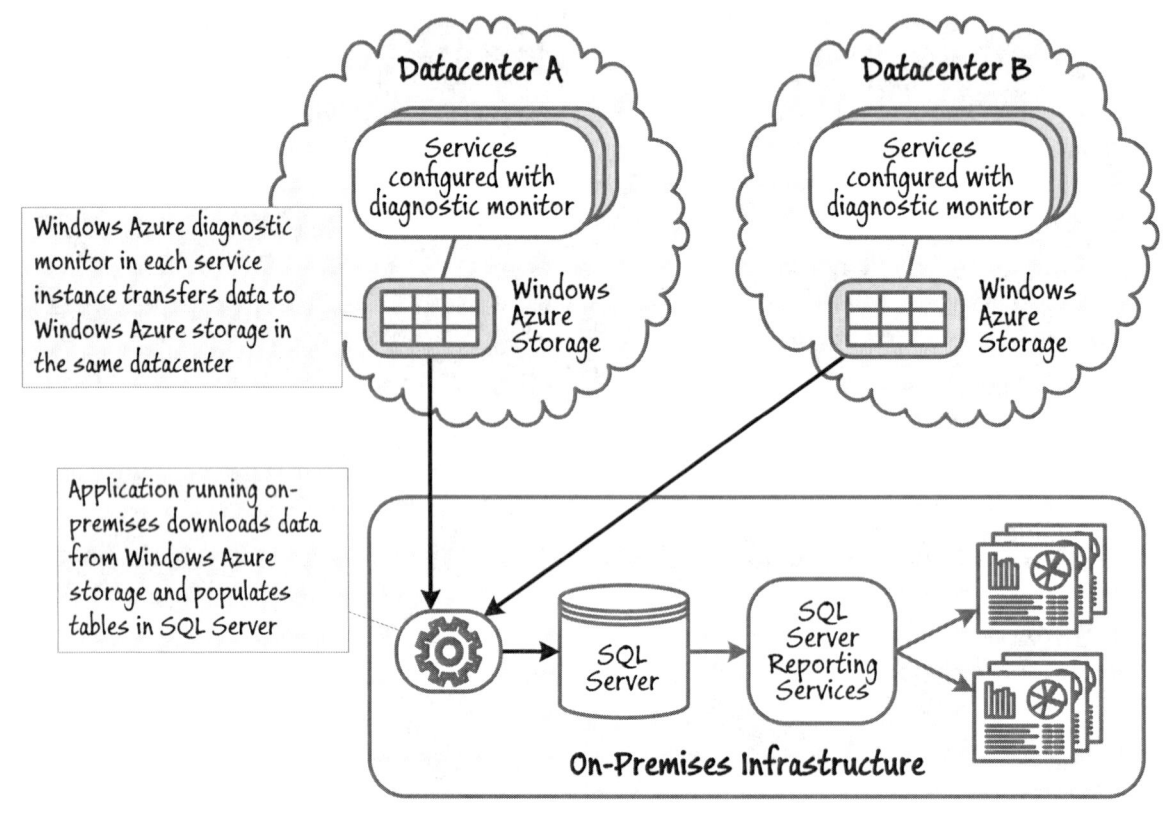

FIGURE 1
Gathering diagnostic data from multiple nodes

An alternative approach is to use a third party solution. Some that were available at the time of writing include the following:
- *Azure Diagnostics Manager* from Cerebrata
- *AzureWatch* from Paraleap Technologies
- *ManageAxis* from Cumulux

- **You need to instrument your web applications and services to identify any potential bottlenecks and capture general information about the health of your solution.**

    Applications running in the cloud can use the Windows Azure Diagnostics APIs to incorporate custom logic that generates diagnostic information, enabling you to instrument your code and monitor the health of your applications by using the performance counters applicable to web and worker roles. You can also define your own custom diagnostics and performance counters. The topic *"Tracing the Flow of Your Windows Azure Application"* provides more information.

    It is also useful to trace and record any exceptions that occur when a web application runs, so that you can analyze the circumstances under which these exceptions arise and if necessary make adjustments to the way in which the application functions. You can make use of programmatic features such as the Microsoft Enterprise Library Exception Handling Application Block to capture and handle exceptions, and you can record information about these exceptions to Windows Azure Diagnostics by using the Enterprise Library Logging Application Block. This data can then be examined by using a tool such as System Center Operations Manager with the Monitoring Pack for Windows Azure, providing a detailed record of the exceptions raised in your application across all nodes, and also generating alerts if any of these exceptions require the intervention of an operator.

    For more information about incorporating the Logging Application Block into a Windows Azure solution, see "Using classic Enterprise Library 5.0 in Windows Azure" on CodePlex.

    For more information about using the Exception Handling Application Block, see "The Exception Handling Application Block."

- **Most of the time your web and worker roles function as expected, but occasionally they run slowly and become unresponsive. At these times you need gather additional detailed diagnostic data to help you to determine the cause of the problems. You need to be able to modify the configuration of the Windows Azure diagnostic monitor on any compute node without stopping and restarting the node.**

    To minimize the overhead associated with logging, only trace logs, infrastructure logs, and IIS logs are captured by default. If you need to examine data from performance counters, Windows® operating system event logs, IIS failed request logs, crash dumps, or other arbitrary logs and files you must enable these items explicitly. An application can dynamically modify the Windows Azure Diagnostics configuration by using the Windows Azure SDK from your applications and services. For more information, see *"How to: Initialize the Windows Azure Diagnostic Monitor and Configure Data Sources."*

You can also configure Windows Azure Diagnostics for a web or worker role remotely by using the Windows Azure SDK. You can follow this approach to implement a custom application running on-premises that connects to a node running a web or worker role, specify the diagnostics to collect, and transfer the diagnostic data periodically to Windows Azure storage. For more information, see *"How to Remotely Change the Diagnostic Monitor Configuration."*

> The Windows Azure diagnostic monitor periodically polls its configuration information, and any changes come into effect after the polling cycle that observes them. The default polling interval is 1 minute. You can modify this interval, but you should not make it too short as you may impact the performance of the diagnostic monitor.

Guidelines for Securing Windows Azure Diagnostic Data

Windows Azure Diagnostics requires that the roles being monitored run with full trust; the **enableNativeCodeExecution** attribute in the service definition file, ServiceDefinition.csdef, for each role must be set to true. This is actually the default value.

The diagnostic information recorded for your system is a sensitive resource and can yield critical information about the structure and security of your system. This information may be useful to an attacker attempting to penetrate your system. Therefore, you should carefully guard the storage accounts that you use to record diagnostic data and ensure that only authorized applications have access to the storage account keys. You should also consider protecting all communications between the Windows Azure storage service and your on-premises applications by using HTTPS.

If you have built on-premises applications or scripts that can dynamically reconfigure the diagnostics configuration for any role, ensure that only trusted personnel with the appropriate authorization can run these applications.

Deploying, Updating, and Restoring Functionality by Using the Windows Azure Service Management API and PowerShell

The Windows Azure Management Portal provides the primary interface for managing Windows Azure subscriptions. You can use this portal to upload applications, and to manage hosted services and storage accounts. However, you can also manage your Windows Azure applications programmatically by using the Windows Azure Service Management API. You can utilize this API to build custom management applications that deploy, configure, and manage your web applications and services. You can also access these APIs through the Windows Azure PowerShell cmdlets; this approach enables you to quickly build scripts that administrators can run to perform common tasks for managing your applications.

The Windows Azure SDK provides tools and utilities to enable a developer to package web and worker roles, and to deploy these packages to Windows Azure. Many of these tools and utilities also employ the Windows Azure Service Management API, and some are invoked by several of the Microsoft Build Engine (MSBuild) tasks and wizards implemented by the Windows Azure templates in Visual Studio.

You can download the Windows Azure PowerShell cmdlets at http://www.windowsazure.com/en-us/manage/downloads/.

> You can download the code for a sample application that provides a client-side command line utility for managing Windows Azure applications and services from *Windows Azure ServiceManagement Sample*.

Guidelines for using the Windows Azure Service Management API and PowerShell

While the Management Portal is a powerful application that enables an administrator to manage and configure Windows Azure services, this is an interactive tool that requires users to have a detailed understanding of the structure of the solution, where the various elements are deployed, and how to configure the security requirements of these elements. It also requires that the user has knowledge of the Windows Live® ID and password associated with the Windows Azure subscription for your organization, and any user who has this information has full authority over your entire solution. If these credentials are disclosed to an attacker or another user with malicious intent, they can easily disrupt your services and damage your business operations.

The following scenarios include suggestions for mitigating these problems:

- **You need to provide controlled access to an operator to enable them to quickly perform everyday tasks such as configuring a role, provisioning services, or starting and stopping role instances. The operator should not require a detailed understanding of the structure of the application, and should not be able to perform tasks other than those explicitly mandated.**

    This is a classic case for using scripts incorporating the Windows Azure PowerShell cmdlets. You can provide a series of scripts that perform the various tasks required, and you can parameterize them to enable the operator to provide any additional details, such as the filename of a package containing a web role to be deployed, or the address of a role instance to be taken offline. This approach also enables you to control the sequence of tasks if the operator needs to perform a complex deployment, involving not just uploading web and worker roles, but also provisioning and managing SQL Azure databases, for example.

To run these scripts, the operator does not need to be provided with the credentials for the Windows Azure subscription. Instead, the security policy enforced by the Windows Azure Service Management API requires that the account that the operator is using to run the scripts is configured with the appropriate management certificate, as described in the section "Guidelines for Securing Management Access to Azure Subscriptions" later in this appendix. The fewer operators that know the credentials necessary for accessing the Windows Azure subscription, the less likely it is that these credentials will be disclosed to an unauthorized third party, inadvertently or otherwise.

Scripting also provides for consistency and repeatability, reducing the chances of human error on the part of the operator, especially when the same series of tasks must be performed across a set of roles and resources hosted in different datacenters.

The disadvantage of this approach is that the scripts must be thoroughly tested, verified, and maintained. Additionally, scripts are not ideal for handling complex logic, such as performing error handling and graceful recovery.

- **You need to provide controlled access to an operator to enable them to quickly perform a potentially complex series of tasks for configuring, deploying, or managing your system. The operator should not require a detailed understanding of the structure of the application, and should not be able to perform tasks other than those explicitly mandated.**

This scenario is an extension of the previous case, except that the operations are more complex and potentially more error-prone. In this scenario, it may be preferable to use the Windows Azure Service Management API directly from a custom application running on-premises. This application can incorporate a high degree of error detection, handling, and retry logic (if appropriate). You can also make the application more interactive, enabling the operator to specify the details of items such as the address of a role to be taken offline, or the filename of a package to deploy, through a graphical user interface with full error checking. A wizard oriented approach is easier to understand and less error prone than expecting the operator to provide a lengthy string of parameters on the command line as is common with the scripted approach.

> A script that creates or updates roles should deploy these roles to the staging environment in one or more datacenters for testing prior to making them available to customers. Switching from the staging to production environment can also be scripted, but should only be performed once testing is complete.

A custom application also enables you to partition the tasks that can be performed by different operators or roles; the application can authenticate the user, and only enable the features and operations relevant to the identity of the user or the role that the user belongs to. However, you should avoid attempting to make the application too complex; keep the features exposed simple to use, and implement meaningful and intelligent default values for items that users must select.

A custom application should audit all operations performed by each user. This audit trail provides a full maintenance and management history of the system, and it must be stored in a secure location.

Of course, the disadvantage of this approach is that, being interactive, such an application cannot easily be used to perform automated routine tasks scheduled to occur at off-peak hours. In this case, the scripted approach or a solution based on a console-mode command-line application may be more appropriate.

- **You are using Systems Center Operations Manager to monitor the health of your system. If a failure is detected in one or more elements, you need to recover from this failure quickly.**

    System Center Operations Manager can raise an alert when a significant event occurs or a performance measure exceeds a specified threshold. You can respond to this alert in a variety of ways, such as notifying an operator or invoking a script. You can exploit this feature to detect the failure of a component in your system and run a PowerShell script that attempts to restart it. For more information, see the topic *"Enable Notification Channels."*

Guidelines for Securing Management Access to Windows Azure Subscriptions

The Windows Azure Service Management API ensures that only authorized applications can perform management operations, by enforcing mutual authentication using management certificates over SSL.

When an on-premises application submits a management request, it must provide the key to a management certificate installed on the computer running the application as part of the request. A management certificate must have a key length of at least 2048 bits and must be installed in the Personal certificate store of the account running the application. This certificate must also include the private key.

> The Windows Azure Service Management API is actually a wrapper around a REST interface; all service management requests are actually transmitted as HTTP REST requests. Therefore, you are not restricted to using Windows applications for building custom management applications; you can use any programming language or environment that is capable of sending HTTP REST requests. For more information, see *"Windows Azure Service Management REST API Reference."*

The same certificate must also be available in the management certificates store in the Windows Azure subscription that manages the web applications and services. You should export the certificate from the on-premises computer as a .cer file without the private key and upload this file to the Management Certificates store by using the Management Portal. For more information about creating management certificates, see the topic *"How to: Manage Management Certificates in Windows Azure."*

> Remember that the Windows Azure SDK includes tools that enable a developer to package web and worker roles, and to deploy these packages to Windows Azure. These tools also require you to specify a management certificate. However, you should be wary of letting developers upload new versions of code to your production site. This is a task that must be performed in a controlled manner and only after the code has been thoroughly tested. For this reason, you should either refrain from provisioning your developers with the management certificates necessary for accessing your Windows Azure subscription, or you should retain a separate Windows Azure subscription (for development purposes) with its own set of management certificates if your developers need to test their own code in the cloud.

More Information

All links in this book are accessible from the book's online bibliography available at: *http://msdn.microsoft.com/en-us/library/hh968447.aspx*.

- Management Portal at *http://windows.azure.com*.
- "Collecting Logging Data by Using Windows Azure Diagnostics" at *http://msdn.microsoft.com/en-us/library/gg433048.aspx*.
- "How to Schedule a Transfer" at *http://msdn.microsoft.com/en-us/library/windowsazure/gg433085.aspx*.
- "How to Perform an On-Demand Transfer" at *http://msdn.microsoft.com/en-us/library/windowsazure/gg433075.aspx*.
- "How to Specify a Storage Account for Transfers" at *http://msdn.microsoft.com/en-us/library/windowsazure/gg433081.aspx*.
- "How to Use the Windows Azure Diagnostics Configuration File" at *http://msdn.microsoft.com/en-us/library/windowsazure/hh411551.aspx*.
- "How to Configure the TraceListener in a Windows Azure Application" at *http://msdn.microsoft.com/en-us/library/hh411522.aspx*.
- "System Center Monitoring Pack for Windows Azure Applications" at *http://pinpoint.microsoft.com/en-us/applications/system-center-monitoring-pack-for-windows-azure-applications-12884907699*.
- Windows Azure Storage Explorer from Neudesic at *http://azurestorageexplorer.codeplex.com/*.
- "How to View Diagnostic Data Stored in Windows Azure Storage" at *http://msdn.microsoft.com/en-us/library/windowsazure/hh411547.aspx*.
- Azure Diagnostics Manager from Cerebrata at *http://www.cerebrata.com/Products/AzureDiagnosticsManager/Default.aspx*.
- AzureWatch from Paraleap Technologies at *http://www.paraleap.com/*.

- "Tracing the Flow of Your Windows Azure Application" at *http://msdn.microsoft.com/en-us/library/windowsazure/hh411529.aspx*.
- "Using classic Enterprise Library 5.0 in Windows Azure" at *http://entlib.codeplex.com/releases/view/75025#DownloadId=336804*.
- "The Exception Handling Application Block" at *http://msdn.microsoft.com/en-us/library/ff664698(v=PandP.50).aspx*.
- "How to: Initialize the Windows Azure Diagnostic Monitor and Configure Data Sources" at *http://msdn.microsoft.com/en-us/library/windowsazure/gg433049.aspx*.
- "How to Remotely Change the Diagnostic Monitor Configuration" at *http://msdn.microsoft.com/en-us/library/windowsazure/gg432992.aspx*.
- Download the code for a sample application that provides a client-side command line utility for managing Windows Azure applications and services from "Windows Azure ServiceManagement Sample" at *http://code.msdn.microsoft.com/windowsazure/Windows-Azure-CSManage-e3f1882c*.
- Download the Windows Azure PowerShell cmdlets from *http://www.windowsazure.com/en-us/manage/downloads/*.
- "Windows Azure Service Management REST API Reference" at *http://msdn.microsoft.com/en-us/library/windowsazure/ee460799.aspx*.
- "Enable Notification Channels" at *http://technet.microsoft.com/en-us/library/dd440882.aspx*.
- "How to: Manage Management Certificates in Windows Azure" at *http://msdn.microsoft.com/en-us/library/windowsazure/gg551721.aspx*.

Index

A

_Layout.cshtml, 53-54
access authentication to service bus queues and topics, 60-61
Access Control Service (ACS) *See* ACS
acknowledgments, xxiii-xxiv
ACS, 238-239
    authentication, 43
    configuration, 47
    unique user IDs, 239
ACSIdentity table, 29
AcsNamespace property, 136-137
actions, 316-317
Active Directory Federation Service (ADFS), 42-43
adapters and connectors for translating and reformatting messages, 91-93
Add method, 115-117
ADFS, 42-43
all data on premises, 18-19
App.config file, 188
applications *See* hybrid applications; Orders application
architecture
    authentication
        with ACS and social identity providers, 236
        implementation, 49
        and sequence in the hybrid Orders application, 46
        service bus endpoints with ACS, 61
        trust chain to support federated identity and SSO, 235
    Autoscaling Application Block, 315-316
    batch processing by Service Bus queue, 274
    bidirectional synchronization
        all databases on-premises and in the cloud, 195
        only databases in the cloud, 196

cache-aside programming pattern, 330
caching static data to reduce network latency, 337
cloud, 8
cloud technology map, 11
compliance system, 106
data
    aggregating and consolidating in the cloud, 215
    physical implementation of synchronization, 28
    replication in the Trey application, 27
    sharing between the cloud and on-premises, 212
    updating in the cache and the authoritative data store, 338
Data Sync SDK for custom synchronization, 222
diagnostic data from multiple nodes, 358
DistributedCacheSessionStateStoreProvider session state provider, 344
Execute method, 121
failover policy, 325
flow of control through a **ServiceBusReceiver-Handler** object, 83
forward-routing messages, 302
hybrid version of the Trey Research solution, 65
load-balancing with multiple receivers, 278
local in-memory caching, 343
messages
    decoupling sender applications from routing logic, 295
    flow for the order processing system, 120
    prioritizing with Service Bus topic and subscriptions, 297
    retrieving and storing session state information, 280

routing to different receivers, 293
TrueFilter logging, 304-307
messaging technology
    for transport partners by Trey, 73
    used by Trey to route orders to the audit log, 101
monitoring approaches for Orders, 171
more information, 107-108, 227
multicasting with Service Bus Relay, 262
newOrderMessageSender instance, 120
on-premises, 6
    authentication at Trey, 41
ordered messages with Service Bus queues, 273
OrderProcessStatus table, 121-122
orders
    locking for processing, 121-124
    processing components in the on-premises application, 64
OrderStatus table, 121
PeekLock to examine messages without dequeueing, 283
on premises
    databases synchronized through the cloud, 199
    master repository with one-way synchronization from the cloud, 197
    master repository with one-way synchronization to the cloud, 198
    resources from a partner organization and roaming computers, 253
    service built with Ruby with Service Bus Relay, 261
    service with HTTP REST requests, 260
publishing on-premises database to the cloud, 214
replication
    data between data centers in the cloud, 213
    database to implement read scale-out, 219
requests and responses through Windows Azure Service Bus Relay, 258
scaling out with Service Bus topic and subscriptions, 300
security architecture of Windows Azure Connect, 254
segregation across the Orders application deployment, 16
Service Bus Relay requests and responses, 258
Service Bus request authentication
    with ACS, 240
    with ADFS and ACS, 242
Service Bus topics and subscriptions limitations to message routing, 304-308
services exposed through Service Bus Relay, 266
sessions
    different state from different datacenters, 345
    to group messages, 279
smart client or service authentication, 237
SQL Azure Data Sync
    to partition and replicate data across branch offices, 218
    to partition and replicate data in the cloud, 217
subscription correlation to deliver response messages, 299
sync group configuration table, 29
synchronization direction for databases participating in Topology D, 210
TCP/SSL, 259
Traffic Manager routing and redirection, 320
transactional and retry mechanism, 113
two-way messaging with response queues and message correlation, 276
updates
    notifications by using a Service Bus topic and subscriptions, 227
    propagation between Azure caches and an authoritative data store, 341
    propagation between Windows Azure caches and replicated data stores, 339
    routing messages, 225
    uncontrolled to replicas of a database, 223
Windows Azure Caching to provide scalability, 330
Windows Azure Traffic Manager, 155
worker role requests through Service Bus Relay, 263
ASP.NET Forms authentication, 42
ASP.NET request validation, 52
asynchronous messages, 284-285
    to a Service Bus queue, 75-77
    from a Service Bus queue, 77-83
    from a topic, 90-91
audience, xviii
audit log data, 21
    orders to the audit log mechanism choice, 100-101
authentication, 39-62, 59
    See also users and requests authentication
    access authentication to service bus queues and topics, 60-61
    ACS, 43
    ACS configuration, 47
    ADFS, 42-43
    ASP.NET Forms authentication, 42
    ASP.NET request validation, 52
    authentication choice, 42-45

Trey's choice, 45
AuthorizeAndRegisterUserAttribute class, 55-56
browser process using ACS and social identity
  providers, 236
choice, 42-45
  Trey's choice, 45
claims-based, 43
  with ACS and ADFS, 45
  with ACS and ADFS by Trey, 45
  with ACS and ADFS for visitor by Trey, 45-59
client, 240-242
combined forms and claims-based, 45
custom authorization attribute, 55-56
custom logon page, 54-55
customer details storage and retrieval, 56-57
ExecuteAction method, 59
federated authentication, 229, 234-235
FindOne method, 57-58
GetDefaultSqlCommandRetryPolicy method, 59
IdentityExtensions class, 51-52
implementation, 48-49
_Layout.cshtml, 53-54
LogOn method, 54-55
more information, 61-62
multiple user IDs, 48
MyOrdersController class, 55
Orders.Website project, 49-51
Orders.Website.Helpers project, 51-52
original on-premises mechanism at Trey Research, 41
public users, 230
retry policy, 58-59
scenario and context, 39-41
sequence in the hybrid Orders application, 46
service bus endpoints with ACS, 61
SingleOrDefault method, 59
smart client or service authentication, 237
summary, 61
Transient Fault Handling Application Block, 59
trust chain that can support federated identity
  and SSO, 235
Views/Shared folder, 53-54
visitor authentication and authorization, 53-54
**visitor authentication and authorizationLayout.
  cshtml**, 53-54
visitors, 42-59, 47
Web.config file, 53
with WIF, 49-52
WIF, 49-52
WsFederationRequestValidator class, 52-53

authorization
  access to Service Bus queues, 232
  rules and rule groups, 244
  service access for non-browser clients, 231
  user actions, 231
AuthorizeAndRegisterUserAttribute class, 55-56
automatic scaling, 145-146
Autoscaler object, 148-151
Autoscaling Application Block, 315-316
  hosting, 147-148
  use guidelines, 317-318
autoscaling rule definitions, 148-151
availability
  See also scalability, availability, and performance
  maximizing for cloud applications, 312
  Windows Azure Traffic Manager, 324-326
Azure
  See also cloud; Windows Azure
  data encryption, 23
  data synchronization, 24-29
  moving to, 7-10
  technologies for routing messages, 292-308
Azure Access Control Service (ACS) *See* ACS
Azure and related technologies, 202
  hybrid application monitoring and managing, 354-364
  maximizing, 313-347
Azure Content Delivery Network (CDN) *See* Content
  Delivery Network (CDN)
Azure Service Bus topics and subscriptions *See* Service Bus;
  Service Bus topics and subscriptions
Azure Service Management API and PowerShell, 360-364
  guidelines, 361

B

batch processing by Service Bus queue, 274
Bharath *See* cloud specialist role (Bharath)
bibliography, xxii
  Orders application deployment to the cloud, 37
  Trey Research, 14
bidirectional synchronization
  across all databases, 195
  only across databases in the cloud, 196
bindings, 268-270
BrokeredMessage class, 271-272
browser authentication process using ACS and social
  identity providers, 236
business logic
  See also cross-boundary communication; cross-
    boundary routing

from message routing, 290
Service Bus topics and subscriptions, 292

C

cache-aside programming pattern, 330
caching
    cached data updating, 331
    and database synching, 159
    functionality, 160-164
    guidelines, 336-346
    local cache, 334
    local in-memory caching, 343
    overview, 313
    static data to reduce network latency in web applications, 337
    web application session state, 335
CachingStrategy class, 161-163
changeSetting action, 317
channel across boundaries with Service Bus queues, 271-287
claims-based authentication, 43
    with ACS and ADFS, 45
        Trey's choice, 45
    process overview, 235-236
CleanupRelyingParties method, 187
clients
    authentication, 240-242
    different session state from different datacenters, 345
cloud
    See also Azure; messaging and communications; Orders application deployment to the cloud; Windows Azure
    complex applications in, 2
    data sharing between on-premises and, 212
    integration, 1-2
    integration challenges, 2-4, 10-12
    and on-premises replication, 194-199
    staged migration to, 12
    technology map, 11
cloud specialist role (Bharath), xxi
combined forms, 45
communication
    See also messaging and communications
    cross-boundary channel, 247
    mechanism choice, 68-71
    with transport partners, 67-100
complex applications, 2

compliance, 104-106
    hosting location, 105
        Trey's choice, 105-106
    system, 106
components in the on-premises application, 64
constraint rules, 314, 316
ContainerBootstrapper class, 164-165
Content Delivery Network (CDN), 157
    overview, 313
contributors and reviewers, xxiii-xxiv
control when receiving messages through a **ServiceBusReceiverHandler** object, 83
copyright, iv
CreateJobProcessors method, 117
cross-boundary communication, 245-287
    See also cross-boundary routing
    asynchronous messages, 284-285
    bindings, 268-270
    BrokeredMessage class, 271-272
    channel across boundaries with Service Bus queues, 271-287
    communications channel across boundaries, 247
    cross-cutting concerns, 248-249
    firewall, 267
    guidelines, 256-263
    interoperability, 249
    message-oriented communications, 250
    message scheduling, expiring, and deferring, 286
    more information, 287
    Network Address Translation (NAT), 247, 259
    on premises
        resources from outside the organization, 246
        services from outside the organization, 246-247
        services from outside with Service Bus Relay, 256-271
    remote procedure call (RPC), 249-250
    responsiveness, 248-249
    security, 248
    Service Bus message, 271-272
    Service Bus queues
        guidelines, 272-283
        security guidelines, 286-287
        sending and receiving guidelines, 283-286
    Service Bus Relay
        guidelines, 256-263
        security guidelines, 264-267
        service naming guidelines, 267-268
    technologies, 249-287

use cases and challenges, 245
    Windows Azure Connect, 251-256
        vs. Service Bus Relay, 270-271
cross-boundary routing, 289-308
    See also cross-boundary communication
    Azure technologies for routing messages, 292-308
    business logic from message routing, 290
        using Service Bus topics and subscriptions, 292
    cross-cutting concerns, 291
    fan-out architecture, 296
    filter rule action, 301, 302-303
    interoperability, 291
    load-leveling, 296
    message routing
        multiple destinations, 291
        technologies, 292-308
    more information, 308
    priority queue, 296
    production line with events based on Service Bus subscriptions, 307
    reliability, 291
    responsiveness and availability, 291
    security, 291
    Service Bus topics and subscriptions
        message routing guidelines, 293-303
        to route messages to multiple destinations guidelines, 304-307
        to route messages to multiple destinations limitations, 308
        security guidelines, 308
    subscription correlation, 298
cross-cutting concerns
    cross-boundary communication, 248-249
    cross-boundary routing, 291
    data replicating, distributing, and synchronizing, 201-202
    hybrid application monitoring and managing, 353-354
    users and requests authentication, 232-233
custom authorization attribute, 55-56
custom logging solution, 173-174
custom logon page, 54-55
custom reporting solution, 31
custom repository, 23
custom service to redirect traffic, 152-153
custom synchronization, 25
custom transactional and retry mechanism, 113
customer data, 20
customer details storage and retrieval, 56-57
Customer table, 29

D

data
    See also deployment
    access security, 201
    all data on premises, 18-19
    cache and the authoritative data store updating, 338
    encryption, 23
    physical implementation of data synchronization, 28
    replication, 27
    reporting data to external partners, 33-36
    retrieving and managing, 159-160
    sharing between the cloud and on-premises, 212
    storage mechanism, 21-23
    synchronization, 24-29
    synchronization architecture, 28
    synchronizing across data sources, 199-201
    Trey Research Orders application, 27
data replicating, distributing, and synchronizing, 193-227
    cloud and on-premises replication, 194-199
    cross-cutting concerns, 201-202
    data access security, 201
    data consistency and application responsiveness, 201
    data synchronizing across data sources, 199-201
    database schema for member databases, 204
    hub database locating and sizing, 207-208
    integrity and reliability, 202
    Service Bus topics and subscription guidelines, 222-227
    SQL Azure Data Sync configuration guidelines, 203-211
    SQL Azure Data Sync Security Model, 220
    SQL Azure Data Sync to replicate and synchronize, 203-220
    SQL Azure Data Sync use guidelines, 211-220
    sync datasets, 203-204
    Sync Framework SDK for custom replication and synchronization, 221-222
    sync groups, 203-204
    sync loop avoidance, 210-211
    synchronization conflict management, 205-206
    synchronization direction for a database, 209
    synchronization schedule for a sync group, 208
    use cases and challenges, 193-194
    Windows Azure and related technologies, 202-227
Data Sync SDK, 222
databases
    existing system, 23
    publishing to the cloud, 214
    schema for member databases, 204

decision recording, 175
deployment
    all data to, 18
    management, 184-190
    solutions, 184-185
    some data, 19
diagnostic data, 181-184
    multiple nodes, 358
diagnostics mechanism configuration, 176-177
diagrams *See* architecture
direct connection over TCP/SSL, 259
DistributedCacheOutputCacheProvider class, 335
DistributedCacheSessionStateStoreProvider session state provider, 334, 344
distribution *See* data replicating, distributing, and synchronizing
DownloadLogs method, 102-103

E

elasticity
    cloud management, 310-311
    control, 144-151
    Enterprise Library Autoscaling Application Block, 314-318
    management choices, 144-146
Electronic Data Interchange (EDI), 68
encryption, 23
Enterprise Library Autoscaling Application Block, 314-318
    overview, 313
Enterprise Library Logging Application Block, 172-173
Execute method, 84, 121-122, 125-127
ExecuteAction method, 59, 86-87
external partners, 33-36
ExternalDataAnalyzer project, 35-36

F

Failover policy, 153, 322
failover policy to achieve maximum availability and functionality, 325
fan-out architecture, 296
federated authentication, 229, 234-235
filter rule action, 301, 302-303
FindAll method, 166
FindOne method, 57-58
firewall, 267
flow of control when receiving messages through a **ServiceBusReceiverHandler** object, 83
forward, xv-xvi
forward-routing messages with filter rule action, 302

G

GetDefaultSqlCommandRetryPolicy method, 59
GetLockedOrders method, 123
GetTokenFromAcs method, 97-98
Guard method, 76
guide
    how to use, xix-xx
    technology map, 12-13
guidelines, 256-263

H

how to use this guide, xix-xx
HTML output caching, 335
hub database locating and sizing, 207-208
hybrid application monitoring and managing, 349-364
    See also Orders monitoring and managing
    Azure and related technologies, 354-364
    Azure Service Management API and PowerShell, 360-364
        guidelines, 361
    cross-cutting concerns, 353-354
    management access to Windows Azure subscriptions guidelines, 363-364
    more information, 364-365
    use cases and challenges, 350-353
    Windows Azure Diagnostics, 355-360
        data security guidelines, 360
        use guidelines, 356-360
hybrid applications, xvii, 1
    and data deployment, 17
    hybrid version of the Trey Research solution, 65

I

ICachingStrategy interface, 160-163
IdentityExtensions class, 51-52
IJob interface, 118
illustrations *See* architecture
integration challenges, 2-4, 10-12
integrity and reliability, 202
interoperability, 233, 249, 291
IOrdersStatistics.cs file, 33
IProductsStore interface, 163-164
IsValidToken method, 99-100
IT professional role (Poe), xxi

J

Jana *See* software architect role (Jana)
job processors, 117-119

L

_Layout.cshtml, 53-54
load-balancing with multiple receivers, 278
load-leveling, 296
local cache, 334
local in-memory caching, 343
location choosing, 17-19
locking for, 121-124
LockOrders method, 122-123
logging
    configuration, 175
    messages with TrueFilter, 304-307
LogOn method, 54-55

M

management access to Windows Azure subscriptions
    guidelines, 363-364
managing *See* hybrid application monitoring and managing;
    Orders monitoring and managing
manual scaling, 145
Markus *See* senior software developer role (Markus)
messages
    See also asynchronous messages; messaging
    adapters and connectors for translating and
        reformatting, 91-93
    flow for the order processing system, 120
    message session state information retrieving and
        storing, 280
    to multiple destinations, 291
    new order message
        creation, 126
        sending, 128
    ordered messages with Service Bus queues, 273
    posting to a topic by Trey, 112-114
    replies correlating, 93-94
    routing
        Azure technologies, 292-308
        to different receivers through a Service Bus
            topic and subscriptions, 293
        update messages through a Service Bus topic
            and subscriptions, 225
    scheduling, expiring, and deferring, 286
    securing, 97-100
    Service Bus topic, 84-87
    Service Bus topic and subscriptions
        prioritization, 297
    ServiceBusReceiverHandler object, 83
    session state information, 280
    technologies, 292-308

messaging
    message-oriented communications, 250
    Trey's technologies for transport partners, 73
    Trey's technology to route orders to the audit
        log, 101
    two-way with response queues and message
        correlation, 276
messaging and communications, 63-108
    adapters and connectors for translating and
        reformatting messages, 91-93
    asynchronous messages
        to a Service Bus queue, 75-77
        from a Service Bus queue, 77-83
        from a topic, 90-91
    communicating with transport partners, 67-100
    communication mechanism choice, 68-71
    compliance, 104-106
        hosting location, 105
        Trey's choice, 105-106
    DownloadLogs method, 102-103
    Electronic Data Interchange (EDI), 68
    Execute method, 84, 93-94
    ExecuteAction method, 86-87
    GetTokenFromAcs method, 97-98
    Guard method, 76
    IsValidToken method, 99-100
    messages
        replies correlation, 93-94
        securing, 97-100
        Service Bus topic, 84-87
    NewOrderJob class, 93-94
    OrderProcessor.cs file, 92
    orders to the audit log, 100-104
        mechanism choice, 100-101
    Orders.Shared project library, 74
    ProcessMessage method, 80-82, 92-93, 104
    ProcessMessages method, 78-79, 90-91
    ReceiveNextMessage method, 79-80
    Run method, 96
    scenario and context, 63-66
    securing message queues, topics, and
        subscriptions, 94-97
    Send method, 76-77, 85-86
    Service Bus queues, 70
    Service Bus topics and subscriptions, 71, 88
    ServiceBusQueue class, 84-85
    ServiceBusReceiverHandler object, 83, 90
    ServiceBusSubscriptionDescription class, 89-90
    ServiceBusTopic class, 96-97

SetupAuditLogListener method, 101-102
SetupServiceBusTopicAndQueue method, 88-89
    transport partners and Trey, 71-100
    web services (pull model), 69
    web services (push model), 68-69
    Windows Azure storage queues, 69-70
Microsoft Active Directory Federation Service *See* ADFS
Microsoft Sync Framework, 25
monitoring
    approaches that Trey considered for the Orders application, 171
    logging, and measuring, 170-184
    logging solution choice, 171
monitoring and managing *See* hybrid application monitoring and managing; Orders monitoring and managing
monitoring endpoints, 321
more information, xxii
    See also bibliography
    architecture, 107-108, 227
    authentication, 61-62
    cross-boundary communication, 287
    cross-boundary routing, 308
    hybrid application monitoring and managing, 364-365
    order processing, 141
    Orders application deployment to the cloud, 37
    Orders monitoring and managing, 190-191
    scalability, availability, and performance, 167
    scalability, availability, and performance maximizing, 347-348
    Trey Research, 14
    users and requests authentication, 244
multicasting, 262
multiple user IDs, 48
MyOrdersController class, 55

N

naming conventions, 267-268
Network Address Translation (NAT), 247, 259
network latency
    and connectivity, 152-156
    connectivity management choices by Trey, 154-156
    reduction for accessing cloud applications, 311
    with Traffic Manager, 318-324
new order message
    creation, 126
    sending, 128
NewOrderJob class, 119-121, 127-128
    messaging and communications, 93-94

newOrderMessageSender instance, 120
no scaling, 144-145

O

on-premise
    all data, 18-19
    architecture, 6
    authentication mechanism at Trey Research, 41
    data sharing between the cloud and, 212
    databases synchronized through the cloud, 199
    master repository with one-way synchronization to the cloud, 198
    original mechanism at Trey Research, 41
    publishing database to the cloud, 214
    resources
        from a partner organization and roaming computers, 253
        from outside the organization, 246
    services
        built with Ruby with Service Bus Relay, 261
        with HTTP REST requests, 260
        from outside with Service Bus Relay, 256-271
OnStart method, 175-176, 178-179
OpenServiceHost method, 33-34
optimistic, with versioning mode, 331
order processing, 109-141
    to a Service Bus topic from the Orders application, 117-131
    AcsNamespace property, 136-137
    Add method, 115-117
    components in the on-premises application, 64
    CreateJobProcessors method, 117
    decoupling from the transport partners' systems, 131-141
    detail recording, 114-117
    Execute method, 125-127
    IJob interface, 118
    job processors, 117-119
    locking for, 121-124
    message flow, 120
    message posting to a topic by Trey, 112-114
    more information, 141
    new order message creation, 126
    new order message sending, 128
    NewOrderJob class, 119-121, 127-128
    order acknowledgment of shipping, 135-139
    OrderProcessor class, 132-133
    OrderProcessStatus tables, 116-117
    Orders application acknowledgment and status messages, 139-141

orders with expired locks, 125
OrderStatus tables, 116-117
posting to the service bus topic, 125
process decoupling from the transport partners'
    systems, 131-141
ProcessMessages method, 133-134
ProcessOrder method, 134-135
reliable send process completion, 129-131
Run method of worker role, 118-119
SendComplete method, 129-130
SendOrderReceived method, 135-136
SendToUpdateStatusQueue method, 138-139
StatusUpdateJob class, 139-141
summary, 141
system message flow, 120
transport partners, 111-141
    receiving and processing, 132-135
TransportPartnerStore class, 125-126
TryUpdateModel method, 114-115
UpdateWithError method, 130-131
worker role, 118
OrderProcessor class, 132-133
OrderProcessor.cs file, 92
OrderProcessStatus table, 116-117, 121-122
orders
    to a Service Bus topic from the Orders
        application, 117-131
    acknowledgment of shipping, 135-139
    to the audit log, 100-104
    to the audit log mechanism, 100-101
        Trey's choice, 101-104
    data, 20-21
    detail recording, 114-117
    with expired locks, 125
    posting to the service bus topic, 125
    receiving and processing in a transport
        partner, 132-135
Orders application, 4, 5-7
    acknowledgment and status messages, 139-141
    monitoring and managing solution choice, 171-174
    monitoring and managing solution choice by
        Trey, 174-184
    Orders application, 188
    original, 6-7
Orders application deployment to the cloud, 15-37
    See also cloud
    all data on premises, 18-19
    application and data deployment, 17
    audit log data, 21

bibliography, 37
custom reporting solution, 31
custom repository, 23
custom synchronization, 25
customer data, 20
data encryption, 23
data replication, 27
data storage mechanism, 21-23
data synchronization architecture, 28
deploying all data to, 18
deploying some data, 19
existing database system, 23
ExternalDataAnalyzer client application, 36
ExternalDataAnalyzer project, 35-36
IOrdersStatistics.cs file, 33
location choosing, 17-19
Microsoft Sync Framework, 25
more information, 37
OpenServiceHost method, 33-34
order data, 20-21
OrderStatistics service, 36
product data, 20
reporting data to external partners, 33-36
reporting solution, 29-36
reporting solution implementation, 29-36
scenario and context, 15-17
SQL Azure, 22
SQL Azure Data Sync, 24
    Trey's choice, 26-29
SQL Azure Reporting Service, 30-31
    Trey's choice, 32
SQL Server Reporting Services, 30
third party reporting solution, 31
third party synchronization, 25
Trey Research
    deployment criteria, 19-21
    reporting criteria, 31
    storage criteria, 23
    synchronization criteria, 26
web.config file, 34-35
Windows Azure storage, 21-22
Orders application4, 5-7
    See also authentication; Orders monitoring and
        managing
Orders monitoring and managing, 169-193
    See also hybrid application monitoring and managing
    App.config file, 188
    Azure configuring with built-in management
        objects, 188

Azure Diagnostics, 172, 174
Azure Management Portal, 184
Azure PowerShell cmdlets library, 185
Azure SDK, 185
Azure Service Management REST API, 185
CleanupRelyingParties method, 187
custom logging solution, 173-174
deployment and management, 184-190
    Trey's choice, 186-190
deployment solution choices, 184-185
    Trey's choice, 185-186
diagnostic data from the cloud, 181-184
Enterprise Library Logging Application Block, 172-173
monitoring, logging, and measuring, 170-184
    solution choice, 171
more information, 190-191
OnStart method, 175-176, 178-179
Orders application, 188
Orders monitoring and managing solution choice, 171-174
    Trey's choice, 174-184
Orders Statistics service, 188-190
ReceiveNextMessage method, 179-181
recording decisions, 175
scenario and context, 169
Service Management Wrapper Library, 186-188
ServiceManagementWrapper object, 186-187
StoreController class, 180-181
summary, 190
third party monitoring solutions, 173
trace message logging, 177-178
trace message writing, 179-181
TraceHelper class, 174-175, 177-178
TraceInformation method, 179-180
TransferLogs method, 181-183
Orders Statistics service, 188-190
Orders.Shared project library, 74
OrderStatistics service, 36
OrderStatus table, 121
    architecture, 121
OrderStatus tables, 116-117
Orders.Website project, 49-51
Orders.Website.Helpers project, 51-52

P

partners
    authenticating corporate users and users from partner organizations, 230-231
    communication with transport partners, 67-100
    external partners, 33-36
    messaging technologies for transport partners by Trey, 73
    resources from a partner organization and roaming computers, 253
    transport partners, 111-141
        Trey's choice, 71-100
PeekLock, 283
performance *See* scalability, availability, and performance
Performance policy, 153, 321
pessimistic, with locking mode, 332-333
physical implementation of data synchronization, 28
Poe *See* IT professional role (Poe)
policies, 321-322
preface, xvii-xxii
prerequisites, xx-xxi
priority queue, 296
processing *See* order processing
ProcessMessage method, 80-82, 92-93, 104
ProcessMessages method, 78-79, 90-91, 133-134
ProcessOrder method, 134-135
product data, 20
production line with events based on Service Bus subscriptions, 307
ProductsStoreWithCache object, 164-165
ProductStore class, 159-160, 163-164, 166

R

reactive rules, 314
ReceiveNextMessage method, 179-181
    messaging and communications, 79-80
relevance of this guide, xviii
reliability
    cross-boundary routing, 291
    send process completion, 129-131
    users and requests authentication, 233
remote procedure call (RPC), 249-250
replication
    data between data centers in the cloud, 213
    a database to implement read scale-out, 219
requests and responses through Azure Service Bus Relay, 258
requirements, xx-xxi
    and challenges, 310-313
response time, 156-166
    optimization, 156-157
        Trey's choice, 158-166
    throughput for cloud applications, 312-313
        with Azure Caching, 327-347

responsiveness, 233, 248-249
    and availability, 291
retry policy, 58-59
reviewers, xxiii-xxiv
roles
    See also cloud specialist role (Bharath); IT
        professional role (Poe); senior software developer
        role (Markus); software architect role (Jana)
    described, xxi
Round Robin policy, 153, 322
Run method, 96
    of worker role, 118-119

S

scalability, availability, and performance, 143-167
    automatic scaling with custom service, 145-146
    automatic scaling with Enterprise Library Autoscaling
        Application Block, 146
    Autoscaler object, 148-151
    Autoscaling Application Block hosting, 147-148
    autoscaling rule definitions, 148-151
    Azure Caching, 156-157
    Azure Caching defining and configuration, 158-159
    Azure Traffic Manager to route customers'
        requests, 153-154
    cache and database synching, 159
    caching functionality, 160-164
    CachingStrategy class, 161-163
    ContainerBootstrapper class, 164-165
    Content Delivery Network (CDN), 157
    custom service to redirect traffic, 152-153
    data retrieving and managing, 159-160
    elasticity control, 144-151
    elasticity management choices, 144-146
    Failover policy, 153
    FindAll method, 166
    ICachingStrategy interface, 160-163
    IProductsStore interface, 163-164
    manual scaling, 145
    more information, 167
    network latency and connectivity, 152-156
        management choices, 152-154
        Trey's choices, 154-156
    no scaling, 144-145
    Performance policy, 153
    ProductsStoreWithCache object, 164-165
    ProductStore class, 159-160, 163-164, 166
    response time, 156-166
        optimization, 156-157
        Trey's choice, 158-166

Round Robin policy, 153
scenario and context, 143-144
ServiceConfiguration.csfg, 158-159
StartDiagnostics method, 151
StoreController class, 166
summary, 167
Traffic Manager, 153-154
Trey's choices, 146-151
scalability, availability, and performance
    maximizing, 309-348
    actions, 316-317
    Autoscaling Application Block, 315
        use guidelines, 317-318
    availability maximizing for cloud applications, 312
        with Azure Traffic Manager, 324-326
    Azure and related technologies, 313-347
    Azure Caching limitations, 346
    Azure Caching provisioning and sizing, 327-329
    Azure Caching security guidelines, 347
    Azure Traffic Manager applications, 318-324
    caching
        data updating, 331
        guidelines, 336-346
        overview, 313
        web application session state, 335
    changeSetting action, 317
    constraint rules, 314, 316
    Content Delivery Network (CDN) overview, 313
    DistributedCacheOutputCacheProvider class, 335
    DistributedCacheSessionStateStoreProvider
        session state provider, 334
    elasticity
        in the cloud management, 310-311
        with Enterprise Library Autoscaling Application
            Block, 314-318
    Enterprise Library Autoscaling Application
        Block, 314-318
        overview, 313
    Failover policy, 322
    HTML output caching, 335
    local cache, 334
    monitoring endpoints, 321
    more information, 347-348
    network latency reduction
        accessing cloud applications, 311
        accessing cloud applications with Traffic
            Manager, 318-324
    optimistic, with versioning mode, 331
    Performance policy, 321
    pessimistic, with locking mode, 332-333

policies, 321-322
reactive rules, 314
requirements and challenges, 310-313
response time and throughput
    for cloud applications, 312-313
    for cloud applications with Azure Caching, 327-347
Round Robin policy, 322
scale action, 317
services that share data with Azure Caching, 329-331
setRange action, 316
Traffic Manager
    limitations, 323-324
    overview, 313
    to reduce network latency guidelines, 323
    request routing, 319-320
    use guidelines, 322-323
scale action, 317
scaling
    automatic, 145-146
    autoscaling rule definitions, 148-151
    no scaling, 144-145
    Service Bus topic and subscriptions, 300
scenario and context, 143-144, 169
    authentication, 39-41
    messaging and communications, 63-66
    Orders application deployment to the cloud, 15-17
scenarios, 1-14
    See also Orders application; Trey Research
security
    cross-boundary communication, 248
    cross-boundary routing, 291
    message queues, topics, and subscriptions, 94-97
    users and requests authentication, 232-233
    Windows Azure Connect, 253-255
security architecture of Windows Azure Connect, 254
segregation across the Orders application deployment, 16
Send method, 76-77
SendComplete method, 129-130
sender applications, 295
SendOrderReceived method, 135-136
SendToUpdateStatusQueue method, 138-139
senior software developer role (Markus), xxi
    GUIDs, 206
Service Bus
    authentication, 95
    authentication and authorization, 239-244
    endpoints and relying parties, 243
    message, 271-272
    queues, 70
    request authentication with ACS, 240
    request authentication with ADFS and ACS, 242
    tokens and token providers, 243
    topics and subscriptions, 71
Service Bus queues, 70
    batch processing, 274
    guidelines, 272-283
    with ordered messages, 273
    security guidelines, 286-287
Service Bus Relay, 240
    guidelines, 256-263
    requests and responses, 258
    service naming guidelines, 267-268
    vs.Windows Azure Connect, 270-271
Service Bus topics and subscriptions, 71, 88
    decoupling a sender application from the message routing logic, 295
    guidelines, 222-227
    message routing
        guidelines, 293-303
        limitations, 304-308
        multiple destinations guidelines, 304-307
        multiple destinations limitations, 308
        updating, 225
    security guidelines, 308
Service Management Wrapper Library, 186-188
ServiceBusQueue class, 84-85
ServiceBusReceiverHandler object, 83, 90
ServiceBusSubscriptionDescription class, 89-90
ServiceBusTopic class, 96-97
ServiceManagementWrapper object, 186-187
services
    exposed through Service Bus Relay, 266
    sharing data with Windows Azure Caching, 329-331
sessions
    different states from different datacenters, 345
    to group messages, 279
setRange action, 316
SetupAuditLogListener method, 101-102
SetupServiceBusTopicAndQueue method, 88-89
SingleOrDefault method, 59
smart client or service authentication, 237
software architect role (Jana), xxi
solutions
    deployment, 184-185
    implementation, 29-36
    reporting, 29-36
    third party monitoring, 173
    third party reporting solution, 31

specifying the synchronization direction for databases participating in Topology D, 210
SQL Azure, 22
SQL Azure Data Sync, 24
    configuration guidelines, 203-211
    to partition and replicate data across branch offices, 218
    to partition and replicate data in the cloud, 217
    to replicate and synchronize, data replicating, distributing, and synchronizing, 203-220
    Trey's choice, 26-29
    use guidelines, 211-220
SQL Azure Data Sync Security Model, 220
SQL Azure Reporting Service, 30-31
    Trey's choice, 32
SQL Server Reporting Services, 30
SSO (single sign-on), 229
staged migration to cloud, 12
StartDiagnostics method, 151
StatusUpdateJob class, 139-141
StoreController class, 166, 180-181
strategy, 5
subscription correlation, 298
    to deliver response messages to a sender, 299
summary
    authentication, 61
    order processing, 141
    Orders monitoring and managing, 190
    scalability, availability, and performance, 167
    Trey Research, 13-14
support, xxii
sync datasets defined, 203-204
Sync Framework SDK for custom replication and synchronization, 221-222
sync groups, 203-204
    configuration table, 29
    defined, 203
sync loop avoidance, 210-211
synchronization
    See also data replicating, distributing, and synchronizing
    bidirectional
        across all databases, 195
        only across databases in the cloud, 196
    conflict management, 205-206
    direction for a database, 209
    schedule for a sync group, 208
system requirements, xx-xxi

T
tables
    authenticating visitors, 47
    diagnostics mechanism configuration, 176-177
    Execute method, 122
    GetLockedOrders method, 123
    LockOrders method, 122-123
    logging configuration, 175
    OrderProcessStatus table, 121-122
    OrderStatus table, 121
    Service Bus authentication, 95
    sync group configuration, 29
    technology map, 13
team, xxiii-xxiv
technology map, 13
third party monitoring solutions, 173
third party reporting solution, 31
third party synchronization, 25
trace messages
    logging, 177-178
    writing, 179-181
TraceHelper class, 174-175, 177-178
TraceInformation method, 179-180
Traffic Manager, 153-154
    limitations, 323-324
    overview, 313
    to reduce network latency guidelines, 323
    request routing, 319-320
    routing and redirection, 320
    use guidelines, 322-323
TransferLogs method, 181-183
Transient Fault Handling Application Block, 59
transport partners, 111-141
    Trey's choice, 71-100
TransportPartnerStore class, 125-126
Trey Research, xvii
    See also Orders application
    bibliography, 14
    choices, elasticity, 146-151
    deployment criteria, 19-21
    more information, 14
    original on-premises authentication mechanism, 41
    reporting criteria, 31
    scenario, 1-14
    storage criteria, 23
    strategy, 5
    summary, 13-14
    synchronization criteria, 26
TryUpdateModel method, 114-115

U

uncontrolled updates, 223
updates
    notifications by Service Bus topic and subscriptions, 227
    propagation between Windows Azure caches and replicated data stores, 339
    uncontrolled, 223
UpdateWithError method, 130-131
URI naming conventions, 267-268
use cases and challenges, 230-232
    cross-boundary communication, 245
    data replicating, distributing, and synchronizing, 193-194
    hybrid application monitoring and managing, 350-353
users and requests authentication, 229-244
    Access Control Service (ACS), 238-239
    ACS and unique user IDs, 239
    authentication
        corporate users and users from partner organizations, 230-231
        public users, 230
    authorization
        access to Service Bus queues, 232
        access to Service Bus Relay endpoints, 232
        rules and rule groups, 244
        service access for non-browser clients, 231
        user actions, 231
    claims-based authentication and authorization technologies, 233-239
    claims-based authentication process overview, 235-236
    client authentication, 240-242
    cross-cutting concerns, 232-233
    federated authentication, 229, 234-235
    interoperability, 233
    more information, 244
    reliability, 233
    responsiveness, 233
    security, 232-233
    Service Bus authentication and authorization, 239-244
    Service Bus endpoints and relying parties, 243
    Service Bus Relay, 240
    Service Bus tokens and token providers, 243
    SSO, 229
    uses cases and challenges, 230-232
    web service requests authorizing, 236-237
    Windows Identity Foundation (WIF), 237-238

V

Views/Shared folder, 53-54
visitor authentication and authorizationLayout.cshtml, 53-54
visitors, 42-59
    authentication and authorization, 53-54

W

web services
    (pull model), 69
    (push model), 68-69
    requests authorizing, 236-237
web.config file, 34-35, 53
who's who, xxi
WIF, 49-52, 237-238
Windows Azure
    See also Azure; cloud
    configuring with built-in management objects, 188
    storage, 21-22
    storage queues, 69-70
    technologies for cross-boundary communication, 249-287
Windows Azure Access Control Service (ACS) *See* ACS
Windows Azure and related technologies, 202-227
Windows Azure Caching, 156-157
    See also caching
    data sharing, 329-331
    limitations, 346
    to provide scalability, 330
    provisioning and sizing, 327-329
Windows Azure Connect, 251-256
    vs. Service Bus Relay, 270-271
Windows Azure Content Delivery Network (CDN) *See* Content Delivery Network (CDN)
Windows Azure Diagnostics, 172, 174
    data security guidelines, 360
    hybrid application monitoring and managing, 355-360
    use guidelines, 356-360
Windows Azure Management Portal, 184
Windows Azure PowerShell cmdlets library, 185
Windows Azure SDK, 185
Windows Azure Service Bus topics and subscriptions *See* Service Bus
Windows Azure Service Management REST API, 185
Windows Azure Traffic Manager, 155
    applications, 318-324
Windows Identity Foundation (WIF) *See* WIF
worker role requests through Service Bus Relay, 263
WsFederationRequestValidator class, 52-53